13th International Workshop on Hadron Physics 2015

Journal of Physics: Conference Series Volume 706

Angra dos Reis, Rio de Janeiro, Brazil
22 – 27 March 2015

ISBN: 978-1-5108-2417-1
ISSN: 1742-6588

Printed from e-media with permission by:

Curran Associates, Inc.
57 Morehouse Lane
Red Hook, NY 12571

Some format issues inherent in the e-media version may also appear in this print version.

Printed by Curran Associates, Inc. (2016)

For permission requests, please contact the Institute of Physics at the address below.

Institute of Physics
Dirac House, Temple Back
Bristol BS1 6BE UK

Phone: 44 1 17 929 7481
Fax: 44 1 17 920 0979

techtracking@iop.org

Additional copies of this publication are available from:

Curran Associates, Inc.
57 Morehouse Lane
Red Hook, NY 12571 USA
Phone: 845-758-0400
Fax: 845-758-2633
Email: curran@proceedings.com
Web: www.proceedings.com

TABLE OF CONTENTS

Journal of Physics: Conference Series **706** (2016) 052001 doi:10.1088/1742-6596/706/5/052001

Telescope performance and image simulations of the balloon-borne coded-mask protoMIRAX experiment

A.V. Penacchioni, J. Braga, M.A. Castro Avila & F. D'Amico

Instituto Nacional de Pesquisas Espaciais (INPE), Av. dos Astronautas, 1758, São José dos Campos, SP, Brazil

E-mail: ana.penacchioni@hotmail.com

Abstract. In this work we present the results of imaging simulations performed with the help of the GEANT4 package for the protoMIRAX hard X-ray balloon experiment. The instrumental background was simulated taking into account the various radiation components and their angular dependence, as well as a detailed mass model of the experiment. We modelled the meridian transits of the Crab Nebula and the Galactic Centre (GC) region during balloon flights in Brazil ($\sim$ -23° of latitude and $\sim$ 40 km of altitude) and introduced the correspondent spectra as inputs to the imaging simulations. We present images of the Crab and of three sources in the GC region: 1E 1740.7-2942, GRS 1758-258 and GX 1+4. The results show that the protoMIRAX experiment is capable of making spectral and timing observations of bright hard X-ray sources as well as important imaging demonstrations that will contribute to the design of the MIRAX satellite mission.

1. Introduction

X and low-energy ($\lesssim$ 1 MeV) γ-ray photons coming from astrophysical sources interact in the Earth's atmosphere through photoelectric effect or Compton scattering, never reaching the surface. However, at usual stratospheric balloon altitudes ($\sim$40 km), a major fraction of the X-ray photons with energy greater than $\sim$30 keV is able to pass through the remaining external atmosphere. This allows the detection and observation of cosmic sources above these energies with instruments carried by balloons. In addition to photons coming directly from the astrophysical sources of interest, radiation from the atmosphere and from other cosmic sources (either point-like or diffuse) in the field of view of an X-ray telescope interacts with the instrument and the balloon platform, generating an intense background. Therefore, detailed predictions of the fluxes and spectra of cosmic sources are crucially dependent upon an accurate knowledge of this background. At balloon altitudes, this is a non-trivial task that has to take into account the several particle and photon fields incident upon the experiment and all the interactions that occur in the detectors and surrounding materials, as well as a detailed mass model of the instrument. By knowing the particle kinds, both primary and secondary, their energy spectra, and by following the secondary emission they produce when they interact, we can simulate the energy depositions on the detectors and identify the contribution of each background component. This is essential to estimate the instrument sensitivity as a function of energy.

Even though protoMIRAX is chiefly a prototype designed to test several subsystems of the MIRAX satellite experiment in a near-space environment, we show in this work that the experiment is also capable of making spectral observations of selected bright hard X-ray sources and perform important imaging demonstrations that will contribute to the design of the MIRAX mission.

In section 2 we describe the protoMIRAX experiment, which will be assembled in a gondola to fly on stratospheric balloons. In section 3 we make a brief description of the nature of the hard X-ray point sources that we have used in our simulations. In section 4 we describe the procedure we took to simulate the instrumental background using GEANT4. In section 5 we show the reconstucted images and discuss their properties. Finally, in section 6 we present the conclusions.

2. The experiment

The protoMIRAX balloon gondola (see Fig. 1) houses the X-ray camera and other subsystems. The camera includes a coded mask based on an extended 4×4 pattern of a 13×13 MURA (1), which is placed 650 mm away from a position-sensitive detector plane. The mask elements are made of lead with 20 mm $\times$ 20 mm in area and 1 mm thickness. Such a design will allow a $\sim 1°45'$ geometrical angular resolution and a $\sim 20 \times 20$ fully-coded FOV. The detector plane comprises 169 square (10 mm $\times$ 10 mm) CZT detectors, 2mm thick, providing a 30 - 200 keV optimum energy coverage.

Figure 1. Design of the protoMIRAX payload, consisting of: (1) the pivot with a decoupling mechanism for momentum dump; (2) the reaction wheel; (3) the elevation step-motor-controlled actuator; (4) star sensors; (5) the MURA-based coded-mask; and (6) the electronics bay. The gondola is $\sim$2.4 m high, 1.4 m $\times$ 1.4 m in cross section, and its total mass is $\sim$600 kg.

3. Emission from the sources

We studied the Crab nebula and three sources in the galactic centre (GC) region: 1E 1740.7-2942, GRS 1758-258 and GX 1+4. The X and low-energy γ-ray photon spectrum of these sources is well described by a power law: $F_0 = AE^{-\gamma}$ [ph cm^2 s^{-1} keV^{-1}]. For the Crab, A = 14.44, γ= 2.169 (2); for 1E, A = 10^{-4}, γ= 1.35 (3); for GRS, A = 4.6 $\times 10^{-5}$, γ = 1.8 (4), and for GX, A = 5.1 $\times 10^{-4}$, γ= 1.9 (5).

The fluxes at each energy are attenuated due to absorption in the atmosphere, so they vary as $F = F_0 e^{-\frac{\mu}{\rho} x sec(z)}$, where $\frac{\mu}{\rho}$ is the absorption coefficient in units of cm^2/g (it depends on the energy) and x = 2.7 g/cm^2 is the atmospheric depth at the balloon altitude ($\sim$42 km). The zenith angle z is given by: $cos(z) = sin(\phi)sin(\delta) + cos(\phi)cos(\delta)cos(t - \alpha)$, where ϕ is the geographic latitude, α is the right ascension and the declination of the source, $H = t - \alpha$ is the hour-angle and t is the sidereal time (LST).

Given the expression for $z(t)$, we can compute the number of photons N coming from the Crab reaching the detector of area S in a time interval T in the detector energy range E_{min}-E_{max}:

$$N = \int_0^T \int_{E_{\min}}^{E_{\max}} \int_S A_0 E^{-\gamma} \exp\left[-\frac{\mu}{\rho}(\mathrm{E})\,\mathrm{x}\sec\mathrm{z}\right] \mathrm{dE\,dt\,dS}. \quad (1)$$

4. Background noise

To compute the protoMIRAX instrumental background at balloon altitudes we made use of the well-known radiation-interaction code GEANT4, developed by CERN (6; 7). We have built a procedure to take into account the main particles (photons, protons, electrons and neutrons) coming from all directions in the atmospheric environment at balloon altitudes, at a latitude of −23°S. Fig. 2 shows each background contribution. Red corresponds to photons, blue to protons, cyan to neutrons, pink to electrons and black rep- resents the sum of all the contributions. The error bars are 1 statistical uncertainties. Fig. 3 shows the spatial distribution of the background across the detector plane. The background is reasonably smooth, with slightly higher count rates near the edges and corners due to geometrical factors.

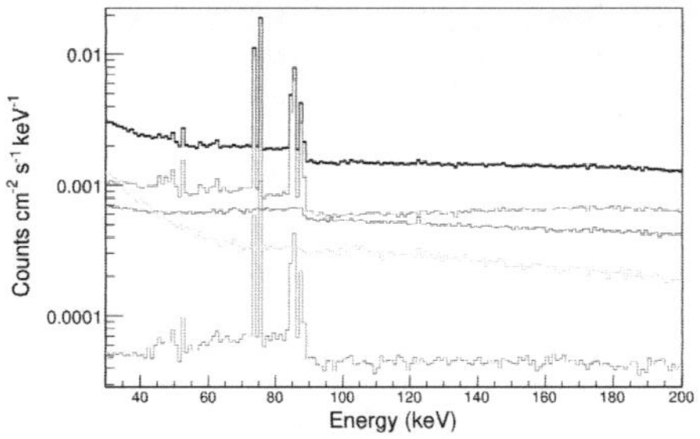

Figure 2. Background spectral contributions estimated for protoMIRAX based on GEANT4 simulations at balloon altitudes at a latitude of −23°S. Red corresponds to photons, blue to protons, cyan to neutrons, pink to electrons and black represents the sum of all the contributions. The error bars are 1σ statistical uncertainties.

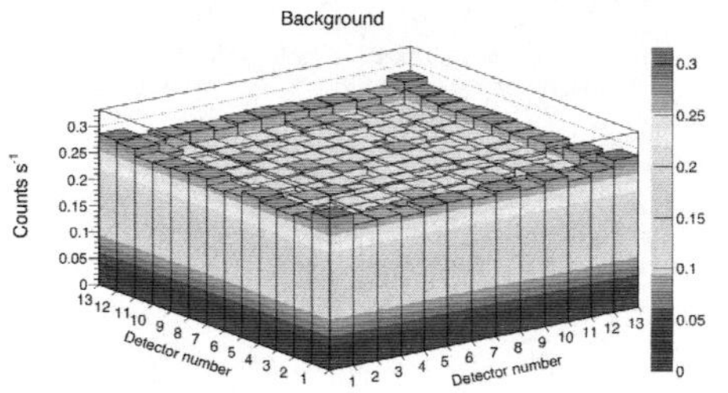

Figure 3. Count distribution of the total background over each pixel on the detection plane in the detector energy range 30-200 keV. The observation time was 8 h at balloon altitudes. Each pixel corresponds to one planar CZT detector (1 cm^2).

5. Image reconstruction and SNR

In a coded-mask experiment, the mask is composed of elements which are transparent (open elements) or opaque (closed elements) to radiation. Photons hitting an open element will pass through the mask and reach a position-sensitive detector plane, and photons hitting a closed element will be absorbed. In this way, a pattern of mask shadows' is projected on the detector plane, which actually corresponds to the superposition of many images of each object (one image for each open element). In order to determine the intensity and position of each observed object in the field of view, a decoding and reprocessing of the shadowgram is needed through the implementation of numerical algorithms (8). Essentially, the reconstructed image is the cross-correlation between the detector count distribution map and a decoding function that mimics the mask pattern (9). The signal-to-noise ratio (SNR) of a peak (i, j) in a reconstructed coded-mask image, which determines the sensitivity of these systems, is given by (1):

$$SNR_{i,j} = \frac{N_S}{\sqrt{N_S + N_T}}, \tag{2}$$

where N_S is the number of net source counts (corresponding to source photons that came through the mask openings and hit the detectors), and N_T is the total number of detector counts, both for the same integration time.

Fig. 4 shows the reconstructed image for the Crab and Fig. 5 shows the reconstructed image for the GC for an integration time of 4h and 8h respectively, in the energy range 30-200 keV (10). The crab is detected at 109σ level, while the SNRs for 1E, GRS and GX are 26, 10 and 9, respectively.

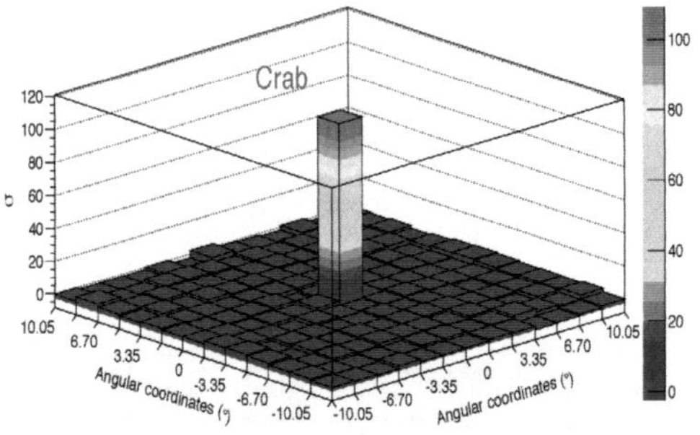

Figure 4. Simulated image of the Crab region as seen by protoMIRAX for 4 hours at balloon altitudes. The Crab is detected at a 109σ level.

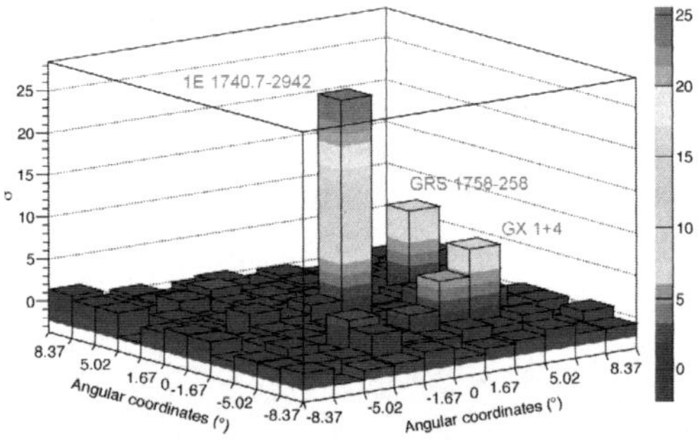

Figure 5. Simulated image of the Galactic Centre region as seen by protoMIRAX for 8 hours. The SNRs for 1E1740.7-2942, GRS1758-258 and GX1+4 are, respectively, 26, 10 and 9. We can see that GX 1+4 counts are distributed in two sky bins.

6. Conclusions

In this work we presented results of simulated images of regions with bright hard X-ray sources with the protoMIRAX balloon experiment. We introduced the results of GEANT4 simulations into which the protoMIRAX experiment will take place, in order to predict the telescope performance during the observation of bright cosmic sources in hard X- rays. We then produced images of the Crab Nebula and a GC field with 3 bright sources: 1E 1740.7-2942, GRS 1758-258 and GX 1+4. We simulated the Crab Nebula image considering a meridian passage of 4 hours, during which the highest source elevation will be 45°. We have shown that we will be able to reach a very good SNR of 109, which will make the Crab a reliable calibration source for the experiment. In the case of the GC, the three sources were imaged together, with 1E 1740.7-2942 in the centre. The results show that we can make observations of crowded fields like the GC with protoMIRAX and provide flux and spectral information without source confusion.

6.1. Acknowledgments

We thank Luiz Reitano, Fernando G. Blanco and Sérgio Amirágile for invaluable technical support. A.V.P. acknowledges the support by the international Cooperation Program CAPES-ICRANET financed by CAPES - Brazilian Federal Agency for Support and Evaluation of Graduate Education within the Ministry of Education of Brazil. We thank FINEP for financial support under Convênio 01.10.0233.00. We also thank CNPq and FAPESP, Brazil, for support under INCT Estudo do Espaço.

References

[1] Gottesman S R and Fenimore E E 1989 *Applied Optics* **28** 4344–4352

[2] Sizun, P et al 2004 *5th INTEGRAL Workshop on the INTEGRAL Universe* (*ESA Special Publication* vol 552) ed Schoenfelder V, Lichti G and Winkler C p 815 (*Preprint* astro-ph/0406058)

[3] Grebenev S A, Pavlinsky M N and Sunyaev R A 1995 *Advances in Space Research* **15** 115

[4] Sunyaev, R et al 1991 *A&A* **247** L29–L32

[5] Dieters S W B, Greenhill J G, Sharma D P, Sood R K, Waldron L and Storey M C 1991 *Advances in Space Research* **11** 35–39

[6] Allison, J et al 2006 *IEEE Transactions on Nuclear Science* **53** 270–278

[7] Agostinelli, S et al 2003 *Nuclear Instruments and Methods in Physics Research A* **506** 250–303

[8] Caroli E, Stephen J B, Di Cocco G, Natalucci L and Spizzichino A 1987 *Space Sci. Rev.* **45** 349–403

[9] Fenimore E E and Cannon T M 1978 *Applied Optics* **17** 337–347

[10] Penacchioni A V, Braga J, Castro M A and D'Amico F 2015 *Journal of High Energy Astrophysics* **5** 22–29 (*Preprint* 1505.05835)

Generalized Scalar Duffin-Kemmer-Petiau Electrodynamics (GSDKP)

R Bufalo, T R Cardoso, A A Nogueira and B M Pimentel

Instituto de Física Teórica (IFT), Universidade Estadual Paulista; Rua Dr. Bento Teobaldo Ferraz 271, Bloco II Barra Funda, CEP 01140-070 São Paulo, SP, Brazil.

E-mail: bufalo@ift.unesp.br, cardoso@ift.unesp.br, nogueira@ift.unesp.br, pimentel@ift.unesp.br

Abstract. The main goal of this work is to investigate the quantum interaction between scalar field and gauge field in the context of Generalized Scalar Duffin-Kemmer-Petiau Electrodynamics (GSDKP) by a quantum theory in the functional approach. The Hamiltonian structure is obtained with the Dirac method and the Faddeev-Senjanovic procedure is established in order to write the transition amplitude in an alternative gauge fixing, known as the non-mixing gauge. As a consequence, the Schwinger-Dyson-Fradkin equations and the Ward-Takahashi-Fradkin identities are obtained.

1. Introduction

The purpose of this work is to study the covariant dynamics of systems with constraints and the quantization procedure in a particular case (GSDKP), so let us make a brief introduction to the subject.

The quantization problem of a classical dynamics was first put in a formal language by Dirac [1] because he observed that the classical dynamics described in the phase space by the observable and Poisson brackets was associated to a quantum dynamics described in the Hilbert space by the operators and commutators\anti-commutators via correspondence principle

$$\{,\}_P \to -i[,]_\pm. \tag{1}$$

Later the existence of constraints in the hamiltonian dynamics led Dirac to extend its mechanical analysis of the phase space by the parenthesis of Dirac and the classification (first class\second class) always being able to make connection to quantum dynamics with the correspondence principle [2],

$$\{,\}_D \to -i[,]_\pm. \tag{2}$$

This is the first look at the connection between classical\quantum dynamics.

The second look begins in a Dirac study [3] about the connection between a dynamics described in the configuration space and its resulting quantum dynamics. In this study we see the emergence of an object very important called transition amplitude

$$Z = \int D\mu \exp[iS], \tag{3}$$

where $D\mu$ is the integration measure and S is the action.

Journal of Physics: Conference Series **706** (2016) 052002 doi:10.1088/1742-6596/706/5/052002

Feynman uses the idea of Dirac to formulate a way to describe the quantum mechanics via path integral [4]. However elegantly Schwinger infers that because the quantum equations in the Heisenberg description preserve its classic form it should be a principle of quantum action [5]. We have now that the amplitude transition, or functional generator is a solution of the functional equation arising from a dynamic

$$[\dot{\varphi} - \{\varphi, H_0 - \varphi J\}_D]|_{\varphi=\frac{\delta}{i\delta J}} Z[J] = 0 \tag{4}$$

where H_0 is the initial hamiltonian.

Continuing, Faddeev explores the integration measure properties $D\mu$ to study connections between classical\quantum dynamics of the physical systems with first-class constraints [6]. Later Senjanovic extends the ideas of Faddeev [7] for second-class constraints. Finally Fradkin and Vilkovisky make an overview of the problem in view of covariant formalisms and its connection with BRST symmetry [8]. See, for more details on quantization of systems with constraints [9].

2. Constraint analysis and transition amplitude

To construct the transition amplitude we will use the Faddeev-Senjanovic [6, 7] analysis where we must first do a short study of constraint.

The Lagrangian density describing the GSDKP is defined by

$$\mathscr{L} = \frac{i}{2}\bar{\psi}\beta^{\mu}\left(\partial_{\mu}\psi\right) - \frac{i}{2}\left(\partial_{\mu}\bar{\psi}\beta^{\mu}\right)\psi - m\bar{\psi}\psi + eA_{\mu}\bar{\psi}\beta^{\mu}\psi - \frac{1}{4}F_{\mu\nu}F^{\mu\nu} + \frac{a^2}{2}\partial^{\mu}F_{\mu\beta}\partial_{\alpha}F^{\alpha\beta} \tag{5}$$

where $F_{\mu\nu} = \partial_{\mu}A_{\nu} - \partial_{\nu}A_{\mu}$ is the usual electromagnetic field-strength tensor and β^{μ} are the DKP matrices that obey the algebra [10]

$$\beta^{\mu}\beta^{\nu}\beta^{\theta} + \beta^{\theta}\beta^{\nu}\beta^{\mu} = \beta^{\mu}\eta^{\nu\theta} + \beta^{\theta}\eta^{\nu\mu}\,. \tag{6}$$

This theory is invariant under local gauge transformations

$$\psi \to e^{i\alpha(x)}\psi, \quad A_{\mu} \to A_{\mu} + \frac{1}{e}\partial_{\mu}\alpha(x)\,. \tag{7}$$

The matter (scalar field, mesons) are described by the Duffin-Kemmer-Petiau (DKP) theory [11], a first-order relativistic theory for the description of spin 0 and spin 1 bosons with a similar form as the Dirac equation, and the radiation (vector field, photons) are described by the Podolsky theory [12]. We can see this gauge theory with U(1) symmetry as an extension of the work on DKP Quantum Electrodynamics [13].

The translational space-time invariance of the Lagrangian density leads us to write the canonical Hamiltonian

$$\begin{aligned} H_c = \int d^3x \Big[&(\partial_0\bar{\psi})\frac{\partial\mathscr{L}}{\partial(\partial_0\bar{\psi})} + \frac{\partial\mathscr{L}}{\partial(\partial_0\psi)}(\partial_0\psi) + \frac{\partial\mathscr{L}}{\partial(\partial_0 A_{\nu})}(\partial_0 A_{\nu}) \\ &- \partial_{\theta}\left(\frac{\partial\mathscr{L}}{\partial(\partial_0\partial_{\theta}A_{\nu})}\right)(\partial_0 A_{\nu}) + \frac{\partial\mathscr{L}}{\partial(\partial_0\partial_{\theta}A_{\nu})}(\partial_{\theta}\partial_0 A_{\nu}) - \mathscr{L}\Big], \end{aligned} \tag{8}$$

where we define the momenta expressions

$$p = \frac{\partial\mathscr{L}}{\partial(\partial_0\bar{\psi})} = -\frac{i}{2}\beta^0\psi, \tag{9}$$

$$\bar{p} = \frac{\partial\mathscr{L}}{\partial(\partial_0\psi)} = \frac{i}{2}\bar{\psi}\beta^0, \tag{10}$$

$$\Pi^{\nu} = F^{\nu 0} + a^2[\eta^{i\nu}\partial_i\partial_{\alpha}F^{\alpha 0} - \partial_0\partial_{\alpha}F^{\alpha\nu}], \tag{11}$$

$$\Phi^{\nu} = a^2[\partial_{\alpha}F^{\alpha\nu} - \eta^{\nu 0}\partial_{\alpha}F^{\alpha 0}]. \tag{12}$$

From the above momentum expressions, we shall study the constraint structure of the theory following Dirac's approach to singular systems [2]. In this approach it is possible to obtain the set of first-class constraints

$$\varphi_1 = \Phi_0 \approx 0, \tag{13}$$
$$\varphi_2 = \Pi_0 - \partial_k \Phi^k \approx 0, \tag{14}$$
$$\varphi_3 = e\bar{\psi}\beta^0\psi - \partial^k \Pi_k \approx 0 \tag{15}$$

and the set of second-class constraints

$$\chi^{(1)} = p + \frac{i}{2}\beta^0\psi \approx 0, \tag{16}$$
$$\bar{\chi}^{(1)} = \bar{p} - \frac{i}{2}\bar{\psi}\beta^0 \approx 0, \tag{17}$$
$$\chi^{(2)} = [1-(\beta^0)^2][i\beta^i\partial_i\psi(x) - m\psi(x) + e\beta^i A_i(x)\psi(x)] \approx 0, \tag{18}$$
$$\bar{\chi}^{(2)} = [-i\partial_i\bar{\psi}(x)\beta^i + m\bar{\psi}(x) - e\bar{\psi}(x)\beta^i A_i(x)][1-(\beta^0)^2] \approx 0. \tag{19}$$

The weak equality $\approx$ is understood in according to Dirac's sense.

With the full set of first-class and second-class constraints determined, we are now in a position to obtain the functional generator. The transition amplitude in the Hamiltonian form is written in the following way:

$$Z = N\int D\mu \exp\left\{i\int d^4x[(\partial_0\bar{\psi})p + \bar{p}(\partial_0\psi) + \Pi^\nu(\partial_0 A_\nu) + \Phi^\nu(\partial_0\Gamma_\nu) - \mathscr{H}_c]\right\} \tag{20}$$

where the canonical Hamiltonian is given by

$$\begin{aligned}\mathscr{H}_c = {} & \Pi_0\Gamma^0 + \Pi_k\Gamma^k + \Phi_k(\partial^k\Gamma_0 - \partial_l F^{lk} + \frac{\Phi^k}{2a^2}) - \frac{i}{2}\bar{\psi}\beta^i\overleftrightarrow{\partial}_i\psi + m\bar{\psi}\psi \\ & - e\bar{\psi}\hat{A}\psi + \frac{1}{4}F_{kj}F^{kj} + \frac{1}{4}(\Gamma_j - \partial_j A_0)^2 - \frac{a^2}{2}(\partial^j\Gamma_j - \partial^j\partial_j A_0)^2,\end{aligned} \tag{21}$$

and the integration measure

$$D\mu = D\Phi^\nu D\Gamma_\nu D\Pi^\mu DA_\mu D\bar{\psi}D\psi D\bar{p}Dp\,\delta(\Theta_l)\det\|\{\Theta_l,\Theta_m\}\|^{\frac{1}{2}}. \tag{22}$$

We have that the complete set of constraints for the GSDKP is

$$\Theta_l = \left\{\chi^{(1)}, \bar{\chi}^{(1)}, \chi^{(2)}, \bar{\chi}^{(2)}, \varphi_1, \varphi_2, \varphi_3, \Sigma_1, \Sigma_2, \Sigma_3\right\}, \tag{23}$$

in which suitable gauge conditions for the first-class constraints are chosen as the generalized radiation conditions [14]

$$\Sigma_1 = \Gamma_0(x) \approx 0, \quad \Sigma_2 = A_0 \approx 0, \quad \Sigma_3 = (1 + a^2\Box)(\vec{\nabla}.\vec{A}) \approx 0. \tag{24}$$

After integrating over the gauge and fermionic momenta, the transition amplitude Z is explicitly written

$$\begin{aligned}Z = N\int & DA_\mu D\bar{\psi}D\psi \det\left\|(1 + a^2\vec{\nabla}^2)\vec{\nabla}^2\right\|\delta((1 + a^2\Box)(\vec{\nabla}.\vec{A})) \\ & \exp[i\int d^4x\{\bar{\psi}(i\beta^\mu\nabla_\mu - m)\psi - \frac{1}{4}F_{\mu\nu}F^{\mu\nu} + \frac{a^2}{2}\partial^\mu F_{\mu\beta}\partial_\alpha F^{\alpha\beta}\}].\end{aligned} \tag{25}$$

Hence, with the Faddeev-Popov-DeWitt ansatz in the non-mixing gauge condition [15]

$$\Omega(A) = \left(1+a^2\Box\right)^{\frac{1}{2}}\partial^\mu A_\mu \tag{26}$$

the transition amplitude can be written in a covariant form

$$Z = N\int DA_\mu D\bar{\psi}D\psi \exp\{i\int d^4x[\bar{\psi}\left(i\beta^\mu\nabla_\mu - m\right)\psi - \frac{1}{4}F_{\mu\nu}F^{\mu\nu} + \frac{a^2}{2}\partial^\mu F_{\mu\beta}\partial_\alpha F^{\alpha\beta} - \frac{1}{2\xi}\left(\partial^\mu A_\mu\right)\left(1+a^2\Box\right)\left(\partial^\mu A_\mu\right)]\}. \tag{27}$$

Although initially Podolsky used the Lorenz gauge fixing condition $\Omega(A) = (\partial_\mu A^\mu)$ to fix the physical degrees of freedom in generalized electrodynamics, lately with a rigorous study of constraint analysis we see that this is not really true [14]. As a consequence, the natural way to fix the physical degrees of freedom was the generalized Lorenz gauge fixing condition $\Omega(A) = (1+a^2\Box)(\partial_\mu A^\mu)$, however it complicates the theory's quantization once it increases the order of the field equation. Today we think that the non-mixing gauge fixing condition [15] $\Omega(A) = \left(1+a^2\Box\right)^{\frac{1}{2}}\partial^\mu A_\mu$ combines with the theory because it preserves the order of the field equation, but it has a pseudo-differential structure.

3. The Schwinger-Dyson-Fradkin equations

The most elegant way of studying the set of field equations in the Heisenberg description is in a functional formulation, consisting in an infinite chain of differential equations that relate different Green's function in a exact manner. This infinite tower of equations is referred to as the Schwinger-Dyson-Fradkin (SDF) equations [17].

3.1. The Schwinger-Dyson-Fradkin equations for the gauge-field propagator

The complete expression of the gauge-field propagator can be determined by means of the functional generator leading to the Schwinger variational equation for the gauge field,

$$\left[\left.\frac{\delta S}{\delta A_\gamma(x)}\right|_{\frac{\delta}{\delta i\eta},\frac{\delta}{\delta i\bar{\eta}},\frac{\delta}{\delta iJ_\mu}} + J^\gamma(x)\right]\mathscr{Z}\left[\eta,\bar{\eta},J_\mu\right] = 0. \tag{28}$$

In terms of the functional $\mathscr{Z}[J] = \exp\{iW[J]\}$ for the connected Green's functions we obtain the equation

$$-J^\gamma(x) = -ie\frac{\delta}{\delta\eta(x)}\beta^\gamma\left(\frac{\delta W}{\delta\bar{\eta}(x)}\right) + e\frac{\delta W}{\delta\eta(x)}\beta^\gamma\frac{\delta W}{\delta\bar{\eta}(x)} + \left[T^{\gamma\mu} + \frac{1}{\xi}L^{\gamma\mu}\right]\left(1+a^2\Box\right)\Box\frac{\delta W}{\delta J^\mu(x)}. \tag{29}$$

The last equation can be interpreted as the complete Podolsky field equation subjected to an external source J^γ. On this equation $T^{\gamma\mu}$ and $L^{\gamma\mu}$ are differential projectors

$$T^{\gamma\mu} + L^{\gamma\mu} = g^{\gamma\mu}, \quad L^{\gamma\mu} = \frac{\partial^\gamma\partial^\mu}{\Box}. \tag{30}$$

In order to obtain the complete gauge-field propagator it proves convenient to introduce also the generating functional for the one particle irreducible (1PI) Green's functions, which is related to W by a functional Legendre transformation

$$\Gamma\left[\psi,\bar{\psi},A_\mu\right] = W\left[\eta,\bar{\eta},J_\mu\right] - \int d^4x\left(\bar{\psi}\eta + \bar{\eta}\psi + A^\mu J_\mu\right). \tag{31}$$

Hence, rewriting (29) in terms of the 1PI functional

$$\frac{\delta^2\Gamma}{\delta A_\nu(y)\,\delta A_\gamma(x)} = \left[T^{\gamma\nu} + \frac{1}{\xi}L^{\gamma\nu}\right]\left(1 + a^2\Box\right)\Box\delta^{(4)}(x,y) + ie^2\int d^4u d^4w Tr\left[\mathscr{S}(x,u;A)\,\beta^\gamma\mathscr{S}(w,x;A)\,\Gamma^\nu(u,w;y)\right]. \tag{32}$$

The second term of (32) can be identified with the polarization operator, $\mathfrak{P}^{\gamma\nu}$,

$$\mathfrak{P}^{\gamma\nu}(x,y) = ie^2\int d^4u d^4w Tr\left[\mathscr{S}(x,u;A)\,\beta^\gamma\mathscr{S}(w,x;A)\,\Gamma^\nu(u,w;y)\right]. \tag{33}$$

The expression for the inverse of the gauge-field complete propagator in momentum representation is

$$\left(\mathfrak{D}^{\gamma\nu}\right)^{-1}(p) = \left(D^{\gamma\nu}\right)^{-1}(p) + \mathfrak{P}^{\gamma\nu}(p). \tag{34}$$

This equation can be viewed as

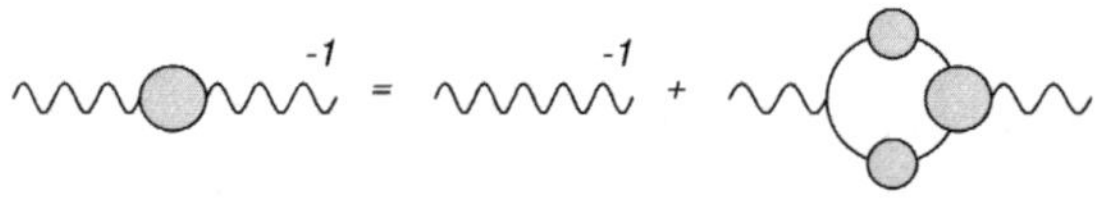

Figure 1. The SDF equation for the gauge-field propagator.

The expression (34) can be solved in order to find

$$i\mathfrak{D}^{\gamma\nu}(p) = -\frac{\left(\eta^{\gamma\nu} - \frac{p^\gamma p^\nu}{p^2}\right)}{\left[\mathfrak{P}(p) - \left(1 - a^2p^2\right)p^2\right]} + \frac{\xi}{p^2\left(1 - a^2p^2\right)}\frac{p^\gamma p^\nu}{p^2}, \tag{35}$$

in which the scalar polarization $\mathfrak{P}(p)$ it is related to the polarization tensor $\mathfrak{P}^{\gamma\nu}(p)$, through the structure

$$\mathfrak{P}^{\gamma\nu}(p) = \left(-p^2\eta^{\gamma\nu} + p^\gamma p^\nu\right)\mathfrak{P}(p). \tag{36}$$

For the free propagator, namely taking $\mathfrak{P}(p) = 0$ on (35) and $a = m_p^{-1}$, one has the expression

$$iD^{\gamma\nu}(p) = \left[\eta^{\gamma\nu} - (1-\xi)\frac{p^\gamma p^\nu}{m_p^2}\right]\left[\frac{1}{p^2} - \frac{1}{p^2 - m_p^2}\right] - (1-\xi)\frac{p^\gamma p^\nu}{\left(p^2\right)^2} \tag{37}$$

3.2. The Schwinger-Dyson-Fradkin equations for the DKP propagator

The expression to the full DKP propagator can be derived starting from Schwinger variational equation

$$\left[\left.\frac{\delta S}{\delta\bar{\psi}(x)}\right|_{\frac{\delta}{\delta i\eta},\frac{\delta}{\delta i\bar{\eta}},\frac{\delta}{\delta iJ_\mu}} + \eta(x)\right]\mathscr{Z}\left[\eta,\bar{\eta},J_\mu\right] = 0. \tag{38}$$

Now, writing the equation (38) in terms of the generating functional W and then differentiating the resulting expression with respect to the source $\eta(y)$, we obtain

$$i\delta^{(4)}(x-y) = -\left[i\beta^\mu\partial_\mu - m + e\beta^\mu\langle A^\mu\rangle - ie\beta^\mu\frac{\delta}{\delta J_\mu(x)}\right]\mathscr{S}(x,y;A). \tag{39}$$

By solving the derivative of the last term we can identify

$$\Sigma(x,z) = ie^2\beta^\mu \int d^4u d^4v \mathfrak{D}^\alpha_\mu(u,x)\mathscr{S}(x,v;A)\Gamma_\alpha(v,z;u), \tag{40}$$

as the DKP self-energy function Σ. Hence, by taking the limit of null sources, we find that

$$i\delta^{(4)}(x-y) = -\left[i\beta^\mu\partial_\mu - m\right]\mathscr{S}(x,y;A) + \int d^4z\Sigma(x,z)\mathscr{S}(z,y;A). \tag{41}$$

In momentum representation the above equation becomes $\mathscr{S}^{-1}(p) = S^{-1}(p) + \Sigma(p)$. The later equation can be viewed as

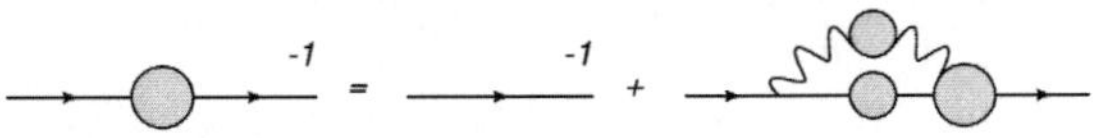

Figure 2. The SDF equation for the scalar propagator.

The equation (39) can formally be written as

$$\mathscr{S}(p) = \frac{i}{\beta^\mu p_\mu - \mathfrak{M}(p)} \tag{42}$$

where we defined the mass operator $\mathfrak{M}$,

$$\mathfrak{M}(p) = m + \Sigma(p). \tag{43}$$

Besides, the expression for the DKP free propagator can be obtained with help of the DKP algebra (6),

$$S(p) = i\frac{1}{m}\left[\frac{\hat{p}(\hat{p}+m)}{(p^2-m^2)} - 1\right], \quad \hat{p} = \beta^\mu p_\mu. \tag{44}$$

3.3. The Schwinger-Dyson-Fradkin equations for the vertex part

The starting point to get the vertex function is the equation (39). In a similar way, we take the derivative of the resulting expression with respect to the field $A_\sigma(z)$, and after some manipulations, we find the following expression for the vertex function

$$i\Gamma^\sigma(q,p;k) = -\beta^\sigma(2\pi)^4\delta(q-p-k) + i\Lambda^\sigma(k,p;q) \tag{45}$$

where we have introduced a new quantity, the vertex part

$$i\Lambda^\sigma(q,p;k) = ie^2\beta^\mu\frac{1}{(2\pi)^4}\int d^4t\mathfrak{D}_{\mu\rho}(t)\mathscr{S}(t+k)\Phi^{\sigma\rho}(t+k,p;q,t) +$$
$$+e^2\beta^\mu\frac{1}{(2\pi)^4}\int d^4p_1d^4p_2\mathfrak{D}_{\mu\rho}(p_1)\mathscr{S}(p_1+k)\Gamma^\sigma((p_1+k),p_2;q)\mathscr{S}(p_2)\Gamma^\rho(p_2,p;p_1), \tag{46}$$

in this expression we have defined the four-point vertex function

$$e^2\Phi^{\sigma\rho}(a,w;z,s) = \frac{\delta^4\Gamma}{\delta A_\rho(s)\delta A_\sigma(z)\delta\psi(w)\delta\bar{\psi}(a)}. \tag{47}$$

From Eq.(46) we clearly see that the three-point vertex function depends on the four-point one.

Diagrammatically, the irreducible vertex part can be visualized

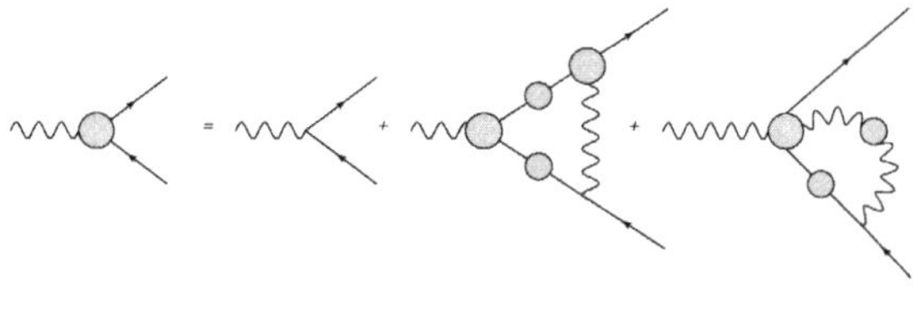

Figure 3. The SDF equation for the vertex function.

4. Ward-Takahashi-Fradkin identities

It is of our understanding that when we formulate the equations of electrodynamics in a covariant way we can describe them by a Lagrangian that has at its cornerstone a gauge symmetry. Classically the matter field (scalar) has a global U(1) gauge symmetry. When we impose a local gauge symmetry we need intermediaries fields of interaction (vector). Therefore when we write the dynamics of this electrodynamics (GSDKP) in a covariant way we have a local gauge symmetry U(1). Similarly we can describe the equations of quantum electrodynamics in a covariant way in the Heisenberg representation. On the other hand when we quantize the theory covariantly using the functional formalism we lost the gauge symmetry, to fix the physical degrees of freedom and define adequately the measure of integration $\int D\mu$. To maintain gauge symmetry we impose it in the functional generator and therefore this generates certain constraints on Green's functions, relationships between them known as Ward-Takahashi-Fradkin identities (WTF) [18]. The gauge symmetry is a consequence of a covariant quantum dynamics, the symmetry is implicit in the quantum equations of motion discussed previously.

The derivation of the WTF identities is formally given in terms of the following identity upon the functional generator

$$\left.\frac{\delta \mathscr{Z}\left[\eta,\bar{\eta},J_\mu\right]}{\delta\alpha(x)}\right|_{\alpha=0}=0. \tag{48}$$

This leads to the equation of motion satisfied by $\mathscr{Z}\left[\eta,\bar{\eta},J_\mu\right]$

$$\left[-i\frac{\Box}{e\xi}\left(1+a^2\Box\right)\partial^\mu\frac{\delta}{\delta J^\mu}-\frac{\delta}{\delta\eta}\eta+\bar{\eta}\frac{\delta}{\delta\bar{\eta}}-\frac{1}{e}\partial^\mu J_\mu\right]\mathscr{Z}=0. \tag{49}$$

Finally, one can get the desired quantum equation of motion for the theory by writing (49) first in terms of W, and then as an expression for the 1PI-generating functional $\Gamma\left[\psi,\bar{\psi},A_\mu\right]$ through the relation (31). Then

$$-i\frac{\Box}{e\xi}\left(1+a^2\Box\right)\partial_x^\mu A_\mu-\bar{\psi}\frac{\delta\Gamma}{\delta\bar{\psi}}+\frac{\delta\Gamma}{\delta\psi}\psi+\frac{1}{e}\partial_x^\mu\frac{\delta\Gamma}{\delta A^\mu}=0. \tag{50}$$

This is the equation that will supply all the WFT identities.

The first identity comes by applying the derivatives of (50) with respect to $\psi(y)$ and $\bar{\psi}(z)$, yielding

$$\partial^\mu\Gamma_\mu(z,y;x)=-\delta(x-z)\Gamma(x,y)+\Gamma(x,z)\delta(x-y). \tag{51}$$

In momentum representation

$$k_\mu\Gamma^\mu\left(p,p',k=p-p'\right)=\mathscr{S}^{-1}\left(p'\right)-\mathscr{S}^{-1}(p). \tag{52}$$

Furthermore, we may consider the limit of this equation as $k\to 0$, in which we find that the vertex part is related to the DKP self-energy function as

$$\Lambda^\mu(p,p,k=0)=-\frac{\partial}{\partial p_\mu}\Sigma^{-1}(p). \tag{53}$$

On the other hand, upon the differentiation of (50) with respect to $A_\nu(y)$, it follows the identity

$$\partial_\mu \Gamma^{\mu\nu}(x,y) = \frac{\Box}{\xi}\left(1+a^2\Box\right)\partial^\nu \delta^{(4)}(x-y) \tag{54}$$

which together with equation (33) implies that

$$k_\mu \mathfrak{P}^{\mu\nu}(k) = 0. \tag{55}$$

5. Conclusions

In this work we have analyzed the covariant dynamics of interaction between scalar particles and generalized photons. The first point to note is that we have implemented the non-mixing gauge in the constraint analysis. The second point to note is that we have used the DKP field to describe the scalar particles. The complete quantum structure of the scalar field, seen by means of SDF diagrams, is exactly as those from $GQED_4$. In this case the same diagram phenomenology for the electromagnetic interaction between scalar or fermionic fields happens. But the DKP fields are described by the DKP algebra (mesonic algebraic structure), while the fermionic field obeys a Clifford algebra (fermionic algebraic structure). This work complements previous studies on covariant quantum dynamics [19] and opens the door to a more complete approach, with the study of renormalization, first radiative corrections and quantum covariant dynamics in equilibrium (Matsubara-Fradkin formalism).

Acknowledgements

R.B. thanks FAPESP for full support, T.R.C. and A.A.N. thank CAPES for full support, B.M.P. thanks CAPES and CNPq for partial suport.

References

[1] Van Der Waerden B L 1967 *Sources of Quantum Mechanics* (North-Holland Publishing Company); Dirac P A M 1958 *The Principle of Quantum Mechanics* (Oxford University Press)
[2] Dirac P A M 1950 *Can. J. Math* **2** 129; Dirac P A M 1964 *Lectures on Quantum Mechanics* (Belter Graduate School of Science)
[3] Dirac P A M *Physikalische Zeitschrift der Sowjetunion* **Band 3 Heft 1** 64 (1933)
[4] Feynman R 2005 *A New Approach to Quantum Theory* (World Scientific)
[5] Schwinger J 2001 *Symbolism of Atomic Measurements* (Springer)
[6] Faddeev L D 1969 *Teor. Mat. Fiz* **1** 3; Faddeev L D 1969 *Theor. Math. Phys* **1** 1
[7] Senjanovic P 1976 *Ann. Phys* **100** 227; Erratum: Senjanovic P 1991 *Ann. Phys* **209** 248
[8] Fradkin E S and Vilkovisky G A 1977 *Quantization of relativistic systems with constraints: Equivalence of canonical and covariant formalisms in quantum theory of gravitational field* (CERN report TH-2332)
[9] Sundermeyer K 1982 *Constrained Dynamics* Lecture Notes in Physics **169** (Springer); Gitman D M and Tyutin I V 1990 *Quantization of Fields with Constraints* (Springer); Rothe H J and Rothe K D 2010 *Classical and quantum dynamics of constrained Hamiltonian systems* (World Scientific)
[10] Lunardi J T , Pimentel B M, Teixeira R G and Valverde J S 2000 *Phys. Lett.* A **268** 165
[11] Duffin R J 1938 *Phy. Rev* **54** 1114; Kemmer N 1939 *Proc. R. Soc. Lond.* A **173** 91; Petiau G 1936 *Académie Royale de Belgique* **16** 3
[12] Bopp F 1940 *Ann. Phys. (Leipzig)* **430** 345; Podolsky B 1942 *Phys. Rev* **62** 68; Podolsky B and Kikuchy C 1944 *Phys. Rev* **65** 228; Podolsky B and Schwed P 1948 *Rev. Mod. Phys* **20** 4
[13] Kinoshita T 1950 *Prog. Theor. Phys* **5** 473; Kinoshita T 1950 *Prog. Theor. Phys* **5** 749 (1950)
[14] Galvão C A P and Pimentel B M 1988 *Can. J. Phys* **66** 460
[15] Chivukula R S, Farzinnia A, Foadi R, and Simmons E H 2010 *Phys. Rev. D* **82** 035015; Bufalo R, Pimentel B M and Soto D E 2014 *Phys. Rev.* D **90** 085012 (2014)
[16] Faddeev L D and Popov V N 1967 *Phys. Lett.* B **25** 29; DeWitt B S 1967 *Phys. Rev* **160** 1113
[17] Nash C 1978 *Relativistic Quantum Fields* (Academic Press Inc, London); E. S. Fradkin 2007 *Selected Papers on Theoretical Physics* (Ed. I.V. Tyutin, Lebedev Institute, Moscow)
[18] Ward J C 1950 *Phys. Rev* **77** 293; Fradkin E S 1955 *Zh. Eksp. Teor. Fiz* **29** 258; Takahashi Y 1957 *Il Nuovo Cimento* **6** 371
[19] Bufalo R, Pimentel B M and Zambrano G E R 2011 *Phys. Rev.* D **83** 045007; Bufalo R, Pimentel B M and Zambrano G E R 2012 *Phys. Rev.* D **86** 125023; Bufalo R and Pimentel B M 2013 *Phys. Rev.* D **88** 125023

A QCDSR calculation of the $J/\psi D_s^* D_s$ strong coupling constant

B. Osório Rodrigues[1], M.E. Bracco[2], M. Chiapparini[1], A. Cerqueira Jr.[3]

[1] Instituto de Física, Universidade do Estado do Rio de Janeiro, Rua São Francisco Xavier 524, 20550-900, Rio de Janeiro, RJ, Brazil
[2] Faculdade de Tecnologia, Universidade do Estado do Rio de Janeiro, Rodovia Presidente Dutra km 298, 27537-000, Resende, RJ, Brazil
[3] Centro Universitário de Volta Redonda, UniFOA, Avenida Paulo Erlei Alves Abrantes 1325, 27240-560, Volta Redonda, RJ, Brazil

E-mail: angelo.cerqueira.cunha@gmail.com

Abstract. In this work, we evaluate the coupling constant and the form factors of the vertex $J/\psi D_s^* D_s$ using the techniques of the QCD sum rules. We consider all the three mesons off shell, resulting in three different form factors. However, despite the different form factors, when extrapolated to the pole of each off-shell meson, we find coupling constants that are in completely agreement each other. The result for the vertex $J/\psi D_s^* D_s$ coupling constant is $g_{J/\psi D_s^* D_s} = 4.30^{+0.42}_{-0.37} GeV^{-}1$.

1. Introduction

The QCD sum rules has been used in the calculation of the coupling constants and form factors of various vertices and a great variety of mesons. Our group has calculated a graet number of vertices as the $D^* D\pi$ [1], $D^* D\rho$ [2], $J/\psi D^* D$ [3] and many others. In recent works using the QCDSR, vertices with the heavy beauty mesons(B, B^*, B_s, B_s^*,...) were object of study as the $B_s^* BK$ [4], $B_s B^* K$ [5], $B_s^* B_s \eta^{(\prime)}$ [6]. Vertices with strange-charmed mesons(D_s, D_s^*, D_{s1},...) has been studied as the $J/\psi D_s D_s$ [7, 8], $J/\psi D_s^* D_s^*$ [8] and others.

The strange-charmed vertices are a subject of a great interest actually in view of the new exotic states detected recently. Some of them with masses that lies just above the $D_s^{(*)} D_s^{(*)}$ mass threshold. Exotic states means that the usual quark-model description as $q\bar{q}$ does not hold for the case of new states. An example of exotic state is the $Y(4140)$, whose its first detection was made by the CDF collaboration [9], it has an observed decay in the pair $J/\psi\phi$. There are some interpretations for its internal structure as the $D_s \bar{D}_s^*$ molecular state [10], $D_s^* \bar{D}_s^*$ molecular state [11, 12, 13, 14], the tetraquark state [15] and a hybrid state [11]. Among these interpretations, we emphasize the molecular $D_s^* \bar{D}_s^*$ hypothesis, so that the decay $Y(4140) \to J/\psi\phi$ can be understood as an intermediate decay as $Y(4140) \to D_s^* \bar{D}_s^* \to J/\psi\phi$. The vertex $J/\psi D_s^* D_s$ is present in this decay, so that a precise knowledge of its form factor and coupling constant may contribute for the understanding of the constitution of the $Y(4140)$ meson.

In this present work, the vertex $J/\psi D_s^* D_s$ is studied by applying the QCDSR formalism. The development of this work consists in evaluating the form factors from the three point correlation

Journal of Physics: Conference Series **706** (2016) 052003 doi:10.1088/1742-6596/706/5/052003

function with quark and gluons degrees of freedom(OPE side) and with hadronic degrees of freedom(phenomenological side); so, the next step is to perform a double Borel transform in both sides and equating them. From the calculated form factor, we obtain the coupling constant by extrapolating its value to the pole of the off-shell meson.

2. Formalism

In this work, we calculate three different three point correlation functions [16], each corresponding to each off-shell meson. The correlation functions corresponds respectively to J/ψ off-shell, D_s^* off-shell and D_s off-shell.

$$\Gamma_{\mu\nu}^{(J/\psi)}(p,p') = \int \langle 0'|T\{j_5^{D_s}(x) j_\mu^{J/\psi\dagger}(y) j_\nu^{D_s^*\dagger}(0)\}|0'\rangle\, e^{ip'x} e^{-iqy} d^4x d^4y, \quad (1)$$

$$\Gamma_{\mu\nu}^{(D_s^*)}(p,p') = \int \langle 0'|T\{j_5^{D_s}(x) j_\nu^{D_s^*\dagger}(y) j_\mu^{J/\psi\dagger}(0)\}|0'\rangle\, e^{ip'x} e^{-iqy} d^4x d^4y, \quad (2)$$

$$\Gamma_{\mu\nu}^{(D_s)}(p,p') = \int \langle 0'|T\{j_\nu^{D_s^*}(x) j_5^{D_s\dagger}(y) j_\mu^{J/\psi\dagger}(0)\}|0'\rangle\, e^{ip'x} e^{-iqy} d^4x d^4y, \quad (3)$$

where $q = p' - p$ is the transferred momentum.

Firstly, these correlation functions are calculated using the quarks and gluons degrees of freedom(OPE side) and using the hadron degrees of freedom(phenomenological side). Then, the form factors and the coupling constants are obtained by equating both sides using the quark-hadron duality, after applying a double Borel transform in both sides.

2.1. The phenomenological side

The effective Lagrangian of $J/\psi D_s^* D_s$ vertex [17, 18] is given by

$$\mathcal{L}_{J/\psi D_s^* D_s} = -g_{J/\psi D_s^* D_s} \varepsilon^{\alpha\beta\gamma\delta} \partial_\alpha \psi_\beta \left(\partial_\gamma D_{s\delta}^{*-} D_s^+ + \partial_\gamma D_{s\delta}^{*+} D_s^- \right), \quad (4)$$

where $\varepsilon^{0123} = +1$ is the Levi-Civita totally antisymmetric tensor.

From this Lagrangian, we obtain the vertices of the hadronic processes that are used to the calculation of the phenomenological side . For the J/ψ, D_s^* and D_s off-shell cases, these vertices are, respectively:

$$\langle D_s^*(p) J/\psi(q)|D_s(p')\rangle = i g_{J/\psi D_s^* D_s}^{(J/\psi)}(q^2) \epsilon_\beta(q,\lambda) \epsilon_\delta(p,\lambda) q_\alpha p_\gamma \varepsilon^{\alpha\beta\gamma\delta}\,, \quad (5)$$

$$\langle J/\psi(p) D_s^*(q)|D_s(p')\rangle = i g_{J/\psi D_s^* D_s}^{(D_s^*)}(q^2) \epsilon_\beta(p,\lambda) \epsilon_\delta(q,\lambda) p_\alpha q_\gamma \varepsilon^{\alpha\beta\gamma\delta}\,, \quad (6)$$

$$\langle J/\psi(p) D_s(q)|D_s^*(p')\rangle = -i g_{J/\psi D_s^* D_s}^{(D_s)}(q^2) \epsilon_\beta(p,\lambda) \epsilon_\delta^*(p',\lambda) p_\alpha p'_\gamma \varepsilon^{\alpha\beta\gamma\delta}\,, \quad (7)$$

where $g_{J/\psi D_s^* D_s}^{(M)}(q^2)$ is the form factor of the $J/\psi D_s^* D_s$ vertex with meson M off-shell ($M = J/\psi, D_s, D_s^*$).

Using each of these vertices in the corresponding correlation function, we can obtain the phenomenological side. It is convenient to make the change of variables $p^2 \to -P^2$, $p'^2 \to -P'^2$ and $q^2 \to -Q^2$. Then, we have the correlation functions of the phenomenological side for each

Journal of Physics: Conference Series **706** (2016) 052003 doi:10.1088/1742-6596/706/5/052003

off-shell case:

$$\Gamma^{phen(J/\psi)}_{\mu\nu} = \frac{C g^{(J/\psi)}_{J/\psi D_s^* D_s}(q^2) p'^\lambda p^\sigma \varepsilon_{\mu\nu\lambda\sigma}}{(p^2+m^2_{D_s^*})(q^2+m^2_{J/\psi})(p'^2+m^2_{D_s})} + h.r.\,, \tag{8}$$

$$\Gamma^{phen(D_s^*)}_{\mu\nu} = \frac{-C g^{(D_s^*)}_{J/\psi D_s^* D_s}(q^2) p'^\lambda p^\sigma \varepsilon_{\mu\nu\lambda\sigma}}{(p^2+m^2_{J/\psi})(q^2+m^2_{D_s^*})(p'^2+m^2_{D_s})} + h.r.\,, \tag{9}$$

$$\Gamma^{phen(D_s)}_{\mu\nu} = \frac{C g^{(D_s)}_{J/\psi D_s^* D_s}(q^2) p'^\lambda p^\sigma \varepsilon_{\mu\nu\lambda\sigma}}{(p^2+m^2_{J/\psi})(q^2+m^2_{D_s})(p'^2+m^2_{D_s^*})} + h.r.\,, \tag{10}$$

where $h.r.$ stands for the contributions of higher resonances and continuum states of each meson and C is defined as:

$$C = \frac{f_{D_s} f_{D_s^*} f_{J/\psi} m^2_{D_s} m_{D_s^*} m_{J/\psi}}{(m_c+m_s)}\,. \tag{11}$$

2.2. The OPE side

In order to obtain the OPE side; in the eqs. (1), (2) and (3), we use the interpolating currents in terms of quark fields: $j^{J/\psi}_\mu = \bar{c}\gamma_\mu c$, $j^{D_s^{*-}}_\nu = \bar{c}\gamma_\nu s$ and $j^{D_s^-}_5 = i\bar{c}\gamma_5 s$. The OPE side is an expansion named as Wilson's Operator Product Expansion. This expansion is dominated by the perturbative term and followed by the non-perturbative contributions:

$$\Gamma^{OPE(M)}_{\mu\nu} = \Gamma^{pert(M)}_{\mu\nu} + \Gamma^{non\text{-}pert(M)}_{\mu\nu}\,, \tag{12}$$

where $\Gamma^{pert(M)}_{\mu\nu}$ is the perturbative term and $\Gamma^{non\text{-}pert(M)}_{\mu\nu}$ are the non-perturbative contributions to the correlation function. Considering the similarities between the $J/\psi D_s^* D_s$ and both $J/\psi D_s D_s$ and $J/\psi D^* D$ vertices, we expect a similar behavior regarding the OPE series as obtained in the two latter works [3, 7].Therefore, it should be adequate to consider non-perturbative contributions up to the mixed quark-gluon condensate:

$$\Gamma^{non\text{-}pert}_{\mu\nu} = \Gamma^{\langle\bar{q}q\rangle}_{\mu\nu} + \Gamma^{m_q\langle\bar{q}q\rangle}_{\mu\nu} + \Gamma^{\langle g^2G^2\rangle}_{\mu\nu} + \Gamma^{\langle\bar{q}g\sigma Gq\rangle}_{\mu\nu} + \Gamma^{m_q\langle\bar{q}g\sigma Gq\rangle}_{\mu\nu}\,. \tag{13}$$

The calculation of eq. (12) regarding the perturbative and the non-perturbative contributions of eq. (13) corresponds to the calculation of the diagrams of fig. 1. The J/ψ off-shell case is the only that has contributions from all the non-perturbative terms as a consequence of the application of the Borel transform. In the D_s and D_s^* off-shell cases the non-perturbative contributions are suppressed because these cases involves the condensates of charm quark that are very small or even zero.

Using the dispersions relations, the perturbative term (fig. 1a) for a given meson M off-shell can be written as:

$$\Gamma^{pert(M)}_{\mu\nu}(p,p') = -\frac{1}{4\pi^2}\int_0^\infty\int_0^\infty \frac{\rho^{pert(M)}_{\mu\nu}(s,u,t)}{(s-p^2)(u-p'^2)} ds du\,, \tag{14}$$

where the spectral density $\rho_{\mu\nu}^{pert(M)}(s,u,t)$ is obtained by the application of Cutkosky's rules. The quantities $s=p^2$, $u=p'^2$ and $t=q^2$ are the Mandelstam variables.

The eq. (14) is the main contributing term of the OPE series in a QCDSR calculation. The spectral density can be parametrized as:

$$\rho_{\mu\nu}^{pert(M)}(s,u,t) = \frac{3}{\sqrt{\lambda}} F^{(M)}(s,u,t) p'^{\lambda} p^{\sigma} \varepsilon_{\mu\nu\lambda\sigma}\,, \tag{15}$$

where $\lambda = (u+s-t)^2 - 4us$ and $F^{(M)}$ is an invariant amplitude. For the J/ψ and D_s off shell, the invariant amplitude can be written as:

$$F^{(J/\psi)} = (m_s - m_c)(A+B) - m_s\,, \tag{16}$$

$$F^{(D_s)} = -F^{(D_s^*)} = (m_c - m_s)B - m_c\,, \tag{17}$$

where

$$A = \left[\frac{k_0}{\sqrt{s}} - \frac{p_0'\overline{|\vec{k}|}\overline{\cos\theta}}{|\vec{p'}|\sqrt{s}}\right], \qquad B = \frac{\overline{|\vec{k}|}\overline{\cos\theta}}{|\vec{p'}|}\,. \tag{18}$$

The quark condensate in fig. 1b corresponds to the first non-perturbative contributions:

$$\Gamma_{\mu\nu}^{\langle\bar{s}s\rangle(J/\psi)} = \frac{-\langle\bar{s}s\rangle p'^{\lambda} p^{\sigma}\varepsilon_{\mu\nu\lambda\sigma}}{(p^2 - m_c^2)(p'^2 - m_c^2)}\,. \tag{19}$$

The expressions for the gluon condensates ($\langle g^2G^2\rangle$) of fig. 1d-1i and mixed quark-gluon condensates ($\langle\bar{s}g\sigma\cdot Gs\rangle$) of fig. 1j-1o for the J/ψ off-shell case can be found in Ref. [23].

2.3. The QCD sum rules

In order to obtain the expressions for the form factors, we make the change of variables $p^2 \to -P^2$, $p'^2 \to -P'^2$ and $q^2 \to -Q^2$ followed by a double Borel transform to both sides of the QCDSR in eqs. (8)-(10) and (12), which involves the transformation: $P^2 \to M^2$ and $P'^2 \to M'^2$, where M and M' are the Borel masses. After that, we equate the phenomenological and OPE sides, taking the quark-hadron duality. So, we obtain the QCDSR expressions for the form factor for each case off-shell:

$$g_{J/\psi D_s^* D_s}^{(J/\psi)}(Q^2) = \frac{-\frac{3}{4\pi^2}\int_{s_{inf}}^{s_0}\int_{u_{inf}}^{u_0}\frac{1}{\sqrt{\lambda}}F^{(J/\psi)}e^{-\frac{s}{M^2}}e^{-\frac{u}{M'^2}}dsdu + \mathcal{BB}\left[\Gamma^{non\text{-}pert}\right]}{\frac{C}{(Q^2+m_{J/\psi}^2)}e^{-m_{D_s^*}^2/M^2}e^{-m_{D_s}^2/M'^2}}\,, \tag{20}$$

$$g_{J/\psi D_s^* D_s}^{(D_s^*)}(Q^2) = \frac{\frac{3}{4\pi^2}\int_{s_{inf}}^{s_0}\int_{u_{inf}}^{u_0}\frac{1}{\sqrt{\lambda}}F^{(D_s^*)}e^{-\frac{s}{M^2}}e^{-\frac{u}{M'^2}}dsdu + \mathcal{BB}\left[\Gamma^{\langle g^2G^2\rangle}\right]}{\frac{C}{(Q^2+m_{D_s^*}^2)}e^{-m_{J/\psi}^2/M^2}e^{-m_{D_s}^2/M'^2}}\,, \tag{21}$$

$$g_{J/\psi D_s^* D_s}^{(D_s)}(Q^2) = \frac{-\frac{3}{4\pi^2}\int_{s_{inf}}^{s_0}\int_{u_{inf}}^{u_0}\frac{1}{\sqrt{\lambda}}F^{(D_s)}e^{-\frac{s}{M^2}}e^{-\frac{u}{M'^2}}dsdu + \mathcal{BB}\left[\Gamma^{\langle g^2G^2\rangle}\right]}{\frac{C}{(Q^2+m_{D_s}^2)}e^{-m_{J/\psi}^2/M^2}e^{-m_{D_s^*}^2/M'^2}}\,, \tag{22}$$

As the QCDSR, the definition for the coupling constant $g_{J/\psi D_s^* D_s}$ is given by:

$$\lim_{Q^2\to -m_M^2} g_{J/\psi D_s^* D_s}^{(M)}(Q^2)\,. \tag{23}$$

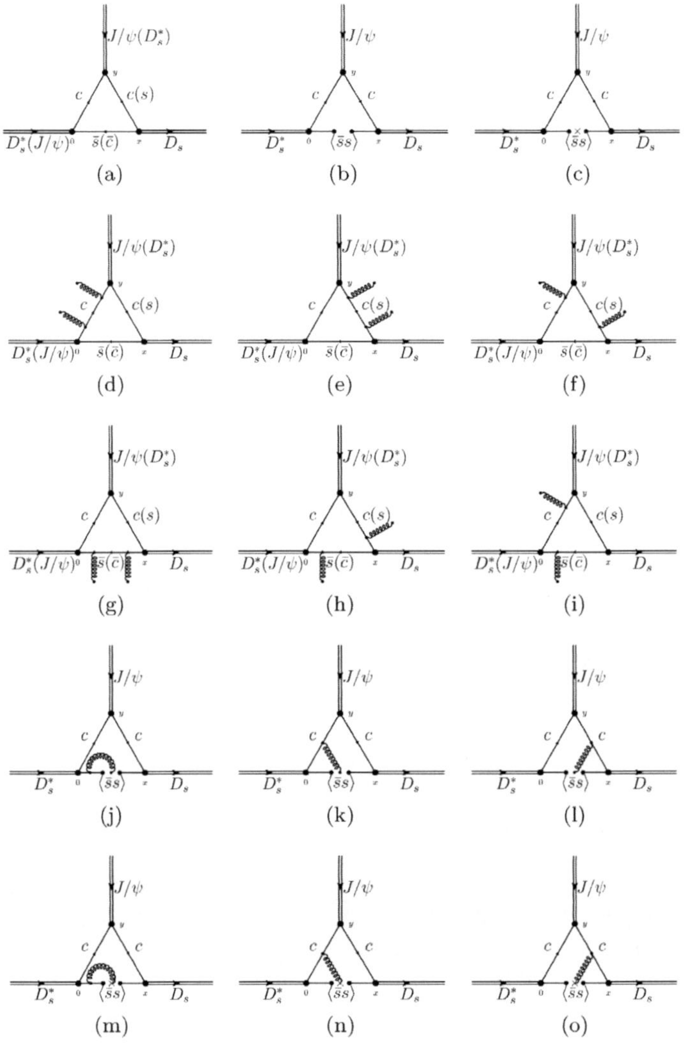

Figure 1: Contributing OPE diagrams for $J/\psi(D_s^*)$ off-shell. The D_s off-shell case corresponds to a permutation of the D_s and D_s^* mesons of the D_s^* off-shell case.

To calculate the coupling constants, it is necessary to extrapolate the results for the form factor to the region of $Q^2 < 0$. From eqs. (20)-(22) we can obtain the coupling constant from three different form factors, one for each meson off-shell. However, the coupling constant must be the same regardless the form factor used for the extrapolation. This condition is used to minimize the uncertainties in the determination of the coupling constant.

3. Results and discussion

The vertex $J/\psi D_s^* D_s$ has three different form factors shown in eqs. (20)-(22) . We perform numerical calculation of these form factors that gives results that must be fitted to an analytical function of Q^2. To minimize the uncertainties, it is required that these three form factors lead to the same coupling constant. This condition is useful to find the Borel masses and continuum thresholds and to reduce the errors.

Table 1: Parametrization of the form factors and numerical results for the coupling constant of this work.

Quantity	J/ψ off-shell	D_s^* off-shell	D_s off-shell
Q^2 (GeV^2)	[0.5, 2.0]	[1.0, 3.0]	[1.0, 4.0]
M^2 (GeV^2)	[7.0, 7.9]	[5.3, 7.3]	[4.9, 5.9]
$g^{(M)}_{J/\psi D_s^* D_s}(Q^2)$	$\frac{A}{B+Q^2}$	$Ae^{-Q^2/B}$	$Ae^{-Q^2/B}$
A	193.4 GeV	2.003 GeV^{-1}	2.330 GeV^{-1}
B (GeV^2)	54.46	6.027	6.122
$g^{(M)}_{J/\psi D_s^* D_s} \pm \sigma$ (GeV^{-1})	$4.31^{+1.21}_{-1.20}$	$4.20^{+1.32}_{-1.12}$	$4.39^{+1.44}_{-1.30}$

Table (1) shows the parameters used in this calculation. The continuum threshold parameters, s_0 and u_0, are defined as $s_0 = (m_i + \Delta_i)^2$ and $u_0 = (m_o + \Delta_o)^2$, where the quantities Δ_i and Δ_o have been determined imposing the most stable Borel window. In order to include the pole and to exclude the contributions from higher resonances and continuum states, the values for $\Delta_{J/\psi}$, $\Delta_{D_s^*}$ and Δ_{D_s} cannot be far from the experimental value (when available) of the distance between the pole and the first excited state [19, 24]. In our analysis, we have found that the best values are $\Delta_{D_s} = 0.6$ GeV and $\Delta_{D_s^*} = \Delta_{J/\psi} = 0.5$ GeV, which leads to a stable Borel windows for the three off-shell cases, as shown in fig. 2 for J/ψ and D_s off-shell. Figures for D_s^* off-shell are omitted as they are very similar to the D_s off-shell case.

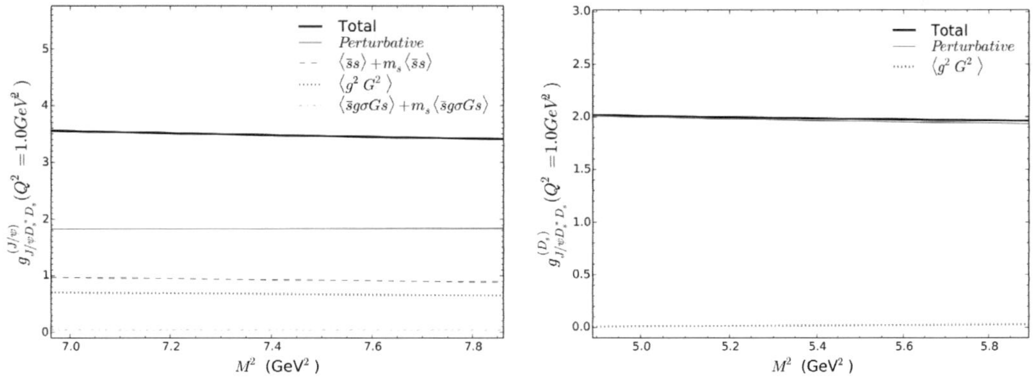

(a) (b)

Figure 2: OPE contributions for J/ψ off-shell (panel **(a)**) and D_s off-shell (panel **(b)**).

Besides the values of the Borel masses and the continuum thresholds, we also need to know the values of decay constants, quark masses, condensates and hadrons and quark masses. These values are: $f_{D_s} = 257.15 \pm 6.1(MeV)$ [19], $f_{D_s^*} = 301 \pm 13(MeV)$ [19], $f_{J/\psi} = 416 \pm 6(MeV)$ [19], $m_c = 1.27^{+0.07}_{-0.09}$ (GeV) [19], $m_s = 101^{+29}_{-21}$ (MeV)[19], $\langle \bar{s}s \rangle = -(290 \pm 15)^3$ (MeV^3)[25], $\langle g^2G^2 \rangle = 0.88 \pm 0.16$(GeV^4)[26], $\langle \bar{s}g\sigma \cdot Gs \rangle = (0.8 \pm 0.2)\langle \bar{s}s \rangle$(GeV^5)[27]. The masses of hadrons of this vertex are $m_{D_s} = 1.968$ GeV, $m_{D_s^*} = 2.112$ GeV and $m_{J/\psi} = 3.097$ GeV [19].

The form factor obtained for the J/ψ off-shell case was well adjusted by a monopolar curve, while for the D_s^* and D_s off-shell cases, the form factors were well adjusted by exponential curves in fig. 4. The coupling constant of the J/ψ off-shell is $g^{(J/\psi)}_{J/\psi D_s^* D_s} \pm \sigma(GeV^{-1}) = 4.31^{+1.21}_{-1.20} GeV^{-1}$.

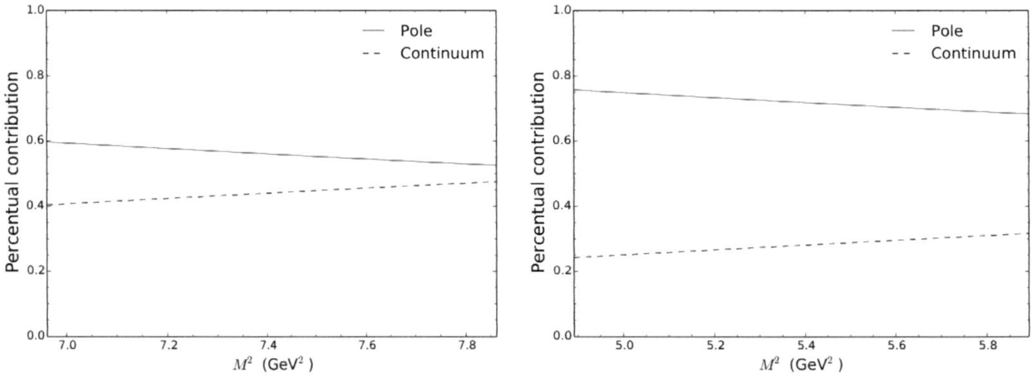

(a) (b)

Figure 3: Pole and continuum contributions for the J/ψ off-shell (panel **(a)**) and for D_s off-shell (panel **(b)**), both at $Q^2 = 1$ GeV^2.

The value found for the D_s^* off-shell case is $g^{(D_s^*)}_{J/\psi D_s^* D_s} \pm \sigma(GeV^{-1}) = 4.20^{+1.32}_{-1.12} GeV^{-1}$ and the coupling constant of the D_s off-shell case is $g^{(D_s)}_{J/\psi D_s^* D_s} \pm \sigma(GeV^{-1}) = 4.39^{+1.44}_{-1.30} GeV^{-1}$.

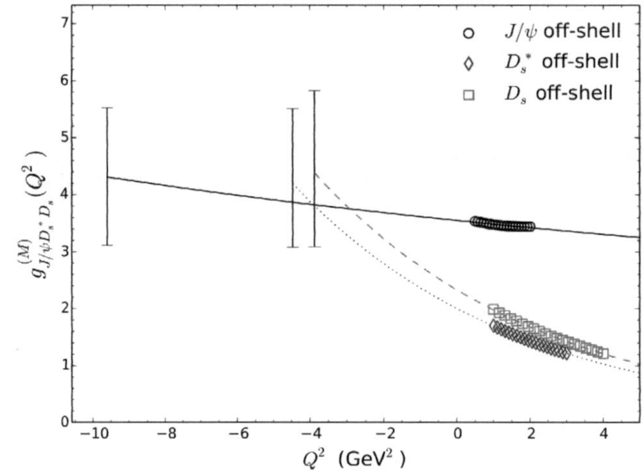

Figure 4: Form factors of the $J/\psi D_s^* D_s$ vertex.

4. Conclusion

In this work we have calculated the $g_{J/\psi D_s^* D_s}$ coupling constant by three different QCD sum rules: one with the D_s meson off-shell, another with the D_s^* meson off-shell and a third one with the J/ψ meson off-shell. This procedure allowed us to reduce the uncertainties related to the method, leading to compatible coupling constants, as seen in fig. 4.

Taking the mean value between the numbers presented in table 1, we obtain the following final result for $g_{J/\psi D_s^* D_s}$:

$$g_{J/\psi D_s^* D_s} = 4.30^{+1.53}_{-1.22}\ \mathrm{GeV}^{-1}\ . \tag{24}$$

This coupling constant was obtained from sum rules that respect the pole dominance over the continuum, the perturbative contribution of the OPE being the dominant one and the form factor stability regarding the Borel mass in the whole Borel window, as shown in fig. 3.

In these results, there is the form factor given by a monopolar parametrization when the heaviest meson (J/ψ) is off-shell, while an exponential one is the case when one of the lightest mesons (D_s or D_s^*) is off-shell. Furthermore, the OPE series presents, in both vertices, the same hierarchy for the contribution of each term, as well as comparable contributions among terms of the same dimension.

We can also compare our result for the $g_{J/\psi D_s^* D_s}$ coupling constant eq. (24) with the results of previous QCDSR works, presented in table 2.

Table 2: Values of coupling constants obtained using different methods.

QCDSR [20] $(GeV)^{-1}$	QCDSR [3] $(GeV)^{-1}$	QCDSR [7]	VMD [21, 22]
$g_{J/\psi D_s^* D_s} = 3.03 \pm 0.62$	$g_{J/\psi D^* D} = 4.0 \pm 0.6$	$g_{J/\psi D_s D_s} = 5.98^{+0.67}_{-0.58}$	$g_{J/\psi DD} = 7.44$

Finally, we can compare our result with the coupling constant $g_{J/\psi DD}$ from the vector meson dominance (VMD) [21, 22] using the SU(3) and HQET relations, which leads to $g_{J/\psi D_s^* D_s} = 3.84$ GeV^{-1}. This coupling is also compatible with the one of eq. (24) within 1σ.

References

[1] Navarra F S, Nielsen M and Bracco M E 2002 *Phys. Rev. D* **65** 037502
[2] Rodrigues B O et al 2011 *Nucl. Phys. A* **852** 127
[3] Matheus R D et al 2005 *Int. J Mod. Phys. E* **14** 555
[4] Cerqueira Jr A, Rodrigues B O and Bracco M E 2012 *Nucl. Phys. A* **874** 130
[5] Cerqueira Jr A, Rodrigues B O, Bracco M and Nielsen M 2015 *Nucl. Phys. A* **936** 45-58
[6] Yazici E et al 2013 *Eur. Phys. J Plus* **128** 113
[7] Rodrigues B O, Bracco M and Chiapparini M 2014 *Nucl. Phys. A* **929** 143
[8] Khosravi R, Janbazi M 2013 *Phys. Rev. D* **87** 016003
[9] Aaltonen T et al (CDF Collaboration) 2009 *Phys. Rev. Lett.* **102** 242002
[10] Qiao, Cong-Feng, Tang and Liang 2014 *EPL* **107** 31001
[11] Mahajan N 2009 *Phys. Lett. B* **679** 228
[12] Albuquerque R M, Bracco M E and Nielsen M 2009 *Phys. Lett. B* **678** 186
[13] Liu X, Zhu S L and Nielsen M 2009 *Phys. Rev. D* **80** 017502
[14] Wang Z G 2009 *Eur. Phys. J* **63** 115
[15] Stancu F 2010 *J Phys. G Nucl. Part. Phys.* **37** 075017
[16] Bracco M E et al 2012 *Prog. Part. Nucl. Phys.* **67** 1019
[17] Liu X, Zhang B and Zhu S L 2007 *Phys. Lett. B* **645** 185
[18] Azevedo R S and Nielsen M 2004 *Phys. Rev. C* **69** 035201
[19] Nakamura K et al 2010 *J Phys. G Nucl. Par. Phys* **37** 1422
[20] Khosravi R, Janbazi M 2013 *Phys. Rev. D* **89** 016001
[21] Oh Y, Song T and Lee S H 2001 *Phys. Rev. C* **63** 034901
[22] Oh Y, Song T and Lee S H 2007 *Phys. Rev. C* **75** 064903
[23] Rodrigues B O, Bracco M E, Chiapparini M and Cerqueira Jr A 2015 *Eur. Phys. J***51** 1-8
[24] Badalian A M and Bakker B L G 2011 *Phys. Rev. D***84** 034006
[25] McNeile C et al 2013 *Phys. Rev. D***87** 034503
[26] Narison S 2012 *Phys. Lett. B***706** 412
[27] Ioffe B 2006 *Prog Part. Nucl. Phys.***56** 232

Gluon Schwinger-Dyson equation in the PT-BFM scheme

A. M. S. Narciso Ferreira and A. C. Aguilar

University of Campinas - UNICAMP, Institute of Physics "Gleb Wataghin",
13083-859 Campinas, SP, Brazil

E-mail: mnarciso@ifi.unicamp.br

Abstract. Schwinger-Dyson equations provide an appropriate framework for tackling non-perturbative QCD phenomena requiring a continuum treatment. However, an inadequate truncation of this tower of integral equations can compromise the symmetries underlying the theory in question. The synthesis of the Pinch Technique and the Background Field method provides a framework where it is possible to devise a self-consistent truncation scheme, exploiting the Ward identities satisfied by the effective Green's functions that emerge. In this work we review how this truncation scheme is implemented, and show that the new series of dressed diagrams for the background gluon propagator organizes itself in characteristic subsets that are *individually transverse*. In addition, we discuss how the Background Quantum identity connects the background gluon propagator with the conventional one, computed in the lattice simulations.

1. Introduction

A variety of interesting QCD phenomena, such as chiral symmetry breaking, formation of bound states, and dynamical mass generation for quarks and gluons occur in the non-perturbative regime of the theory [1, 2, 3]. In particular, it is well know that the symmetries of the QCD Lagrangian prohibit a gluon mass term, and therefore its appearance can only happen through purely dynamical effects of non-perturbative nature [4]. Those effects must be triggered by a subtle mechanism which will guarantee that the local gauge invariance remains intact [3].

The most widely used continuum approach for the study of non-perturbative QCD is based on the Schwinger-Dyson equations (SDEs). The SDEs consist in an infinite system of coupled integral equations connecting all Green's functions of the theory. A major difficulty in dealing with this system is related to the fact that it is necessary to truncate the series at some level. In general, a good truncation scheme is considered to be one where a tractable subset of these equations is selected and, at the same time, the underlying symmetries of the theory are not compromised. Most importantly, the gauge invariance must be kept intact at all stages of any truncated treatment [5, 6, 7].

Here we are particularly interested in the non-perturbative behavior of the gluon self-energy. The advent of a self-consistent truncation scheme that preserves the transversality of the gluon self-energy is a highly non-trivial task, mainly because the fully dressed vertices appearing in the standard gluon SDE satisfy complicated Slavnov-Taylor identities (STIs). As a result, the transversality of the gluon self-energy emerges only after the contributions of all diagrams appearing in Fig. 1 have been included [3].

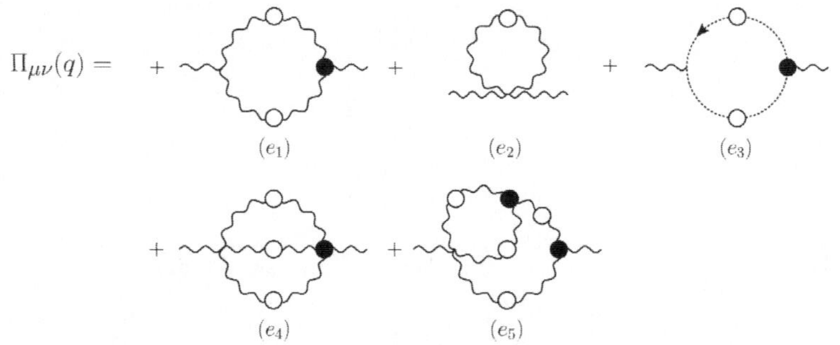

Figure 1. The conventional gluon self-energy. Wavy lines with white blobs represent full gluon propagators, while dashed lines with white blobs are full quantum ghost ones. The black blobs represent full vertices.

The formulation of the SDE based on the synthesis of the Pinch Technique (PT) [4, 5, 8] and the Background Field method (BFM) [9], denoted simply as PT-BFM scheme [3, 6, 7], provides a suitable framework for eliminating this drawback. Essentially, this special framework allows the construction of a new SDE series for the gluon self-energy, where the dressed diagrams are organized in a natural way into independently transverse subgroups, thus leading to a transversality preserving truncation scheme [5, 6, 7]. The reason why such a truncation becomes possible may be traced back to the fact that the PT-BFM Green's functions satisfy Abelian Ward identities (WIs), in contradistinction to the ghost-infested STIs satisfied by the conventional functions; for more details see Ref. [5].

In this work we will review the main results obtained in Ref. [6], where the PT-BFM truncation scheme for the gluon SDE was first presented. To do that, in the Section 2 we introduce all the WIs satisfied by the vertices appearing in the PT-BFM gluon SDE. Then, in Section 3 we show how the transversality in blocks emerges within this formalism. In Section 4 we discuss how the background quantum identity (BQI) [10] connects the two distinct gluon propagators: *i.e.* the conventional and the PT-BFM one, appearing in the resulting gluon SDE [5, 6, 7]. Finally, in Section 5 we summarize our conclusions.

2. The PT-BFM formalism and the WIs

In the PT-BFM framework one distinguishes between three types of gluon propagators, depending on the nature of the the external gluon legs [5]. The legs can be formed either by a background (B) or by a quantum (Q) gluon. More specifically, the BB gluon propagator, to be denoted $\widehat{\Delta}_{\mu\nu}(q)$, is formed by two B gluons. On the other hand, the BQ propagator, $\widetilde{\Delta}_{\mu\nu}(q)$, is composed by one Q and other B gluon fields. Finally, the conventional propagator, $\Delta_{\mu\nu}(q)$, is formed by QQ gluons.

In general covariant R_ξ gauges, the BB gluon propagator is defined as

$$\widehat{\Delta}_{\mu\nu}(q) = -i\left[P_{\mu\nu}(q)\widehat{\Delta}(q^2) + \xi\frac{q_\mu q_\nu}{q^4}\right], \tag{1}$$

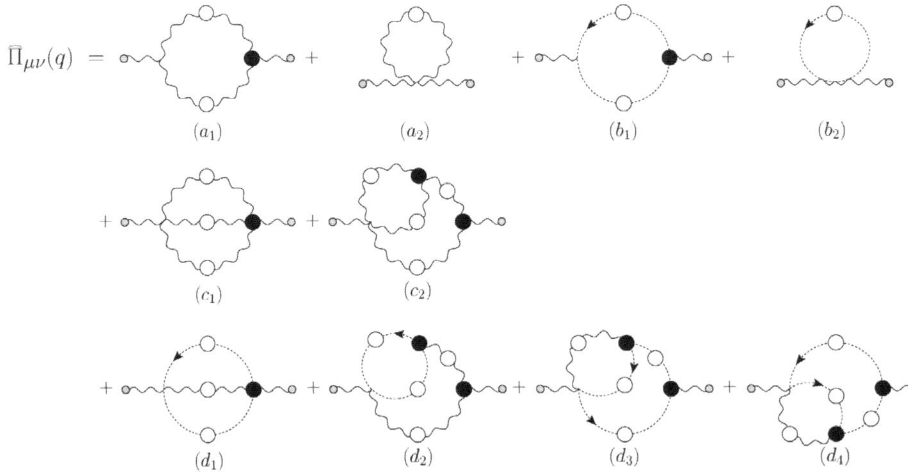

Figure 2. The PT-BFM gluon self-energy. Wavy lines with white blobs represent full quantum gluon propagators, while dashed lines with white blobs are full quantum ghost ones. External lines ending with a small gray circle are background gluons. The black blobs represent full vertices.

where the transverse projector reads

$$P_{\mu\nu}(q) = g_{\mu\nu} - \frac{q_\mu q_\nu}{q^2}. \tag{2}$$

The scalar co-factor $\widehat{\Delta}(q^2)$ is related to the all-order BB self-energy $\widehat{\Pi}_{\mu\nu}(q)$ by

$$\widehat{\Pi}_{\mu\nu}(q) = P_{\mu\nu}(q)\widehat{\Pi}(q^2); \qquad \widehat{\Delta}^{-1}(q^2) = q^2 + i\widehat{\Pi}(q^2), \tag{3}$$

where the diagrammatic representation of $\widehat{\Pi}_{\mu\nu}(q)$ is given in Fig. 2.

On the other hand, the definition for the conventional propagator $\Delta_{\mu\nu}(q)$ can be obtained by just omitting the hats in the above equations. However, in this case the diagrammatic representation of the self-energy $\Pi_{\mu\nu}(q)$ is given in Fig. 1.

Let us now concentrate on the structure of the PT-BFM self-energy given in Fig. 2. We notice that $\widehat{\Pi}_{\mu\nu}(q)$ is built out of ten dressed diagrams, to be grouped into the following four categories: (a) one-loop dressed gluonic diagrams containing gluons only [diagrams (a_1) and (a_2)], (b) one-loop dressed diagrams with internal ghost lines [(b_1) and (b_2)], (c) two-loop dressed gluonic diagrams [(c_1) and (c_2)], and finally, (d) the two-loop dressed ghost contribution [(d_1), (d_2), (d_3), (d_4)].

Here we will show that the aforementioned groups of diagrams are separately transverse [6]. In other words, we will demonstrate that for each subgroup

$$q^\mu \, \widehat{\Pi}^{ab}_{\mu\nu}\Big|_{(i)} = 0, \tag{4}$$

where q^μ is the gluon momentum, and $i = a$, b, c or d refers to each group of diagrams appearing in Fig. 2.

Notice that $\Pi_{\mu\nu}(q)$ of Fig. 2 contains vertices of the type BQQ, $B\bar{c}c$, $BQQQ$, and $BQ\bar{c}c$, to be denoted $\mathbf{\Gamma}_{\mu\alpha\beta}$, $\mathbf{\Gamma}_{\mu}$, $\mathbf{\Gamma}_{\mu\nu\alpha\beta}$ and $\mathbf{\Gamma}_{\nu\alpha\beta}$, respectively. The trilinear vertices satisfy the following

WIs [3, 5, 6]

$$
\begin{aligned}
q_1^\mu \widetilde{\mathbf{\Gamma}}^{abc}_{\mu\alpha\beta}(q_1,q_2,q_3) &= g f^{abc}\left[\Delta^{-1}_{\alpha\beta}(q_2) - \Delta^{-1}_{\alpha\beta}(q_3)\right],\\
q_1^\mu \widetilde{\mathbf{\Gamma}}^{acb}_{\mu}(q_1,q_2,q_3) &= g f^{abc}\left[D^{-1}(q_2) - D^{-1}(q_3)\right],
\end{aligned} \tag{5}
$$

where in the RHS of these expressions appear the inverse of the all-order QQ gluon, $\Delta_{\alpha\beta}(q)$, and ghost, $D(q)$, propagators.

Similarly, the four-point Green's functions, $\widetilde{\mathbf{\Gamma}}_{\mu\nu\alpha\beta}$ and $\widetilde{\mathbf{\Gamma}}_{\mu\nu}$, satisfy the WIs [3, 5, 6]

$$
\begin{aligned}
q_1^\mu \widetilde{\mathbf{\Gamma}}^{abcd}_{\mu\nu\alpha\beta}(q_1,q_2,q_3,q_4) &= ig f^{abx}\mathbf{\Gamma}^{cdx}_{\alpha\beta\nu}(q_3,q_4,q_1+q_2)\\
&\quad + ig f^{acx}\mathbf{\Gamma}^{dbx}_{\beta\nu\alpha}(q_4,q_2,q_1+q_3)\\
&\quad + ig f^{adx}\mathbf{\Gamma}^{bcx}_{\nu\alpha\beta}(q_2,q_3,q_1+q_4),
\end{aligned} \tag{6}
$$

and

$$
\begin{aligned}
q_1^\mu \widetilde{\mathbf{\Gamma}}^{cdba}_{\mu\nu}(q_1,q_2,q_3,q_4) &= -ig f^{cdx}\mathbf{\Gamma}^{axb}_{\nu}(q_4,q_2+q_1,q_3)\\
&\quad - ig f^{cbx}\mathbf{\Gamma}^{adx}_{\nu}(q_4,q_2,q_3+q_1)\\
&\quad - ig f^{cax}\mathbf{\Gamma}^{xdb}_{\nu}(q_4+q_1,q_2,q_3),
\end{aligned} \tag{7}
$$

where on the RHS we have the all-order conventional QQQ ($\Gamma_{\nu\alpha\beta}$) and $Q\bar{c}c$ (Γ_ν) vertices.

3. Transversality Property of the gluon SDE

Now we are in position to show how the separation of the diagrams into independently transverse groups emerges. To that end, we will exploit the rearrangements that the above WIs enforce, when the vertices in the diagrams are contracted by the external gluon momentum [6].

Let us start with the contraction of diagram (a_1) by q^ν; with the help of the first line of Eq. (5) we can write

$$
\begin{aligned}
q^\nu\, \widehat{\Pi}^{ab}_{\mu\nu}\Big|_{(a_1)} &= \frac{1}{2}\int_k \widetilde{\Gamma}^{aex}_{\mu\alpha\beta}\Delta^{\alpha\rho}_{ee'}(k)\left[q^\nu \widetilde{\mathbf{\Gamma}}^{be'x'}_{\nu\rho\sigma}\right]\Delta^{\beta\sigma}_{xx'}(k+q)\\
&= \frac{1}{2}\int_k \widetilde{\Gamma}^{aex}_{\mu\alpha\beta}\Delta^{\alpha\rho}_{ee'}(k) g f^{be'x'}\left[\Delta^{-1}_{\rho\sigma}(k+q) - \Delta^{-1}_{\rho\sigma}(k)\right]\Delta^{\beta\sigma}_{xx'}(k+q),
\end{aligned} \tag{8}
$$

where the brackets have been employed to emphasize where the use of the WI takes place. In addition, we have introduced the compact notation $\int_k \equiv \int \mathrm{d}^d k/(2\pi)^d$, where $d = 4-\epsilon$ is the dimension of space-time, appearing in the dimensional regularization.

To proceed further, we use into Eq. (8) the Feynman rule for the bare PT-BFM three gluon vertex, and recalling that $f^{aex}f^{bex} = C_A\delta^{ab}$ we obtain

$$
\begin{aligned}
q^\nu\, \widehat{\Pi}^{ab}_{\mu\nu}\Big|_{(a_1)} &= \frac{1}{2}g^2 C_A\delta^{ab}\int_k \left\{\left[k - q - \frac{1}{\xi_Q}(k+q)\right]_\gamma g_{\alpha\mu} + \left[q + (k+q) - \frac{k}{\xi_Q}\right]_\alpha g_{\gamma\mu}\right.\\
&\quad \left. + \left[-(k+q) - k\right]_\mu g_{\gamma\alpha}\right\} \times \left[\Delta^{\gamma\alpha}(k) - \Delta^{\gamma\alpha}(k+q)\right].
\end{aligned} \tag{9}
$$

Performing the shift $k \to k+q$, it is straightforward to show that $\int_{k+q}\Delta^{\gamma\alpha}(k+q) = \int_k \Delta^{\gamma\alpha}(k)$, which leads to

$$
q^\nu\, \widehat{\Pi}^{ab}_{\mu\nu}\Big|_{(a_1)} = C_A g^2 \delta^{ab} q_\mu \int_k \Delta^\rho_\rho(k). \tag{10}
$$

Journal of Physics: Conference Series **706** (2016) 052004 doi:10.1088/1742-6596/706/5/052004

Then, a direct calculation of the contraction of diagram (a_2) by q^ν yields

$$q^\nu\, \widehat{\Pi}^{ab}_{\mu\nu}\Big|_{(a_2)} = \frac{1}{2}q^\nu \int_k \widetilde{\Gamma}^{abex}_{\mu\nu\alpha\beta}\Delta^{\alpha\beta}_{ex}(k) = -C_A g^2 \delta^{ab} q_\mu \int_k \Delta^\rho_\rho(k). \tag{11}$$

Thus, from Eqs. (10) and (11) follows that the subgroup (a) obeys

$$q^\nu\, \widehat{\Pi}^{ab}_{\mu\nu}(q)\Big|_{(a)} = q^\nu \left(\widehat{\Pi}^{ab}_{\mu\nu}\Big|_{(a_1)} + \widehat{\Pi}^{ab}_{\mu\nu}\Big|_{(a_2)} \right) = 0, \tag{12}$$

which implies that the sum of the diagrams (a_1) and (a_2) is independently transverse.

Now let us turn our attention to the diagrams of the group (b). In this case the proof of the transversality uses the second line of Eq. (5). Following the same steps performed in the previous case, one finds that

$$q^\nu\, \widehat{\Pi}^{ab}_{\mu\nu}\Big|_{(b_1)} = -\int_k \widetilde{\Gamma}^{aex}_\mu D_{ee'}(k)\left[q^\nu \widetilde{\mathbf{\Gamma}}^{be'x'}_\nu\right] D_{xx'}(k+q) = 2C_A g^2 \delta^{ab} q_\mu \int_k D(k), \tag{13}$$

and

$$q^\nu\, \widehat{\Pi}^{ab}_{\mu\nu}\Big|_{(b_2)} = -q^\nu \int_k \widetilde{\Gamma}^{abex}_{\mu\nu} D_{ex} = -2C_A g^2 \delta^{ab} q_\mu \int_k D(k). \tag{14}$$

Then, we find that the combination of diagrams (b) vanishes, *i.e*

$$q^\nu\, \widehat{\Pi}^{ab}_{\mu\nu}(q)\Big|_{(b)} = q^\nu \left(\widehat{\Pi}^{ab}_{\mu\nu}\Big|_{(b_1)} + \widehat{\Pi}^{ab}_{\mu\nu}\Big|_{(b_2)} \right) = 0, \tag{15}$$

which proves the transversality of the subgroup (b).

The analysis of the two-loop dressed diagrams is slightly more difficult. We begin with group (c). The contraction of the diagram (c_1) by q^ν triggers the WIs of the Eq. (6), where $q_1 = -q$, $q_2 = l + q$, $q_3 = -k - l$ and $q_4 = k$, yieldings the result

$$\begin{aligned} q^\nu\, \widehat{\Pi}^{ab}_{\mu\nu}\Big|_{(c_1)} &= \frac{1}{6}\int_k\int_l \widetilde{\Gamma}^{acex}_{\mu\alpha\beta\gamma}\Delta^{\alpha\alpha'}_{cc'}(k)\Delta^{\beta\beta'}_{ee'}(k+l)\Delta^{\gamma\gamma'}_{xx'}(l+q)\left[q^\nu \widetilde{\mathbf{\Gamma}}^{bx'e'c'}_{\nu\gamma'\beta'\alpha'}\right] \\ &= \frac{ig}{6}\int_k\int_l \widetilde{\Gamma}^{acex}_{\mu\alpha\beta\gamma}\Delta^{\alpha\alpha'}(k)\Delta^{\beta\beta'}(k+l)\Delta^{\gamma\gamma'}(l+q)\left[f^{bxi}\mathbf{\Gamma}^{eci}_{\beta'\alpha'\gamma'}(-k-l,k,l)\right. \\ &\quad \left. + f^{bei}\mathbf{\Gamma}^{cxi}_{\alpha'\gamma'\beta'}(k,l+q,-q-k-l) + f^{bci}\mathbf{\Gamma}^{xei}_{\gamma'\beta'\alpha'}(l+q,-k-l,k-q)\right]. \end{aligned} \tag{16}$$

At this point, since $\widetilde{\Gamma}^{acex}_{\mu\alpha\beta\gamma}$, does not depend on the momenta, we can shift the integration momenta for the different terms as in the derivations above. Then, with an appropriate relabeling of the dummy indices, it is possible to show that all three terms contribute equally [6], leading to

$$q^\nu\, \widehat{\Pi}^{ab}_{\mu\nu}\Big|_{(c_1)} = \frac{1}{2}igf^{bxi}\int_k\int_l \widetilde{\Gamma}^{acex}_{\mu\alpha\beta\gamma}\Delta^{\alpha\alpha'}(k)\Delta^{\beta\beta'}(k+l)\Delta^{\gamma\gamma'}(l+q)\mathbf{\Gamma}^{eci}_{\beta'\alpha'\gamma'}(-k-l,k,l). \tag{17}$$

On the other hand, for the diagram (c_2) we invoke again the first WI of Eq. (5), which lead us to

$$\begin{aligned} q^\nu\,\widehat{\Pi}^{ab}_{\mu\nu}\Big|_{(c_2)} &= \frac{1}{2}\int_k\int_l \widetilde{\Gamma}^{acex}_{\mu\alpha\beta\gamma}\Delta^{\alpha\alpha'}_{cc'}(k)\Delta^{\beta\beta'}_{ee'}(k+l)\mathbf{\Gamma}^{ne'c'}_{\sigma\beta'\alpha'}\Delta^{\sigma\sigma'}_{nn'}(l)\left[q^\nu\widetilde{\mathbf{\Gamma}}^{bx'n'}_{\nu\gamma'\sigma'}\right]\Delta^{\gamma\gamma'}_{xx'}(l+q)\\ &= -\frac{1}{2}igf^{bxi}\int_k\int_l \widetilde{\Gamma}^{acex}_{\mu\alpha\beta\gamma}\Delta^{\alpha\alpha'}(k)\Delta^{\beta\beta'}(k+l)\mathbf{\Gamma}^{iec}_{\gamma'\beta'\alpha'}(l,-k-l,k)\Delta^{\gamma\gamma'}(l+q)\\ &\quad+\frac{1}{2}igf^{bxi}\int_k\int_l \widetilde{\Gamma}^{acex}_{\mu\alpha\beta\gamma}\Delta^{\alpha\alpha'}(k)\Delta^{\beta\beta'}(k+l)\mathbf{\Gamma}^{iec}_{\gamma'\beta'\alpha'}(l,-k-l,k)\Delta^{\gamma\gamma'}(l). \end{aligned} \tag{18}$$

The last term is independent of q and must vanish identically, since the free Lorentz index μ cannot be saturated. Then, observing that the conventional three-gluon vertex inside the integral is Bose symmetric, $\mathbf{\Gamma}^{eci}_{\beta'\alpha'\gamma'}(-k-l,k,l) = \mathbf{\Gamma}^{iec}_{\gamma'\beta'\alpha'}(l,-k-l,k)$, we conclude that

$$q^\nu\,\widehat{\Pi}^{ab}_{\mu\nu}(q)\Big|_{(c)} = q^\nu\left(\widehat{\Pi}^{ab}_{\mu\nu}\Big|_{(c_1)} + \widehat{\Pi}^{ab}_{\mu\nu}\Big|_{(c_2)}\right) = 0. \tag{19}$$

Finally, let us analyze the transversality of group (d). For the calculation of the divergence of the graph (d_1) we use the WI of Eq. (7), obtaining

$$\begin{aligned} q^\nu\,\widehat{\Pi}^{ab}_{\mu\nu}\Big|_{(d_1)} &= -\int_k\int_l \widetilde{\Gamma}^{acex}_{\mu\alpha}D_{cc'}(k+l)\Delta^{\alpha\beta}_{ee'}(k)D_{xx'}(l-q)\\ &\quad\times\left[q^\nu\widetilde{\mathbf{\Gamma}}^{bx'e'c'}_{\nu\beta}(-q,q-l,-k,l+k)\right]\\ &= ig\int_k\int_l \widetilde{\Gamma}^{acex}_{\mu\alpha}D(k+l)\Delta^{\alpha\beta}(k)D(l-q)\left[f^{ebi}\mathbf{\Gamma}^{bxi}_{\beta}(l+k,-l,-k)\right.\\ &\quad\left.+f^{bei}\mathbf{\Gamma}^{cxi}_{\beta}(l+k,q-l,-k-q)+f^{bci}\mathbf{\Gamma}^{ixe}_{\beta}(k+l-q,q-l,-k)\right]. \end{aligned} \tag{20}$$

Each of these three terms cancels one of the remaining three diagrams. Here, we will show only the cancellation of diagram (d_2). Using once again the first WI of Eq. (5), one obtains

$$\begin{aligned} q^\nu\,\widehat{\Pi}^{ab}_{\mu\nu}\Big|_{(d_2)} &= -\int_k\int_l \widetilde{\Gamma}^{acex}_{\mu\alpha}D_{cc'}(k+l)D_{ee'}(l-q)\mathbf{\Gamma}^{e'nc'}_{\beta}(q-l,-k-q,k+l)\\ &\quad\times\Delta^{\beta\beta'}_{nn'}(k+q)\left[q^\nu\widetilde{\mathbf{\Gamma}}^{bx'n'}_{\nu\alpha'\beta'}(-q,-k,k+q)\right]\Delta^{\alpha\alpha'}_{xx'}(k)\\ &= -igf^{bxi}\int_k\int_l \widetilde{\Gamma}^{acex}_{\mu\alpha}D(k+l)D(l-q)\mathbf{\Gamma}^{eic}_{\beta}(q-l,-k-q,k+l)\\ &\quad\times\left[\Delta^{\alpha\beta}(k)-\Delta^{\alpha\beta}(k+q)\right]. \end{aligned} \tag{21}$$

Now, shifting the second integral by $k\to k+q$ and $l-q\to l$, we again obtain a q-independent integral with a free Lorentz index, which cannot be saturated. Then, the remaining integral cancel against the second term of the Eq. (20).

Similarly, one can show the cancellation of graphs (d_3) and (d_4) against the remaining terms of Eq. (20) [6]. Finally, this cancellation lead us to the conclusion that

$$q^\nu\,\widehat{\Pi}^{ab}_{\mu\nu}(q)\Big|_{(d)} = q^\nu\left(\widehat{\Pi}^{ab}_{\mu\nu}\Big|_{(d_1)} + \widehat{\Pi}^{ab}_{\mu\nu}\Big|_{(d_2)} + \widehat{\Pi}^{ab}_{\mu\nu}\Big|_{(d_3)} + \widehat{\Pi}^{ab}_{\mu\nu}\Big|_{(d_4)}\right) = 0, \tag{22}$$

i.e. the subgroup (d) is also transverse.

Journal of Physics: Conference Series **706** (2016) 052004 doi:10.1088/1742-6596/706/5/052004

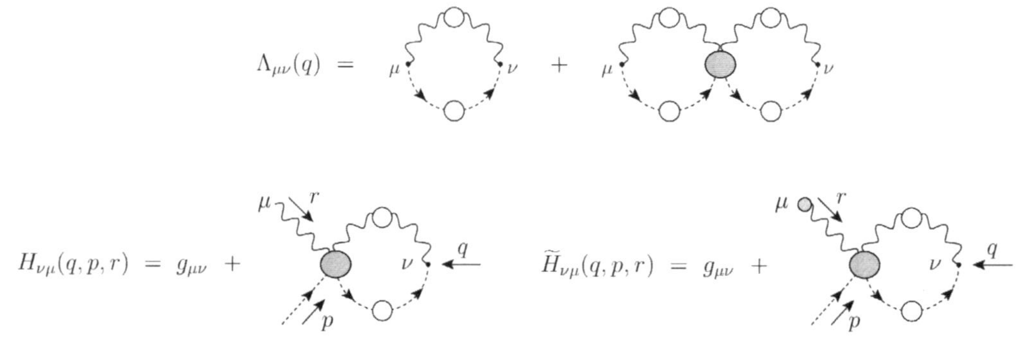

Figure 3. Diagrammatic representation of the auxiliary functions $\Lambda_{\mu\nu}$ and $H_{\mu\nu}$.

Then, it follows directly from Eqs. (12), (15), (19), and (22) that the full PT-BFM self-energy can be written as

$$\widehat{\Pi}(q^2)P_{\mu\nu}(q) = \left[\widehat{\Pi}(q^2)\Big|_{(a)} + \widehat{\Pi}(q^2)\Big|_{(b)} + \widehat{\Pi}(q^2)\Big|_{(c)} + \widehat{\Pi}(q^2)\Big|_{(d)}\right] P_{\mu\nu}(q)\,, \tag{23}$$

where we have omitted the color indices.

As announced, the four groups of diagrams defining the PT-BFM gluon self-energy are independently transverse. In addition, we showed that the gluonic and ghost contributions are independently transverse. Moreover, we also demonstrated that the transversality properties of the one-loop and two-loop dressed diagrams do not mix.

4. A dynamical equation for the PT-BFM gluon propagator

As mentioned previously the PT-BFM SDE given in the Fig. 2, mixes the BB gluon propagator, $\widehat{\Delta}_{\mu\nu}(q)$, with the conventional gluon propagator, $\Delta_{\mu\nu}(q)$, appearing in the internal lines of the diagrams. In order to convert, this equation into a *dynamical* integral equation, it is necessary to translate $\Delta_{\mu\nu}$ into $\widehat{\Delta}_{\mu\nu}$ or vice-versa. As has been explained in the literature [3, 5, 6], we can accomplish this task by employing the so-called Background Quantum Identities (BQIs), which relate the PT-BFM Green's function with the conventional ones [10].

In the case of the gluon propagators, the corresponding BQI states that

$$\widehat{\Delta}(q) = \left[1 + G(q^2)\right]^2 \Delta(q). \tag{24}$$

The function $G(q^2)$ that appears in the above equation is defined as the scalar co-factor of $g_{\mu\nu}$ in the Lorentz decomposition of a special two-point function $\Lambda_{\mu\nu}(q)$, defined diagrammatically in the Fig. 4 [5, 6, 7]. More specifically, we can write

$$\begin{aligned}\Lambda_{\mu\nu}(q) &= \int_k H^{(0)}_{\mu\alpha} D(k)\Delta^{\alpha\beta}(k+q)H_{\nu\beta}(k+q,-k,-q) \\ &= g_{\mu\nu}G(q^2) + q_\mu q_\nu L(q^2),\end{aligned} \tag{25}$$

where $H^{(0)}_{\mu\nu} = -igg_{\mu\nu}$. Its fully-dressed $H_{\alpha\beta}$ is related to the conventional gluon-ghost vertex $\mathbf{\Gamma}_\beta$ by the following STI [11]

$$q^\alpha H_{\alpha\beta}(p,r,q) = \mathbf{\Gamma}_\beta(p,r,q). \tag{26}$$

Journal of Physics: Conference Series **706** (2016) 052004 doi:10.1088/1742-6596/706/5/052004

Using Eqs. (3), (23) and (24), we can rewrite the SDE exclusively in terms of the conventional propagator in the following way [6]

$$\Delta^{-1}(q^2) = \frac{q^2 + i\widehat{\Pi}(q^2)}{[1+G]^2}\,, \tag{27}$$

since the diagrams that define $\widehat{\Pi}(q^2)$ involve only conventional propagators, $\Delta(q^2)$.

Notice that the PT-BFM formalism provides a systematic way of truncating the gluon SDE without compromising the gauge invariance of the equation [3, 6, 7]. For example, the simplest truncation scheme would be one where we retain only the one-loop gluonic contribution, *i.e.* the diagrams of the group (a). A improved version of this truncation is the case where the ghost effects at one-loop dressed are also included, *i.e.* retain diagrams of the group (a) and (b). Then, we can improve further by including the two purely gluonic two-loop dressed diagrams of the group (c). Finally, the gluon SDE will be treated thoroughly when the group (d), is finally entirely included into the game.

5. Conclusion

In this presentation we have reviewed how the transversality properties of the gluon SDE emerge in blocks within the PT-BFM truncation scheme. In particular, we have shown that the formulation of this SDE in the context of this formalism furnishes considerable advantages, because it allows for a systematic truncation that respects manifestly, and at every step the gauge-invariance, expressed by the transversality of the gluon self-energy. We also reviewed that one may use the SDE for $\widehat{\Delta}(q)$, take advantage of its improved truncation properties, and then convert it to an equivalent equation for $\Delta(q)$ (the conventional gluon propagator simulated on the lattice) by means of the BQIs. It is important to mention that different versions of truncations, within this scheme, have been studied in the last few years [3, 6, 7, 12], and all results found so far corroborate the notion of an IR finite, massive gluon propagator.

Acknowledgments

The authors thank the organizers of the XIII International Workshop on Hadron Physics for their hospitality. This work was developed under the support of the National Council for Scientific and Technological Development - CNPq (AMSNF and ACA) and São Paulo Research Foundation - FAPESP (ACA).

References

[1] C. D. Roberts, arXiv:1509.02925 [nucl-th].
[2] C. D. Roberts and A. G. Williams, Prog. Part. Nucl. Phys. **33**, 477 (1994).
[3] J. Papavassiliou, J. Phys. Conf. Ser. **631**, no. 1, 012006 (2015).
[4] J. M. Cornwall, Phys. Rev. D **26**, 1453 (1982).
[5] D. Binosi and J. Papavassiliou, Phys. Rept. **479**, 1 (2009).
[6] A. C. Aguilar and J. Papavassiliou, JHEP **0612**, 012 (2006).
[7] A. C. Aguilar, D. Binosi and J. Papavassiliou, Phys. Rev. D **78**, 025010 (2008).
[8] D. Binosi and J. Papavassiliou, Phys. Rev. D **66**, 111901 (2002).
[9] L. F. Abbott, Nucl. Phys. B **185**, 189 (1981).
[10] D. Binosi and J. Papavassiliou, Phys. Rev. D **66**, 025024 (2002).
[11] P. Pascual and R .Tarrach, *QCD: Renormalization for the practitioner* (Springer-Verlag, Heidelberg, 1984).
[12] A. C. Aguilar, D. Binosi and J. Papavassiliou, Phys. Rev. D **86**, 014032 (2012).

Heavy flavor electron R_{AA} and v_2 in event-by-event relativistic hydrodynamics

Caio A G Prado[1], Mauro R Cosentino[1,2], Marcelo G Munhoz[1], Jorge Noronha[1,3] and Alexandre A P Suaide[1]

[1] Instituto de Física, Universidade de São Paulo, C.P. 66318, 05315-970 São Paulo, SP, Brazil
[2] Centro de Ciências Naturais e Humanas, Universidade Federal do ABC, Av. dos Estados 5001, Bairro Santa Terezina, 09210-508 Santo André, SP, Brazil
[3] Department of Physics, Columbia University, 538 West 120th Street, New York, NY 10027, USA

E-mail: caio.prado@usp.br mr.cosentino@gmail.com munhoz@if.usp.br noronha@if.usp.br suaide@if.usp.br

Abstract. In this work we investigate how event-by-event hydrodynamics fluctuations affect the nuclear suppression factor and elliptic flow of heavy flavor mesons and non-photonic electrons. We use a 2D+1 Lagrangian ideal hydrodynamic code [1, 2] on an event-by-event basis in order to compute local temperature and flow profiles. Using a strong coupling inspired energy loss parametrization [3] on top of the evolving space-time energy density distributions we are able to propagate the heavy quarks inside the medium until the freeze-out temperature is reached and a PYTHIA [4] modeling of hadronization takes place. The resulting D^0 and heavy-flavor electron yield is compared with recent experimental data for R_{AA} and v_2 from the STAR and PHENIX collaborations [5–7]. In addition we present preditions for the higher order Fourier harmonic coefficients $v_3(p_T)$ of heavy-flavor electrons at RHIC's $\sqrt{S_{NN}} = 200\,\text{GeV}$ collisions.

1. Introduction

Central Au+Au collisions at RHIC and Pb+Pb at LHC exhibit a strong particle suppression when compared to p+p collisions as well as anisotropic flow. The suppression is usually related with jet quenching or energy loss of partons inside the quark-gluon-plasma (QGP) whereas the flow might be due to lumps of higher density inside the medium due to initial fluctuations. Furthermore while the QGP dynamics may affect differently the expansion of these high-density spots in comparison with the rest of the plasma it can affect higher harmonic orders of anisotropic flow such as v_3.

We aim to study these effects of the medium on the heavy quarks suppression during their evolution inside the QGP. In order to achieve this goal we performed computational simulations of the evolution of heavy quarks. The simulation consists of a sampled quark drawn from an initial condition energy density profile. The quark is then evolved inside the medium while the medium itself expands. The initial p_T distribution is given by First-Order Next-to-Leading Logs (FONLL [8]) calculation. After evolution quarks reach the freeze-out temperature and hadronize. The resulting meson decays into electrons we can analyse and compare with experimental data of electron R_{AA} and v_2.

2. Simulation

This simulation takes for granted the factorization on QCD in order to follow a modular paradigm. This allows one to replace part of the calculation without affecting the other ones, for instance, one could change the energy loss model in order to study a different one. We can summarize the simulation with the following modules:

- Initial Conditions;
- Hydrodynamics;
- Energy loss model;
- Fragmentation;
- Meson decay.

Also, the simulation only takes into account the effects of the medium over the quark probes and not the other way around.

2.1. Initial Conditions

Initial conditions are constructed using a modified version of the PHOBOS Glauber Monte Carlo [9] code with default parameters, and selecting the appropriate centrality range. From the nucleons participants in the collision given by the PHOBOS simulation we can create an energy density profile for central rapidity by attributing gaussian distributions for each wounded nucleon and binary collision [10] with width given in terms of the inelastic nucleon-nucleon cross section [9]. The normalization factor which arises from this method is calculated by matching the maximum initial average temperature with literature values [11–14].

2.2. Hydrodynamics

The evolution of the medium is performed by an implementation of the Smoothed Particles Hydrodynamics (SPH) [1, 15] algorithm. We use a longitudinal expansion with boost invariance [16, 17] and set the initial transverse velocity to zero while also considering it to be independent of the rapidity. In addition we assume the baryon chemical potential to be zero. The equation of state EOS s95n-v1 [18] is used in this work [2].

The evolution starts from $\tau = 1.0\,\mathrm{fm/c}$ and we set the freeze-out temperature to $140\,\mathrm{MeV}$ [11, 12]. The smoothing SPH parameter is set to $h = 0.3\,\mathrm{fm}$ in order to perform the simulation in a doable time while still preserving the important structures of the initial conditions. The evolution goes until the complete decoupling of the particles in the medium following the Cooper-Frye prescription [19]. The hydrodynamics is performed separately from the quark evolution.

2.3. Energy Loss Model

We draw quarks from the medium using the initial energy density profile as probability distribution for the initial position. From a FONLL calculation [8, 20, 21] we set the initial momentum of the quark with uniformly distributed φ direction.

The quarks are propagated by a numerical integration of the energy loss over step distance. The integration is performed locally in the medium frame so Lorentz transformations are performed every step and this accounts for some quarks gaining energy while being pushed by the medium.

The parametrization used in these simulations for the energy loss is given by:

$$\frac{\mathrm{d}E}{\mathrm{d}x} \propto v\gamma(v)T^2, \tag{1}$$

where v is the quark velocity, $\gamma(v)$ is the Lorentz factor and T is the medium temperature. A similar expression is obtained by using AdS/CFT correspondence and a classical test string approximation to calculate the drag force on an external quark moving in a thermal plasma [3] and a Langevin simulation to describe the diffusion dynamics of heavy quarks in QGP [22].

The scale factor of the parametrization is fitted against experimental data, we use D^0 meson R_{AA} for the charm quarks factor and the electron R_{AA} for the bottom one.

2.4. Fragmentation, decay and output

After the propagation is performed and the quarks have reached the freeze-out temperature we use Peterson fragmentation function in order to hadronize the quarks which then decay into electrons using the semi-leptonic channels through PYTHIA8 [4].

The final output is the electron distribution after all quarks have hadronized and decayed. By analysing these electron spectra we can obtain heavy flavor electron R_{AA} and v_2 in order to compare the results with experimental data.

3. Results

We now present some results that were obtained with the simulation. Figures 1 and 2 show the electron spectra for RHIC and LHC energies comparing both bottom and charm flavors with and without the energy loss. The ratio between them is the nuclear modification factor R_{AA}.

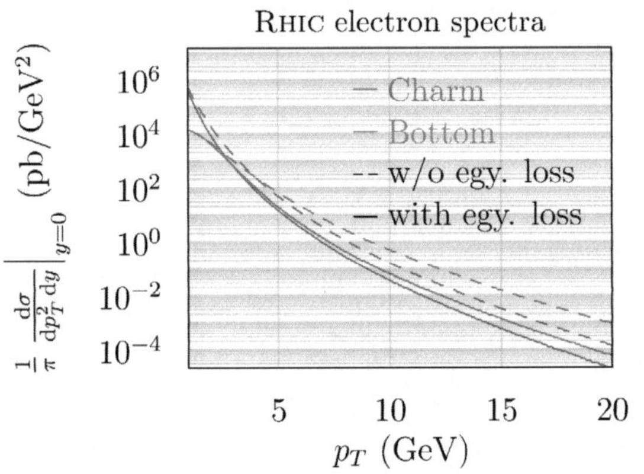

Figure 1. Electron spectra for RHIC ($\sqrt{S_{NN}} = 200\,\text{GeV}$) energy comparing charm and bottom flavors.

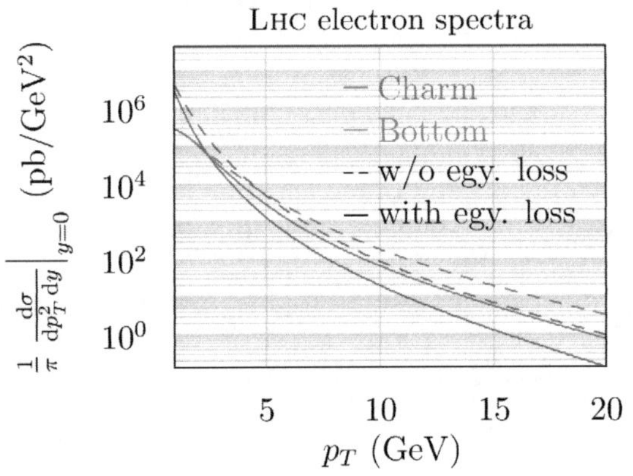

Figure 2. Electron spectra for LHC ($\sqrt{S_{NN}} = 2.76\,\text{TeV}$) energy comparing charm and bottom flavors.

We show the resulting R_{AA} calculations for both RHIC and LHC energies in Figures 3 and 4. The simulation is performed separately for bottom and charm quark and in order to obtain the full R_{AA} spectra we combine both results weighted by the FONLL cross section calculations. The spectra are compared with experimental data.

Also, in Figures 5 and 6 flow calculations for the RHIC energy. The results for v_2 are below the data which might be due to the energy loss model which we want to improve in order to obtain better results, also the p_T range of the data is fairly limited in comparison with the available range for R_{AA} spectra.

4. Conclusions

We implemented a 2D+1 Langrangian ideal hydrodynamic code on an event-by-event basis and evolved heavy quarks inside the expanding medium in order to obtain electron p_T and φ

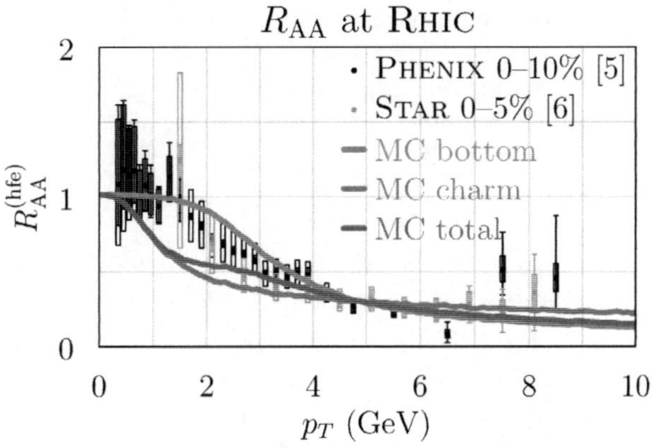

Figure 3. Electron nuclear modification factor for each quark flavor and the total one in comparison with experiment results for RHIC's energy 200 GeV

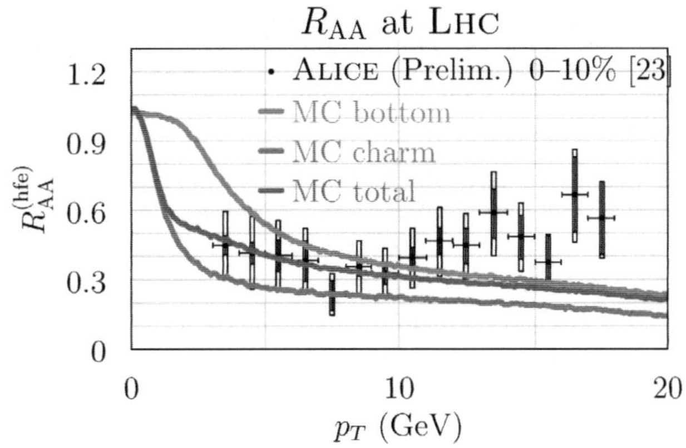

Figure 4. Electron nuclear modification factor for each quark flavor and the total one in comparison with experiment results for LHC's energy 2.76 TeV.

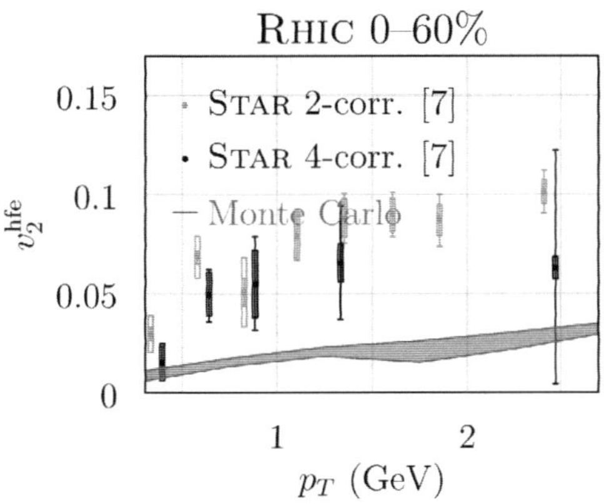

Figure 5. Calculation of electron v_2 for RHIC energy.

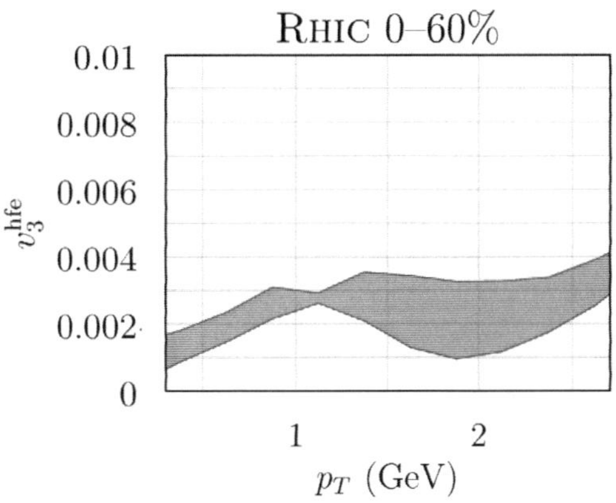

Figure 6. Calculation of electron v_3 for RHIC energy.

spectra so we could calculate the nuclear modification factor R_{AA} and the Fourier harmonic coefficients. Our framework is intentionally very modular so one could use it to test different models for energy loss, hydrodynamics or fragmentation, allowing for a comprehensive study of different aspects of the QGP phenomenology. We obtained R_{AA} spectra for both RHIC and LHC energies.

References

[1] Andrade R P G and Noronha J 2013 *Phys. Rev. C* **88** 1–13 ISSN 0556-2813 URL `http://link.aps.org/doi/10.1103/PhysRevC.88.034909`

[2] Andrade R P G, Noronha J and Denicol G S 2014 *Phys. Rev. C* **90** 1–11 ISSN 0556-2813 URL `http://link.aps.org/doi/10.1103/PhysRevC.90.024914`

[3] Gubser S S 2006 *Phys. Rev. D* **74** 126005 ISSN 1550-7998 URL `http://link.aps.org/doi/10.1103/PhysRevD.74.126005`

[4] Sjostrand T, Mrenna S and Skands P 2008 *Comput. Phys. Commun.* **178** 852–867 ISSN 00104655 (*Preprint* 0710.3820) URL `http://linkinghub.elsevier.com/retrieve/pii/S0010465508000441`

[5] PHENIX Collaboration 2007 *Phys. Rev. Lett.* **98** 172301 ISSN 0031-9007 URL `http://link.aps.org/doi/10.1103/PhysRevLett.98.172301`

[6] STAR Collaboration 2007 *Phys. Rev. Lett.* **98** 192301 ISSN 0031-9007 URL `http://www.ncbi.nlm.nih.gov/pubmed/17677616`

[7] STAR Collaboration 2014 *Arxiv Prepr. arXiv1405.6348* 10 (*Preprint* 1405.6348)

[8] Cacciari M, Nason P and Vogt R 2005 *Phys. Rev. Lett.* **95** 1–4 ISSN 0031-9007 URL `http://link.aps.org/doi/10.1103/PhysRevLett.95.122001`

[9] Loizides C, Nagle J and Steinberg P 2014 *Arxiv Prepr. 1408.2549* 6 (*Preprint* 1408.2549) URL `http://arxiv.org/abs/1408.2549`

[10] Nonaka C and Bass S a 2007 *Eur. Phys. J. C* **49** 97–102 ISSN 14346044 (*Preprint* 0607018)

[11] Luzum M and Romatschke P 2009 *Phys. Rev. Lett.* **103** 262302 ISSN 0031-9007 URL `http://link.aps.org/doi/10.1103/PhysRevLett.103.262302`

[12] Luzum M, Gombeaud C and Ollitrault J Y 2010 *Phys. Rev. C - Nucl. Phys.* **81** 1–5 ISSN 05562813 (*Preprint* 1004.2024)

[13] Roy V and Chaudhuri a K 2011 *Phys. Lett. Sect. B Nucl. Elem. Part. High-Energy Phys.* **703** 313–317 ISSN 03702693 (*Preprint* 1103.2870) URL `http://dx.doi.org/10.1016/j.physletb.2011.08.006`

[14] Bhatt J R, Mishra H and Sreekanth V 2011 *Phys. Lett. Sect. B Nucl. Elem. Part. High-Energy Phys.* **704** 486–489 ISSN 03702693 (*Preprint* 1103.4333) URL `http://dx.doi.org/10.1016/j.physletb.2011.09.052`

[15] Aguiar C E, Kodama T, Osada T and Hama Y 2000 *J. Phys. G Nucl. Part. Phys.* **75** 14 (*Preprint* 0006239) URL `http://arxiv.org/abs/hep-ph/0006239`

[16] Hwa R C 1974 *Phys. Rev. D* **10** 2260–2268 ISSN 05562821

[17] Bjorken J D 1983 *Phys. Rev. D* **27** 140–151 ISSN 05562821 URL `http://link.aps.org/doi/10.1103/PhysRevD.27.140`

[18] Huovinen P and Petreczky P 2010 *Nucl. Phys. A* **837** 26–53 ISSN 03759474 (*Preprint* 0912.2541) URL `http://dx.doi.org/10.1016/j.nuclphysa.2010.02.015`

[19] Cooper F and Frye G 1974 *Phys. Rev. D* **10** 186–189 ISSN 0556-2821 URL `http://link.aps.org/doi/10.1103/PhysRevD.10.186`

[20] Cacciari M, Greco M and Nason P 1998 *J. High Energy Phys.* **1998** 007–007 ISSN 1029-8479

[21] Cacciari M and Frixione S 2001 *J. High Energy Phys.* **2001** 006–006 URL http://iopscience.iop.org/1126-6708/2001/03/006

[22] Akamatsu Y, Hatsuda T and Hirano T 2009 *Phys. Rev. C* **79** 054907 ISSN 0556-2813 URL http://link.aps.org/doi/10.1103/PhysRevC.79.054907

[23] de Godoy D M and ALICE Collaboration 2014 Elliptic azimuthal anisotropy of heavy-flavour decay electrons in Pb-Pb collisions at SNN = 2.76 TeV measured with ALICE *AIP Conf. Proc.* vol 1625 pp 226–229 ISBN 9780735412620 URL http://scitation.aip.org/content/aip/proceeding/aipcp/10.1063/1.4901799

Energy-dependent dipole form factor in a QCD-inspired model

C.A.S. Bahia, M. Broilo, E.G.S. Luna

Instituto de Física, Universidade Federal do Rio Grande do Sul, Caixa Postal 15051, 91501-970, Porto Alegre, RS, Brazil

E-mail: andre.bahia@ufrgs.br,mateus.broilo@ufrgs.br,luna@if.ufrgs.br

Abstract. We consider the effect of an energy-dependent dipole form factor in the high-energy behavior of the forward amplitude. The connection between the semihard parton-level dynamics and the hadron-hadron scattering is established by an eikonal QCD-based model. Our results for the proton-proton (pp) and antiproton-proton ($\bar{p}p$) total cross sections, $\sigma_{tot}^{pp,\bar{p}p}(s)$, obtained using the CTEQ6L1 parton distribution function, are consistent with the recent data from the TOTEM experiment.

1. Introduction

The total cross section $\sigma_{tot}(s)$ is a fundamental quantity in collisions of strongly interacting particles. At present one of the main theoretical approaches for the description of σ_{tot} is the QCD-inspired formalism [1, 2, 3, 4]. In this approach the energy dependence of the total cross section is obtained from QCD via an eikonal formulation. The high-energy dependence of σ_{tot} is driven by the rapid increase in gluon density at small-x. In this work we explore the non-perturbative dynamics of QCD in order to describe, in both pp and $\bar{p}p$ channels, the total cross sections $\sigma_{tot}^{pp,\bar{p}p}(s)$ and the ratios of the real to imaginary part of the forward scattering amplitude, $\rho^{pp,\bar{p}p}(s)$. In our calculations we introduce an energy-dependent dipole form factor which represents the overlap density for the partons at impact parameter b. The behavior of these forward quantities is derived from the QCD parton model using standard parton-parton elementary processes and an updated set of gluon distribution, namely the CTEQ6L1 set. In our analysis we have included recent measurements of pp total cross sections at the LHC by the TOTEM Collaboration [5]. This work is organized as follows: in the next section we introduce our QCD-based model where the onset of the dominance of semihard partons is managed by the dynamical gluon mass. We investigate a form factor which the spatial distribution of gluons changes with the energy and introduce integral dispersion relations to connect the real and imaginary parts of eikonals. In Section III we present our results and in Section IV we draw our conclusions.

2. The QCD-inspired eikonal model

In our model the increase of $\sigma_{tot}^{pp,\bar{p}p}(s)$ is associated with elementary semihard processes in the hadrons. In the eikonal representation $\sigma_{tot}(s)$ and $\rho(s)$ are given by

$$\sigma_{tot}(s) = 4\pi \int_0^{\infty} b\,db\,[1 - e^{-\chi_R(s,b)} \cos\chi_I(s,b)] \quad (1)$$

and

$$\rho(s) = \frac{-\int_0^\infty b\, db\, e^{-\chi_R(s,b)} \sin\chi_I(s,b)}{\int_0^\infty b\, db\, [1 - e^{-\chi_R(s,b)} \cos\chi_I(s,b)]}, \tag{2}$$

respectively, where s is the square of the total center-of-mass energy, b is the impact parameter, and $\chi(s,b) = \mathrm{Re}\,\chi(s,b) + i\mathrm{Im}\,\chi(s,b) \equiv \chi_R(s,b) + i\chi_I(s,b)$ is the (complex) eikonal function. The eikonal functions for pp and $\bar{p}p$ scatterings are the sum of the soft and semihard (SH) parton interactions in the hadron-hadron collision,

$$\chi(s,b) = \chi_{soft}(s,b) + \chi_{SH}(s,b). \tag{3}$$

It follows from the QCD parton model that the real part of the eikonal, $\mathrm{Re}\,\chi_{SH}(s,b)$, can be factored as

$$\mathrm{Re}\,\chi_{SH}(s,b) = \frac{1}{2} W_{SH}(b)\, \sigma_{QCD}(s), \tag{4}$$

where $W_{SH}(b)$ is an overlap density for the partons at impact parameter space b,

$$W_{SH}(b) = \int d^2b'\, \rho_A(|\mathbf{b} - \mathbf{b}'|)\, \rho_B(b'), \tag{5}$$

and $\sigma_{QCD}(s)$ is the usual QCD cross section

$$\begin{aligned} \sigma_{QCD}(s) &= \sum_{ij} \frac{1}{1+\delta_{ij}} \int_0^1 dx_1 \int_0^1 dx_2 \int_{Q^2_{min}}^\infty d|\hat{t}| \frac{d\hat{\sigma}_{ij}}{d|\hat{t}|}(\hat{s},\hat{t}) \\ &\times f_{i/A}(x_1,|\hat{t}|) f_{j/B}(x_2,|\hat{t}|)\, \Theta\left(\frac{\hat{s}}{2} - |\hat{t}|\right), \end{aligned} \tag{6}$$

with $|\hat{t}| \equiv Q^2$, $\hat{s} = x_1 x_2 s$ and $i,j = q, \bar{q}, g$.

The eikonal functions for pp and $\bar{p}p$ scatterings are written as $\chi_{pp}^{\bar{p}p}(s,b) = \chi^+(s,b) \pm \chi^-(s,b)$, with $\chi^+(s,b) = \chi^+_{soft}(s,b) + \chi^+_{SH}(s,b)$ and $\chi^-(s,b) = \chi^-_{soft}(s,b) + \chi^-_{SH}(s,b)$. Since in the parton model $\chi^-_{SH}(s,b)$ decreases rapidly with increasing s, the difference between pp and $\bar{p}p$ cross sections is due only to the different weighting of the quark-antiquark (valence) annihilation cross sections in the two channels. Thus the crossing-odd eikonal $\chi^-(s,b)$ receives no contribution from semihard processes at high energies, and it is sufficient to take $\chi_{SH}(s,b) = \chi^+_{SH}(s,b)$ and, consequently, $\chi^-(s,b) = \chi^-_{soft}(s,b)$. The connection between the real and imaginary parts of the crossing-even eikonal $\chi^+(s,b)$ is obtained by means of dispersion relation

$$\mathrm{Im}\,\chi^+(s,b) = -\frac{2s}{\pi} \mathcal{P} \int_0^\infty ds' \frac{\mathrm{Re}\,\chi^+(s',b)}{s'^2 - s^2}, \tag{7}$$

valid at $s \gg 1$ GeV^2. The soft term is given by

$$\chi^+_{soft}(s,b) = \frac{1}{2} W^+_{soft}(b;\mu^+_{soft}) \left[A' + \frac{B'}{(s/s_0)^\gamma} e^{i\pi\gamma/2} - iC' \left[\ln\left(\frac{s}{s_0}\right) - i\frac{\pi}{2}\right]\right], \tag{8}$$

where $\sqrt{s_0} \equiv 5$ GeV and A', B', C', γ and μ^+_{soft} are fitting parameters. The odd eikonal $\chi^-(s,b)$ is given by

$$\chi^-(s,b) = \frac{1}{2} W^-_{soft}(b;\mu^-_{soft})\, D' \frac{e^{-i\pi/4}}{\sqrt{s}}, \tag{9}$$

where $\mu^-_{soft} \equiv 0.5$ GeV and D' is also a fitting parameter.

In the expression (6) the term $d\hat{\sigma}_{ij}/d|\hat{t}|$ is the differential cross section for ij scattering ($i, j = q, \bar{q}, g$). We select parton-parton processes containing at least one gluon in the initial state, i.e., we consider the processes

$$\frac{d\hat{\sigma}}{d\hat{t}}(gg \to gg) = \frac{9\pi\bar{\alpha}_s^2}{2\hat{s}^2}\left(3 - \frac{\hat{t}\hat{u}}{\hat{s}^2} - \frac{\hat{s}\hat{u}}{\hat{t}^2} - \frac{\hat{t}\hat{s}}{\hat{u}^2}\right), \tag{10}$$

$$\frac{d\hat{\sigma}}{d\hat{t}}(qg \to qg) = \frac{\pi\bar{\alpha}_s^2}{\hat{s}^2}\left(\hat{s}^2 + \hat{u}^2\right)\left(\frac{1}{\hat{t}^2} - \frac{4}{9\hat{s}\hat{u}}\right), \tag{11}$$

$$\frac{d\hat{\sigma}}{d\hat{t}}(gg \to \bar{q}q) = \frac{3\pi\bar{\alpha}_s^2}{8\hat{s}^2}\left(\hat{t}^2 + \hat{u}^2\right)\left(\frac{4}{9\hat{t}\hat{u}} - \frac{1}{\hat{s}^2}\right), \tag{12}$$

where

$$\bar{\alpha}_s(Q^2) = \frac{4\pi}{\beta_0 \ln\left[(Q^2 + 4M_g^2(Q^2))/\Lambda^2\right]}, \tag{13}$$

$$M_g^2(Q^2) = m_g^2\left[\frac{\ln\left(\frac{Q^2+4m_g{}^2}{\Lambda^2}\right)}{\ln\left(\frac{4m_g{}^2}{\Lambda^2}\right)}\right]^{-12/11}; \tag{14}$$

here $\bar{\alpha}_s(Q^2)$ is a non-perturbative QCD effective charge, obtained by Cornwall through the use of the pinch technique in order to derive a gauge invariant Schwinger-Dyson equation for the gluon propagator [6].

In the case of semihard gluons we consider the possibility of a "broadening" of the spatial distribution. Our assumption suggests an increase of the average gluon radius when $\sqrt{s}$ increases, and can be properly implemented using the energy-dependent dipole form factor

$$G_{SH}^{(d)}(s, k_\perp; \nu_{SH}) = \left(\frac{\nu_{SH}^2}{k_\perp^2 + \nu_{SH}^2}\right)^2, \tag{15}$$

where $\nu_{SH} = \nu_1 - \nu_2 \ln(\frac{s}{s_0})$, with $\sqrt{s_0} \equiv 5$ GeV. Here ν_1 and ν_2 are constants to be fitted. From equation (5) we have

$$\begin{aligned} W_{SH}(s, b; \nu_{SH}) &= \frac{1}{2\pi}\int_0^\infty dk_\perp\, k_\perp\, J_0(k_\perp b)\, [G_{SH}^{(d)}(s, k_\perp; \nu_{SH}]^2 \\ &= \frac{\nu_{SH}^2}{96\pi}(\nu_{SH} b)^3 K_3(\nu_{SH} b). \end{aligned} \tag{16}$$

Since semihard interactions dominate at high energies, we consider an energy-dependence only in the case of $W_{SH}(s, b; \nu_{SH})$. In this way the soft overlap densities $W^+_{soft}(b; \mu^+_{soft})$ and $W^-_{soft}(b; \mu^-_{soft})$ are static, i.e., μ^+_{soft} and μ^-_{soft} are not energy dependent.

3. Results

First, in order to determine the model parameters, we set the value of the gluon mass scale to $m_g = 400$ MeV. This choice for the mass scale is not only consistent to our LO procedure, but is also the one usually obtained in other calculations of strongly interacting processes [2, 3, 7, 8]. Next, we carry out a global fit to high-energy forward pp and $\bar{p}p$ scattering data above $\sqrt{s} = 10$ GeV, namely the total cross section $\sigma_{tot}^{pp,\bar{p}p}$ and the ratio of the real to imaginary part of the forward scattering amplitude $\rho^{pp,\bar{p}p}$. We use data sets compiled and analyzed by the Particle Data Group [9] as well as the recent data at LHC from the TOTEM Collaboration, with the statistic and systematic errors added in quadrature. We include in the dataset the first estimate for the ρ parameter made by the TOTEM Collaboration in their ρ-independent measurement at $\sqrt{s} = 7$ TeV, namely $\rho^{pp} = 0.145 \pm 0.091$ [5].

The values of the fitted parameters are given in the legend of Figure 1. The result $\chi^2/DOF = 1.062$ for the fit was obtained for 154 degrees of freedom. The results of the fit to σ_{tot} and ρ for both pp and $\bar{p}p$ channels are displayed in Figure 1, together with the experimental data. The curves depicted in Figure 1 were all calculated using the cutoff $Q_{min} = 1.3$ GeV, the value of the CTEQ6 fixed initial scale Q_0.

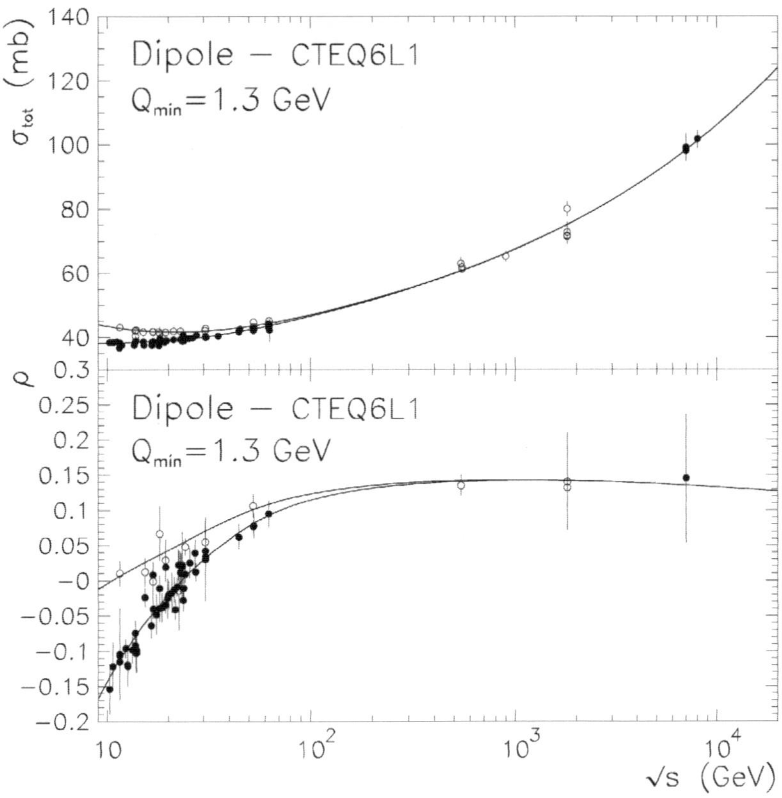

Figure 1. The total cross section (upper) and the ratio of the real to imaginary part of the forward scattering amplitude (lower), for pp (•) and $\bar{p}p$ (○). The values of the fitted parameters are: $\nu_1 = 2.770 \pm 0.865$ GeV, $\nu_2 = (7.860 \pm 5.444) \times 10^{-2}$ GeV, $A' = 108.9 \pm 8.6$ GeV^{-1}, $B' = 30.19 \pm 15.78$ GeV^{-1}, $C' = 1.260 \pm 0.437$ GeV^{-1}, $\gamma = 0.719 \pm 0.200$, $\mu^{+}_{soft} = 0.457 \pm 0.209$ GeV and $D' = 21.73 \pm 2.26$ GeV^{-1}.

4. Conclusions

In this work we have investigated infrared effects on semihard parton-parton interactions. We have studied pp and $\bar{p}p$ scattering in the LHC energy region with the assumption that the increase of their total cross sections arises exclusively from semihard interactions. In the calculation of $\sigma_{tot}^{pp,\bar{p}p}$ and $\rho^{pp,\bar{p}p}$ we have considered the phenomenological implication of an energy-dependent form factor for semihard partons. We introduce integral dispersion relations to connect the real and imaginary parts of eikonals with energy-dependent form factors. In our analysis we have included the recent data at LHC from the TOTEM Collaboration. Our results show that very good descriptions of $\sigma_{tot}^{pp,\bar{p}p}$ and $\rho^{pp,\bar{p}p}$ data are obtained by constraining the value of the cutoff at $Q_{min} = 1.3$ GeV. The χ^2/DOF for the global fit was 1.062 for 154 degrees of freedom. This good statistical result shows that our eikonal model, where non-perturbative effects are naturally included via a QCD effective charge, is well suited for detailed predictions of the forward quantities to be measured at higher energies. In the semihard sector we have considered a form factor in which the average gluon radius increases with $\sqrt{s}$. With this assumption we have obtained another form in which the eikonal can be factored into the QCD parton model, namely $\mathrm{Re}\,\chi_{SH}(s,b) = \frac{1}{2}\,W_{SH}(s,b)\,\mathrm{Re}\,\sigma_{QCD}(s)$. The imaginary part of this *semi-factorizable* eikonal was obtained by means of appropriate integral dispersion relations which take into account eikonals with energy-dependent form factors.

At the moment we are analyzing the effects of different updated sets of parton distributions on the forward quantities, namely CTEQ6L and MSTW, investigating the uncertainty on these forward observables coming from the uncertainties associated with the dynamical mass scale and the parton distribution functions, and exploring a new ansatz for the energy-dependent form factor.

References

[1] L. Durand and H. Pi, Phys. Rev. Lett. **58**, 303 (1987); Phys. Rev. D **38**, 78 (1988); **40**, 1436 (1989).
[2] E.G.S. Luna, A.F. Martini, M.J. Menon, A. Mihara, and A.A. Natale, Phys. Rev. D **72**, 034019 (2005); E.G.S. Luna, Phys. Lett. B **641**, 171 (2006); E.G.S. Luna and A.A. Natale, Phys. Rev. D **73**, 074019 (2006); E.G.S. Luna, Braz. J. Phys. **37**, 84 (2007); D.A. Fagundes, E.G.S. Luna, M.J. Menon, and A.A. Natale, Nucl. Phys. A **886**, 48 (2012); E.G.S. Luna and P.C. Beggio, Nucl. Phys. A **929**, 230 (2014).
[3] E.G.S. Luna, A.L. dos Santos, and A.A. Natale, Phys. Lett. B **698**, 52 (2011).
[4] A. Corsetti, A. Grau, G. Pancheri, and Y.N. Srivastava, Phys. Lett. B **382**, 282 (1996); A. Grau, G. Pancheri, and Y.N. Srivastava, Phys. Rev. D **60**, 114020 (1999); R.M. Godbole, A. Grau, G. Pancheri, and Y.N. Srivastava, Phys. Rev. D **72**, 076001 (2005); A. Achilli *et al.*, Phys. Lett. B **659**, 137(2008); A. Grau, R.M. Godbole, G. Pancheri, and Y.N. Srivastava, Phys. Lett. B **682**, 55 (2009); G. Pancheri, D.A. Fagundes, A. Grau, S. Pacetti, and Y.N. Srivastava, arXiv:1301.2925 [hep-ph].
[5] G. Antchev *et al.*, EPL **96**, 21002 (2011); G. Antchev *et al.*, EPL **101**, 21002 (2013); G. Antchev *et al.*, EPL **101**, 21004 (2013); G. Antchev *et al.*, Phys. Rev. Lett. **111**, 012001 (2013).
[6] J. M. Cornwall, Phys. Rev. D **22**, 1452 (1980); D **26**, 1453 (1982).
[7] A. Doff, E.G.S. Luna, and A.A. Natale, Phys. Rev. D **88**, 055008 (2013).
[8] E.G.S. Luna and A.A. Natale, J. Phys. G **42**, 015003 (2015).
[9] K.A. Olive *et al.*, Chin. Phys. C **38**, 090001 (2014).

Mass generation and the problem of seagull divergences

C. T. Figueiredo and A. C. Aguilar
University of Campinas - UNICAMP, Institute of Physics “Gleb Wataghin”,
13083-859 Campinas, SP, Brazil
E-mail: ctfigue@ifi.unicamp.br

Abstract. The gluon mass generation is a purely non-perturbative effect, and the natural framework to study it in the continuum are the Schwinger-Dyson equations (SDEs) of the theory. At the level of the SDEs the generation of such a mass is associated with the existence of infrared finite solutions for the gluon propagator. From the theoretical point of view, the dynamical gluon mass generation has been traditionally plagued with seagull divergences. In this work, we will review how such divergences can be eliminated completely by virtue of a characteristic identity, valid in dimensional regularization. As a pedagogical example, we will first discuss in the context of scalar QED how it is possible to eliminate all seagull divergences, by triggering the aforementioned special identity, which enforces the masslessness of the photon. Then, we will discuss what happens in QCD and present an Ansatz for the three gluon vertex, which completely eliminates all seagull divergences and at same time allows for the possibility of a dynamical gluon mass generation.

1. Introduction

The dynamical gluon mass generation has been the subject of numerous studies in the past few years [1–6]. Being a non-perturbative effect, the most adequate formalism to study it in the continuum are the QCD Schwinger-Dyson equations (SDEs). Specifically, we will work in the framework provided by the synthesis of the pinch technique (PT) [1, 3, 4] with the background field method (BFM) [7], known in the literature as the PT-BFM scheme.

The most crucial theoretical ingredient for obtaining out of the SDEs an infrared finite gluon propagator, without interfering with the gauge invariance of the theory, is the existence of a set of special vertices endowed with massless poles. These poles act as composite, longitudinally coupled Nambu-Goldstone bosons, which at the same time trigger the Schwinger mechanism in QCD and maintain the gauge invariance of the theory. Specifically, the form of the Ward identities (WIs) and the Slavnov-Taylor identities (STIs) are kept intact before and after the dynamical generation of a gluon mass. It is interesting to mention that a series of secondary effects related to the existence of a gluon mass can be observed in the QCD dynamics [8]. For example, it is expected that the QCD effective charge “freezes” at a finite value in the infrared due the presence of the momentum dependent gluon mass in the argument of the leading logarithm [1, 9].

However, one of the QCD long-standing problems is how to generate a dynamical gluon mass without the appearance of seagull divergences, *i.e.* divergent integrals of the type $\int_k \Delta(k)$ and

$\int_k k^2\Delta^2(k)$. Here, we will follow the same steps presented in the Ref. [5] and we will review how these divergences can be eliminated by virtue of a special identity, valid in dimensional regularization.

These proceedings is organized as follows. In Section 2, we present the main ingredients and the crucial relations necessary for the generation of a dynamical gluon mass through the Schwinger mechanism. We also briefly mention how the seagull divergences come into play. In Section 3, we explore a didactic example, within the context of scalar QED, showing how it is possible to eliminate all seagull divergences by triggering the aforementioned characteristic identity. In addition, we discuss how this identity is instrumental for ensuring the masslessness of the photon in the absence of bound-state poles, *i.e.* when the Schwinger mechanism is not triggered. In Section 4, we put together all theoretical ingredients introduced in the the previous sections, in order to construct an Ansatz for the QCD three gluon vertex. The aforementioned Ansatz completely eliminates all seagull divergences and at the same time allows for the possibility of a dynamical gluon mass generation. Finally, in Section 5 we summarize our conclusions.

2. Gluon Mass Generation and Seagull Divergences

The gluon propagator, $\Delta_{\mu\nu}(q)$, can be written in R_ξ-type gauges as

$$\Delta_{\mu\nu}(q) = -i\left[P_{\mu\nu}(q)\Delta(q^2) + \xi\,\frac{q_\mu q_\nu}{q^4}\right], \tag{1}$$

where ξ is the gauge fixing parameter and $P_{\mu\nu}(q)$ is the transverse projector, given by

$$P_{\mu\nu}(q) = g_{\mu\nu} - \frac{q_\nu q_\mu}{q^2}\,. \tag{2}$$

The scalar factor $\Delta(q^2)$ can be obtained from

$$\Delta^{-1}(q^2) = q^2 + i\Pi(q^2)\,, \tag{3}$$

where $\Pi_{\mu\nu}(q) = P_{\mu\nu}(q)\,\Pi(q^2)$ is the gluon self-energy. It is interesting to mention that, a long time ago, Schwinger pointed out that if the theory has sufficient strength to produce a vacuum polarization of the gauge bosons with a pole at zero momentum transfer, the vector meson becomes massive, without breaking gauge symmetry [10]. The dimensionless vacuum polarization, $\mathbf{\Pi}(q^2)$, is usually defined as $\Pi(q^2) = q^2\mathbf{\Pi}(q^2)$. Thus, it is clear that if $\mathbf{\Pi}(q^2)$ has a pole at $q^2 = 0$ with positive residue μ^2, then

$$\Delta^{-1}(q^2) = q^2 + \mu^2\,, \tag{4}$$

in Euclidean space. Since $\Delta^{-1}(0) = \mu^2$, the gluon becomes massive, even though no mass term is permitted at the level of the fundamental Lagrangian. We note that the existence of this required pole for $\mathbf{\Pi}(q^2)$ occurs due to purely dynamical reasons, in such a way that the dynamically generated mass for the vector meson does not interfere with gauge invariance [11–13].

In order to obtain the momentum dependence of the dynamical gluon mass, we must study the SDE that governs the gluon self-energy. The Schwinger mechanism is integrated into the SDE through the three-gluon vertex. However, despite the use of a three-gluon vertex containing massless poles and satisfying the correct WI, the value of the gluon propagator at zero momentum is expressed in terms of formally divergent seagull-type contributions of the form

$$\Delta^{-1}(0) = c_1\int_k \Delta(k) + c_2\int_k k^2\Delta^2(k)\,. \tag{5}$$

where $\int_k \equiv \mu^{2\epsilon}(2\pi)^{-d}\int d^dk$ and $d = 4 - \epsilon$, in dimensional regularization (DR).

The downside of the presence of such divergent integrals is that they need to be regularized. It turns out that the regularization procedure employed in the literature furnishes finite but scheme-dependent gluon masses [3].

Therefore, the full realization of gluon mass generation preserving all symmetries of the theory is an extremely subtle task. Basically, it demands that our Ansatz for the three gluon vertex has the following simultaneous features (i) contains massless poles, (ii) respect the corresponding WIs, and (iii) preserves the cancellation that eliminates all seagull divergences of the type of Eq. (5).

3. Scalar QED

Let us first analyze some of the basic issues regarding the appearance and cancellation of seagull divergent integrals in the context of scalar QED, which has a much simpler structure than QCD. We start by showing in Fig. 1 the diagrams contributing to the SDE for the photon of scalar QED.

Using the WIs satisfied by the vertices appearing in Fig. 1, it is straightforward to demonstrate that the SDE may be truncated "loop-wise" [5], so that

$$\begin{aligned} q^\mu \Pi_{\mu\nu}(q)\big|_{[(d_1)+(d_2)]} &= 0\,, \\ q^\mu \Pi_{\mu\nu}(q)\big|_{[(d_3)+(d_4)+(d_5)]} &= 0\,. \end{aligned} \tag{6}$$

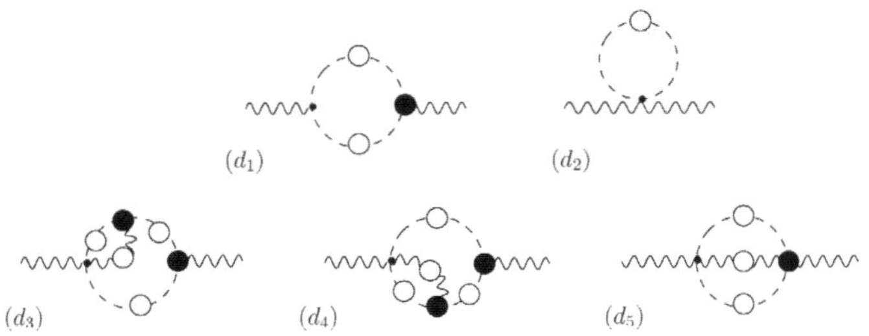

Figure 1. Diagrams contributing to the SDE for the photon self energy in scalar QED [5].

In the Fig. 2 we show the terms contributing to the SDE for the photon self-energy at the "one-loop dressed" level, which can be written as

$$\Pi_{\mu\nu}(q) = e^2 \int_k \Gamma^{(0)}_\mu \mathcal{D}(k)\,\mathcal{D}(k+q)\Gamma_\nu + e^2 \int_k \Gamma^{(0)}_{\mu\nu}\,\mathcal{D}(k)\,, \tag{7}$$

where $\mathcal{D}(k)$ is the full propagator of the scalar field and Γ_ν is the full photon-scalar vertex, whose tree-level expression is

$$\Gamma^{(0)}_\nu = -i(2k+q)_\mu\,. \tag{8}$$

In addition, the bare quadrilinear photon-scalar vertex appearing in the diagram (d_2) is expressed as $\Gamma^{(0)}_{\mu\nu} = 2ig_{\mu\nu}$.

The Abelian WI for the photon-scalar vertex Γ_μ reads

$$q^\mu \Gamma_\mu = \mathcal{D}^{-1}(k+q) - \mathcal{D}^{-1}(k)\,. \tag{9}$$

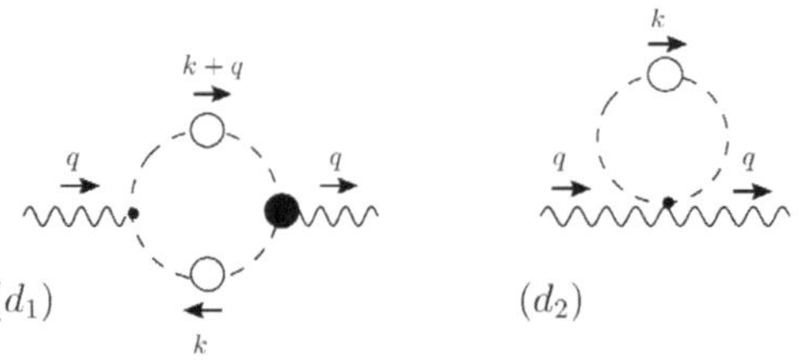

Figure 2. The "one-loop dressed" contribution to the SDE for the photon self-energy in scalar QED.

With the help of the WI, one can easily show that $q^\nu \Pi_{\mu\nu}(q) = 0$. Therefore, $\Pi_{\mu\nu}(q)$ is transverse and assumes the form $\Pi_{\mu\nu}(q) = \Pi(q^2)P_{\mu\nu}(q)$. Thus, it follows that the scalar $\Pi(q^2)$ is given by

$$\Pi(q^2) = \frac{-2ie^2}{d-1}\left[\int_k \mathcal{D}(k)\,\mathcal{D}(k+q)k^\mu\,\Gamma_\mu - d\int_k \mathcal{D}(k)\right]. \tag{10}$$

In addition, we can write the one-loop expression for $\Pi(q^2)$ as being

$$\Pi^{(1)}(q^2) = \frac{-ie^2}{d-1}\left[\int_k (4k^2 - q^2)\,\mathcal{D}_0(k)\,\mathcal{D}_0(k+q) - 2d\int_k \mathcal{D}_0(k)\right], \tag{11}$$

where $\mathcal{D}_0(k) = (k^2 - m^2)^{-1}$. Now, we take the limit $q \to 0$, which leads to

$$\Pi^{(1)}(0) = \frac{-4ie^2}{d-1}\left[\int_k k^2\,\mathcal{D}_0^2(k) - \frac{d}{2}\int_k \mathcal{D}_0(k)\right]. \tag{12}$$

Using spherical coordinates in the Euclidean space, it can be demonstrated that [5]

$$\int_k k^2\,\frac{\partial\,\mathcal{D}_0(k)}{\partial k^2} = -\frac{d}{2}\int_k \mathcal{D}_0(k)\,. \tag{13}$$

Consequently, we have $\Pi^{(1)}(0) = 0$, as expected, given that the photon remains massless perturbatively. It is important to mention that the same result can be obtained using the following exact identity in DR:

$$\int_k \frac{k^2}{(k^2-m^2)^2} = \frac{d}{2}\int_k \frac{1}{k^2-m^2}\,. \tag{14}$$

The equation above can be verified applying the standard integration rules of DR [14].

Now, let us analyze the full case given by Eq. (7). Evidently, in order to preserve the transversality of $\Pi_{\mu\nu}(q)$ any Ansatz for Γ_μ should satisfy the WI of Eq. (9). A suitable form is the one proposed by Ball and Chiu [15] given by

$$\Gamma_\mu = \frac{(2k+q)_\mu}{(k+q)^2 - k^2}\left[\mathcal{D}^{-1}(k+q) - \mathcal{D}^{-1}(k)\right] + A(k,q)\left[(k+q)\cdot q\,k_\mu - k\cdot q\,(k+q)_\mu\right], \tag{15}$$

where $A(k,q)$ is finite as $q \to 0$.

The term proportional to $A(k,q)$, when inserted into Eq. (10), yields a contribution that vanishes as $q \to 0$. Since we are interested in the limit of vanishing q, only the the first term of Γ_μ contributes, producing

$$\Pi(0) = \frac{4ie^2}{d-1}\left[\int_k k^2 \frac{\partial \mathcal{D}(k)}{\partial k^2} + \frac{d}{2}\int_k \mathcal{D}(k)\right], \tag{16}$$

Certainly, we must have $\Pi(0) = 0$, since we have not applied Schwinger's mechanism, *i.e.*, no dynamical poles were introduced in the Ansatz of Eq. (15). Therefore, we must have

$$\int_k k^2 \frac{\partial \mathcal{D}(k)}{\partial k^2} = -\frac{d}{2}\int_k \mathcal{D}(k), \tag{17}$$

which is the non-perturbative generalization of Eq. (13). The demonstration of this identity follows along the same lines as that for Eq. (13).

We can now try another Ansatz for Γ_μ with massless poles, which will not trigger anymore the Eq. (17). Let us consider the vertex given by

$$\Gamma_\mu = \Gamma_\mu^{(0)} + \frac{q_\mu}{q^2}\left[\Sigma(k+q) - \Sigma(k)\right], \tag{18}$$

where $\Sigma(k)$ is the self-energy of the scalar field and $\mathcal{D}^{-1}(k) = k^2 - m^2 + \Sigma(k)$. Thus,

$$\Gamma_\mu = \left\{\Gamma_\mu^{(0)} - \frac{q_\mu}{q^2}\left[(k+q)^2 - k^2\right]\right\} + \frac{q_\mu}{q^2}\left[\mathcal{D}^{-1}(k+q) - \mathcal{D}^{-1}(k)\right], \tag{19}$$

which satisfies the WI of Eq. (9), but contains massless poles. Substituting Γ_μ of Eq. (19) into Eq. (10), one finds

$$\Pi(0) = ie^2\left[2\int_k \mathcal{D}(k) - \int_k k^2 \mathcal{D}^2(k)\right], \tag{20}$$

which possess the general form of Eq. (5). Hence, we end up with a non-zero $\Delta^{-1}(0)$, but expressed in terms of seagull divergences.

In the next section, then, we will look for an Ansatz, in the case of a pure Yang-Mills theory, that adds non-perturbative pole terms to the vertex of Eq. (15), but also preserves the cancellation implemented by Eq. (17), in order to obtain a finite $\Delta^{-1}(0)$.

4. Finite Gluon Mass Generation

Let us now turn to the case of pure QCD and analyze the SDE of the background gluon propagator within the PT-BFM framework. The advantage of working with the PT-BFM formalism, it is that the transversality of the gluon self-energy is preserved at every step, allowing for a gauge-invariant truncation of the SD series [2, 4]. Here, we will study only the "one-loop dressed" gluonic part of the SDE diagrammatically represented in the Fig. 3. In addition, the Feynman rules for the two tree-level vertices appearing in the PT-BFM formulation are presented in Fig. 4.

To reduce the algebraic complexity of the problem, we set $\Delta_{\alpha\beta} \to -ig_{\alpha\beta}\Delta$. In order to be consistent and maintain the transversality of the gluon self-energy, $\widehat{\Pi}_{\mu\nu}(q)$, we write the WI satisfied by $\widetilde{\Gamma}_{\nu\alpha\beta}$ as

$$q^\nu \widetilde{\Gamma}_{\nu\alpha\beta} = \left[\Delta^{-1}(k+q) - \Delta^{-1}(k)\right] g_{\alpha\beta}\,. \tag{21}$$

Journal of Physics: Conference Series **706** (2016) 052007 doi:10.1088/1742-6596/706/5/052007

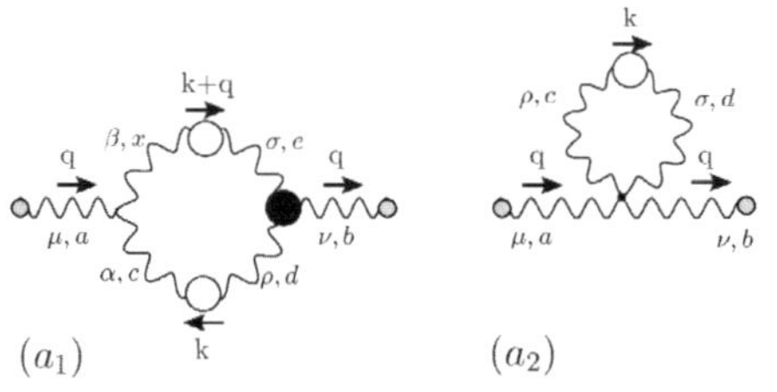

Figure 3. The "one-loop dressed" gluonic graphs for the SDE of the (background) gluon self-energy.

Then, it can be demonstrated that the mathematical expression for the SDE of Fig. 3 is

$$\widehat{\Delta}^{-1}(q) = q^2 + \frac{iq^2 C_A}{2(d-1)} \left[\int_k \widetilde{\Gamma}^{(0)}_{\mu\alpha\beta} \Delta(k)\Delta(k+q)\widetilde{\Gamma}^{\mu\alpha\beta} + 2d^2 \int_k \Delta(k) \right] , \tag{22}$$

where C_A is the Casimir eigenvalue of the adjoint representation, $\widetilde{\Gamma}^{(0)}_{\mu\alpha\beta}(q, k, -k-q)$ is the bare BFM three-gluon vertex in the Feynman gauge (see Fig. 4), and $\widetilde{\Gamma}_{\mu\alpha\beta}$ represents its fully-dressed version.

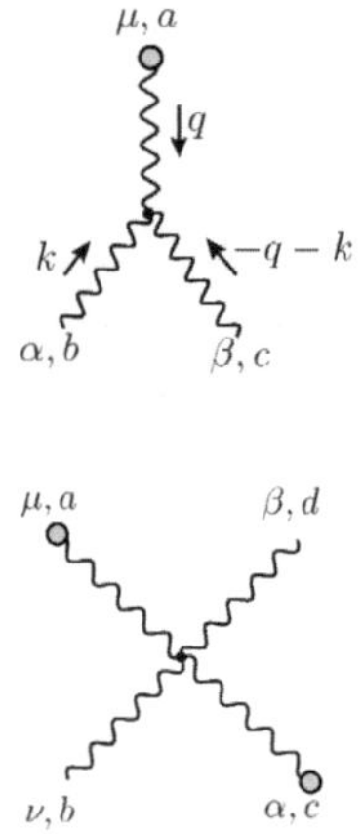

$$g\, f^{abc} \left[(2k+q)_\mu g_{\alpha\beta} + 2q_\beta g_{\mu\alpha} - 2q_\alpha g_{\mu\beta} \right] ,$$

$$\begin{aligned} -ig^2 \Big[& f^{abx} f^{xcd} (g_{\mu\alpha} g_{\nu\beta} - g_{\mu\beta} g_{\nu\alpha} + g_{\mu\nu} g_{\alpha\beta}) \\ & + f^{adx} f^{xbc} (g_{\mu\nu} g_{\alpha\beta} - g_{\mu\alpha} g_{\nu\beta} + g_{\mu\beta} g_{\nu\alpha}) \\ & + f^{acx} f^{xbd} (g_{\mu\nu} g_{\alpha\beta} - g_{\mu\beta} g_{\nu\alpha}) \Big] . \end{aligned}$$

Figure 4. The trilinear and quadrilinear gluon vertices in the Feynman gauge of PT-BFM [5].

The function $\widehat{\Delta}(q)$ appearing on the lhs of Eq. (22) is related to the conventional $\Delta(q)$ through

$$\widehat{\Delta}(q)[1 + G(q^2)]^2 = \Delta(q) , \tag{23}$$

where $G(q^2)$ is the auxiliary two-point function of the PT-BFM framework [4,16–18]. To simplify our work, we set $G(q^2) = 0$, which does not alter the essential features we intend to check. Therefore, we assume $\Delta(q) = \Delta(q)$, so that we can suppress the "hats" from now on.

We can now write $\Delta^{-1}(q)$ (in Minkowski space) as

$$\Delta^{-1}(q) = q^2 J(q) - \widetilde{m}^2(q)\,. \tag{24}$$

Then, an adequate Ansatz for $\widetilde{\Gamma}_{\nu\alpha\beta}$ is

$$i\widetilde{\Gamma}_{\nu\alpha\beta} = \left[\frac{(k+q)^2 J(k+q) - k^2 J(k)}{(k+q)^2 - k^2}\right] \widetilde{\Gamma}^{(0)}_{\nu\alpha\beta} + V_{\mu\alpha\beta}\,, \tag{25}$$

where $V_{\mu\alpha\beta}$ contains the non-perturbative contributions due to bound-state poles associated with the Schwinger mechanism. Note that, in order for $\widetilde{\Gamma}_{\nu\alpha\beta}$ to satisfy the WI of Eq. (21), we should have

$$q^\mu V_{\mu\alpha\beta} = \left[\widetilde{m}^2(k) - \widetilde{m}^2(k+q)\right] g_{\alpha\beta}\,. \tag{26}$$

The Ansatz of Eq. (25) is somewhat analogous to that of Eq. (15). The first term presents the correct structure for obtaining the derivative term of Eq. (17), once we insert it into Eq. (22). In addition, the rhs of Eq. (17) is obtained from the second term of the rhs of Eq. (22). Therefore, we can eliminate the seagull divergence, and $V_{\mu\alpha\beta}$ can produce an IR-finite gluon propagator, $\Delta^{-1}(0) = -\widetilde{m}^2(0)$.

Moreover, $V_{\mu\alpha\beta}$ can be decomposed in a transverse and a longitudinal piece, such that

$$V_{\mu\alpha\beta} = V^l_{\mu\alpha\beta} + V^t_{\mu\alpha\beta}\,, \tag{27}$$

where

$$V^l_{\mu\alpha\beta} = \frac{q_\mu}{q^2}\left[\widetilde{m}^2(k) - \widetilde{m}^2(k+q)\right] g_{\alpha\beta}\,, \tag{28}$$

and

$$q^\mu V^t_{\mu\alpha\beta} = 0\,. \tag{29}$$

Thus, we can write the vertex of Eq. (25) as

$$i\widetilde{\Gamma}_{\mu\alpha\beta} = \left[\frac{\Delta^{-1}(k+q) - \Delta^{-1}(k)}{(k+q)^2 - k^2}\right] \widetilde{\Gamma}^{(0)}_{\mu\alpha\beta} + \overline{V}_{\mu\alpha\beta}\,, \tag{30}$$

with

$$\overline{V}_{\mu\alpha\beta} = V_{\mu\alpha\beta} + (2k+q)_\mu \left[\frac{\widetilde{m}^2(k+q) - \widetilde{m}^2(k)}{(k+q)^2 - k^2}\right] g_{\alpha\beta}\,. \tag{31}$$

This last form of $\widetilde{\Gamma}_{\mu\alpha\beta}$ allows for the immediate use of the identity of Eq. (17). Hence, the final non-perturbative effective vertex $\overline{V}_{\mu\alpha\beta}$ is transverse:

$$q^\mu \overline{V}_{\mu\alpha\beta} = 0\,. \tag{32}$$

Therefore in Eq. (30), we were able to propose an Ansatz for the vertex with three important characteristics:

(i) It satisfies the WIs of Eq. (21), ensuring the transversality of the gluon self-energy;

(ii) The pole term contained in $V_{\mu\alpha\beta}$ allows for a non-vanishing $\Delta^{-1}(0)$;

(iii) It triggers Eq. (17), which cancels out all the the seagull integrals.

Consequently, the resulting $\Delta^{-1}(0)$ is both non-vanishing and finite, according to our goal.

5. Conclusions

In this work, we have demonstrated the possibility of eliminating the problem of seagull divergences in the generation of a dynamical gluon mass by means of an insightful Ansatz for the three gluon vertex. This Ansatz is given in Eq. (30) and its form mimics that of the Ball and Chiu vertex in the case of scalar QED. Specifically, the Ansatz in question was chosen so that it would trigger the seagull identity of Eq. (17), valid in dimensional regularization. Therefore, this method allows for the dynamical generation of a non-vanishing but finite gluon mass free of seagull divergences.

Acknowledgments

We would like to thank the organizers of the XIII Hadron Physics for the pleasant workshop. The research of ACA is supported by National Council for Scientific and Technological Development-CNPq under the grant 306537/20 12-5 and project 473260/2012-3, and by São Paulo Research Foundation - FAPESP through the project 2012/15643-1. The work of CTF is supported by FAPESP under the grant 2014/16247-8.

References

[1] J. M. Cornwall, Phys. Rev. D **26**, 1453 (1982).
[2] A. C. Aguilar and J. Papavassiliou, JHEP **0612**, 012 (2006).
[3] A. C. Aguilar, D. Binosi and J. Papavassiliou, Phys. Rev. D **78**, 025010 (2008).
[4] D. Binosi and J. Papavassiliou, Phys. Rept. **479**, 1 (2009).
[5] A. C. Aguilar and J. Papavassiliou, Phys. Rev. D **81**, 034003 (2010).
[6] D. Binosi, D. Ibanez and J. Papavassiliou, Phys. Rev. D **86**, 085033 (2012).
[7] L. F. Abbott, Nucl. Phys. B **185**, 189 (1981).
[8] A. C. Aguilar, A. Mihara and A. A. Natale, Phys. Rev. D **65**, 054011 (2002).
[9] A. C. Aguilar, D. Binosi, J. Papavassiliou and J. Rodriguez-Quintero, Phys. Rev. D **80**, 085018 (2009).
[10] J. S. Schwinger, Phys. Rev. **128**, 2425 (1962).
[11] R. Jackiw and K. Johnson, Phys. Rev. D **8**, 2386 (1973).
[12] J. M. Cornwall and R. E. Norton, Phys. Rev. D **8**, 3338 (1973).
[13] E. Eichten and F. Feinberg, Phys. Rev. D **10**, 3254 (1974).
[14] M. E. Peskin and D. V. Schroeder, "An Introduction to quantum field theory", Reading, USA: Addison-Wesley (1995) 842 p.
[15] J. S.Ball and T. W. Chiu, Phys. Rev. D **22**, 2542 (1980).
[16] D. Binosi and J. Papavassiliou, Phys. Rev. D **66**, 025024 (2002).
[17] D. Binosi and J. Papavassiliou, Phys. Rev. D **66**, 111901 (2002).
[18] D. Binosi and J. Papavassiliou, J. Phys. G **30**, 203 (2004).

The pion electromagnetic structure with self-energy

Clayton Santos Mello[a], J. P. B. C. de Melo[b] and T. Frederico[a]

[a]Instituto Tecnológico de Aeronáutica, DCTA, 12.228-900 São José dos Campos, SP, Brazil.
[b]Laboratório de Física Teórica e Computacional - LFTC Universidade Cruzeiro do Sul, 01506-000, São Paulo, SP, Brazil

E-mail: claytonsantosmello@yahoo.com.br

Abstract. We study the electromagnetic structure of the pion in terms of the quantum cromodynamic (QCD) model on the Breit-frame. We calculated the observables, such as the electromagnetic form factor. The priori to have a calculation covariant need to get the valence term of the eletromagnetic form factor. We use the usual formalism in quantum field theory (QFT) and light-front quantum field theory (LFQFT) in order to test the properties of form factor in nonperturbative QCD. In this particular case, the form factor can be obtained using the pion Light-Front (LF) wave function including self-energy from Lattice-QCD. Specifically, these calculations was performed in LF formalism. We consider a quark-antiquark vertex model having a quark self-energy. Also we can use other models to compare the pion electromagnetic form factor with different wave function and to observe the degree of agreement between them.

1. Introduction

The dynamics of the internal structure of hadrons affects their observable properties, and the electromagnetic form factor of hadrons is an example of such an observable [1, 2, 3, 4, 5, 6, 7, 8, 9, 10]. The interaction of a virtual photon with a meson probes its internal structure and dynamics through the meson form factor [11, 12, 13, 14]. The study of their form factors will thus allow us to extract information about the nonperturbative dynamics of its constituents [9, 14]. Specifically in this paper we treat only the pion meson internal structure. The theoretical prediction of electromagnetic form factor $F_\pi(q^2)$ at experimentally accessible q^2, below say 10 GeV^2, is a nontrivial task since the complex nonperturbative physics of confinement [1], dynamical chiral symmetry breaking (DCSB), and bound state structure are highly dependent on the modelling of the strong coupling regime that is not reachable using perturbative-QCD.

For our model we propose the inclusion of the self-energy of the quark at the quark-antiquark vertex $\Gamma_\pi(k)$ in the chiral limit [15], where the pion mass $m_\pi = 0$. This account was used only to define this vertex. Since the calculation of the form factor was conducted in the Breit-frame the vertex must be out of the chiral limit to satisfy the momentum conservation. The form factor was derived from the Feynman triangular diagram at impulse approximation. So we have only one pion elastic form factor [9].

The valence wave function of the pion in our model comes from the projection of the Bethe-Salpeter amplitude in the Light-Front (LF), and integrated in k^- (energy in LF). The four-vector momentum space in the LF, is defined as, k^μ, where $\mu = -, +, \perp$, energy, longitudinal and transversal momenta respectively.

In the configurations space we have x^μ, where $\mu = -, +, \perp$, longitudinal position, time and transversal direction respectively [16]. Here we use the valence wave function for the purpose of showing only the pion form factor of valence whose is the square modulus of the wave function less the quark-photon vertex $\Gamma^\mu(k,P)$. It will be constructed from Ward-Takahashi identity (WTI) [17] in order to ensure the momentum conservation for the electromagnetic current J^μ. In order to minimize the zero-mode terms in LF [18] we calculate only the current component J^+. Since we have the electromagnetic current of the pion we can get the electromagnetic form factor, as explained in the following sections.

2. Pion and its constituent quarks

Dynamical chiral symmetry breaking (DCSB) is one of the most important properties of low energy QCD [19], and its breaking pattern has profound impact on phenomenological quantities, e.g. the appearance of pseudoscalar Goldstone bosons [20] and the non-degeneracy of chiral partners. The spontaneous breaking of chiral symmetry is a remarkable feature of QCD because it cannot be derived directly from the Lagrangian [19] it is related to the nontrivial structure of the QCD vacuum, characterised by strong condensates of quarks and gluons [15]. This is quite different from the explicit symmetry breaking, which is put in by hand through the finite quark masses, and appears in a similar way through the Higgs mechanism. There are two important consequences of the spontaneous breaking of chiral symmetry. The first one is that the valence quarks acquire a dynamical or constituent mass through their interactions with the collective excitations of the QCD vacuum that is much larger than the seed mass present in the Lagrangian. The second one is the appearance of a triplet of pseudoscalar mesons of low mass (π^+, π^-, π^0) which represent the associated Goldstone bosons [21].

The prominent role played by the pion as the Goldstone boson of spontaneously broken chiral symmetry has its impact on the low-energy structure of hadrons through pion cloud effects in the quark propagation [22]. In full QCD there are hadron contributions to the fully dressed quark-gluon vertex. These effects are generated by the inclusion of dynamical sea quarks in the quark-gluon interaction, and are therefore only present in the unquenched case. It is the aim of this paper to introduce these pion cloud effects into the quark propagator through quark mass function, and then all the way up into the meson Bethe-Salpeter amplitude (BSA) and the pion electromagnetic form factor.

3. Self-energy into pion-quark-antiquark vertex

Due to the inclusion of quark self-energy at vertex we have a pseudoscalar pion vertex it is given by [15]:

$$\Gamma_\pi(k) = i\gamma^5 M(k) \ ; \ \text{where } M(k) = m_0 - \frac{m^3}{k^2 - \lambda^2 - i\epsilon} \ , \tag{1}$$

where γ^5 is the Dirac matrix, k is the relative momentum between the constituent quarks of the pion. $M(k)$ is the quark mass function that has been obtained from Schwinger-Dyson equation (SDE) solutions [5, 23] who was able to fit the Lattice-QCD calculations [24, 25, 26]. The $m_0 = 0.014\ GeV$ is the current quark mass and $m^3 = 0.189\ GeV^3$ and $\lambda^2 = 0.639\ GeV^2$ are the Lattice-QCD parameters [6]. The quark propagator also contains the quark mass function due to the presence of the self-energy in the legs into pion-quark-antiquark vertex. So we have to the quark propagator in the LF coordinates:

$$S(k) \ = \ \frac{\not{k} + M(k)}{k^2 - M^2(k) + i\epsilon} = \frac{\not{k}_{\rm on} + M(k)}{k^2 - M^2(k) + i\epsilon} + \frac{\gamma^+}{2k^+} \ , \tag{2}$$

In the expression above, the $\not{k}_{\rm on}$ subscript indicates the quark is on-shell.

And that has been obtained by the separation $\frac{\gamma^+}{2k^+}$ instant term in LF. The $k^-_{\rm on} = \frac{k_\perp^2+M^2(k)}{k^+}$, it is on-shell energy. The four-vector in LF is defined for the usual coordinates as: $k^\pm = k^0 \pm k^3$ and $k_\perp = (k^1, k^2)$ the same for the Dirac matrices. The consequence of this is that the dot product $k^2 = k^+k^- - k_\perp^2$ [16]. The increase $i\epsilon$ allows us to delocate the poles that contribute to integration on k^- via Cauchy's theorem [18]. This calculation makes the relative time between quarks to be eliminated [9].

3.1. Quark-photon vertex

For the Dirac structure of the electromagnetic current, we make use of Ward-Takahashi identity [17]. So we can extract the quark-photon vertex as follows:

$$\begin{aligned} q_\mu \Gamma^\mu(k; P, P') &= S^{-1}(P'-k) - S^{-1}(P-k) \\ \Gamma^\mu(k; P, P') &= \gamma^\mu + \Lambda^\mu(k; P, P')\ , \end{aligned} \tag{3}$$

where the correlation function, $\Lambda^\mu(k, P)$, is given by:

$$\Lambda^\mu(k; P, P') = \frac{m^3\,(2k - P' - P)^\mu}{[(P'-k)^2 - \lambda^2 + i\epsilon]\,[(P-k)^2 - \lambda^2 + i\epsilon]}\ . \tag{4}$$

From the Feynman triangular diagram, we obtain the pion electromagnetic current J^μ:

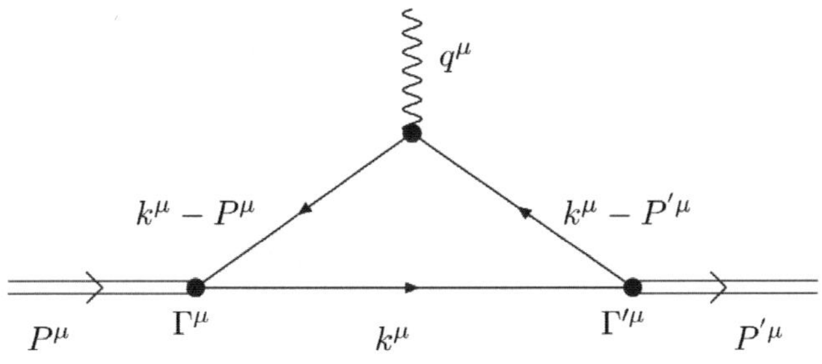

Figure 1: Triangular diagram in impulse approximation.

$$\begin{aligned} J^\mu &= i\frac{N^2 N_c}{f_\pi^2}\int \frac{d^4k}{(2\pi)^4} Tr\left[S(k)\Gamma_\pi(P'-k)S(P'-k)\Gamma^\mu(k;P,P')S(P-k)\Gamma_\pi(P-k)\right] \\ J^\mu &= i\frac{N^2 N_c}{f_\pi^2}\int \frac{d^4k}{(2\pi)^4}\frac{Tr\left[\mathcal{O}^\mu(k;P,P')\right]}{[k^2 - M^2(k) + i\epsilon]} M(P'-k)M(P-k) \\ & \frac{1}{[(P-k)^2 - M^2(P-k) + i\epsilon]\,[(P'-k)^2 - M^2(P'-k) + i\epsilon]}\ . \end{aligned} \tag{5}$$

Now we might writte the Dirac structure of this way:

$$\begin{aligned} \mathcal{O}^\mu(k;P,P') &= (\not{k} + M(k))\,\gamma^5\,(\not{P}' - \not{k} + M(P'-k))\,\Gamma^\mu(k;P,P') \\ & (\not{P} - \not{k} + M(P-k))\,\gamma^5\ , \end{aligned} \tag{6}$$

where $q = P' - P$, and using the Breit frame that is the transfered momentum in x-direction $q = (0, q^x, 0, 0)$. To the total initial momentum of the pion it is $P = (P^0, -q^x/2, 0, 0)$ and total final momentum is $P' = (P^0, q^x/2, 0, 0)$. We also made use of Drell-Yan condition, $q^+ = 0$, when there no transfer momentum on longitudinal direction [18, 27, 28]. The Dirac trace structure, is writing as,

$$\begin{aligned}
\mathcal{P}_n^+(k,P,P') &= -4m^6\left[k^+(a_n-b_n-c_n)+(b_nP^++c_nP'^+)\right] \\
&+ 4m_0m^3\left[(c_nP'^++b_nP^+)a_n+(-2k^++P^++P'^+)b_nc_n\right] \\
&+ m_0\left[4(k_x^2+m_0^2)+2(P'^-P^++P^-P'^+)+q^2\right]\times \\
&\times a_n(2k^+-P^+-P'^+)m^3 \\
&+ a_nb_nc_n\{[-4[k_n^-(k^+-P^+)(k^+-P'^+) \\
&+ k_x^2(P^++P'^+)]-2k_x(P'^+-P^+)q \\
&+ k^+(4k_x^2+q^2)]+4m_0^2[k^+-P^+-P'^+]\} \\
&+ \{[-2k^+P'^-a_nc_n+2k_xq(c_n-b_n)a_n-2k^+P^-a_nb_n \\
&- 4k_x^2(a_n(b_n+c_n)-b_nc_n)+2k^+b_nc_n(P'^-+P^-) \\
&- 4m_0^2(a_n(b_n+c_n)+b_nc_n) \\
&+ b_nc_n(-2P'^-P^+-2P^-P'^+-q^2)]\,m^3 \\
&+ 2k_n^-\left[(-P'^+a_nc_n-P^+a_nb_n)+2k^+(a_n(b_n+c_n)-b_nc_n)\right. \\
&+ \left.(P^++P'^+)b_nc_n\right]m^3-4m^6\left[m^3-m_0(a_n+b_n+c_n)\right] \\
&- 4m_0k_n^-k^+a_nb_nc_n\}\,\frac{(2k^+-P^+-P'^+)m^3}{b_nc_n} \\
Tr[\mathcal{O}_n^+(k;P,P')] &= \frac{\mathcal{P}_n^+(k;P,P')}{a_nb_nc_n}\,. \qquad (7)
\end{aligned}$$

Thus we have the electromagnetic current this way:

$$\begin{aligned}
J^+ &= i\frac{N^2N_c}{f_\pi^2}\int\frac{d^4k}{(2\pi)^4}\frac{m^6Tr\left[\mathcal{O}^+(k;P,P')\right]a_n^2b_n^2c_n^2}{(a_n^2(k^2+i\epsilon)-(m_0a_n-m^3)^2))\,b_nc_n} \\
&\quad\frac{1}{(b_n^2((P-k)^2+i\epsilon)-(m_0b_n-m^3)^2))} \\
&\quad\frac{1}{(c_n^2((P'-k)^2+i\epsilon)-(m_0c_n-m^3)^2))} \\
&= i\frac{N^2N_c}{f_\pi^2}\int\frac{d^4k}{(2\pi)^4}\frac{m^6\mathcal{P}^+(k;P,P')(y-\lambda^2+i\epsilon)}{(y-y_1)(y-y_2)(y-y_3)} \\
&\quad\frac{1}{(z-z_1)(z-z_2)(z-z_3)(w-w_1)(w-w_2)(w-w_3)}\,. \qquad (8)
\end{aligned}$$

Now we have written the quark mass function below, in order to obtain an expression for the pion electromagnetic form factor:

$$\begin{aligned}
M(k) &= m_0-\frac{m^3}{a_n}\,,\ a_n=k^+\left(k_n^--\frac{f_a-i\epsilon}{k^+}\right), \\
M(P-k) &= m_0-\frac{m^3}{b_n}\,,\ b_n=(P^+-k^+)\left(P^--k_n^--\frac{f_b-i\epsilon}{(P^+-k^+)}\right), \\
M(P'-k) &= m_0-\frac{m^3}{c_n}\,,\ c_n=(P'^+-k^+)\left(P'^--k_n^--\frac{f_c-i\epsilon}{(P'^+-k^+)}\right), \qquad (9)
\end{aligned}$$

where:

$$\begin{aligned}
f_1 &= k_\perp^2 + Re[y_1]\ ,\ f_2 = k_\perp^2 + Re[y_2]\ ,\ f_3 = k_\perp^2 + Re[y_3]\ ,\ f_a = k_\perp^2 + \lambda^2 \\
f_4 &= (P-k)_\perp^2 + Re[z_1]\ ,\ f_5 = (P'-k)_\perp^2 + Re[z_2]\ ,\ f_6 = (P-k)_\perp^2 + Re[z_3] \\
f_7 &= (P'-k)_\perp^2 + Re[w_1]\ ,\ f_8 = (P-k)_\perp^2 + Re[w_2]\ , \\
f_9 &= (P'-k)_\perp^2 + Re[w_3]\ ,\ f_b = (P-k)_\perp^2 + \lambda^2\ ,\ f_c = (P'-k)_\perp^2 + \lambda^2\ . \qquad (10)
\end{aligned}$$

where $z_1 = w_1 = y_1$, $z_2 = w_2 = y_2$ and $z_3 = w_3 = y_3$ are the denominator roots of the electromagnetic current and they are derived of propagators when including the self-energy in legs at pion vertex. We identify the poles in k^- when we perform a change of variable $k^2 = y$, $(P-k)^2 = z$ and $(P'-k)^2 = w$, and the roots are:

$$\begin{aligned}
e &= -6m^3 m_0 - 2\lambda^2 m_0^2 + m_0^4\ , \\
f &= -18\lambda^2 m^3 m_0 - 6\lambda^4 m_0^2 + 18m^3 m_0^3 + 6\lambda^2 m_0^4 - 2m_0^6\ , \\
g &= 36\lambda^2 m^9 m_0 + 8\lambda^4 m^6 m_0^2 - 4m^9 m_0^3 - 4\lambda^2 m^6 m_0^4\ , \\
d &= \sqrt[3]{2\lambda^6 - 27m^6 + 3\sqrt{3}\sqrt{-4\lambda^6 m^6 + 27m^{12} + g} + f}\ , \\
y_1 &= \frac{2}{3}\lambda^2 + \frac{m_0^2}{3} - \frac{\sqrt[3]{2}}{3d}\left(\lambda^4 + e\right) - \frac{d}{3\sqrt[3]{2}} - i\epsilon\ , \\
y_2 &= \frac{2}{3}\lambda^2 + \frac{m_0^2}{3} + \frac{(1+i\sqrt{3})}{3\sqrt[3]{4}d}\left(\lambda^4 + e\right) + \frac{(1-i\sqrt{3})d}{6\sqrt[3]{2}} - i\epsilon\ , \\
y_3 &= \frac{2}{3}\lambda^2 + \frac{m_0^2}{3} + \frac{(1-i\sqrt{3})}{3\sqrt[3]{4}d}\left(\lambda^4 + e\right) + \frac{(1+i\sqrt{3})d}{6\sqrt[3]{2}} - i\epsilon\ . \qquad (11)
\end{aligned}$$

The electromagnetic current,J^+, is also, writting like,

$$J^+ = i\frac{N^2 N_c}{f_\pi^2}\int \frac{d^2k_\perp dk^+ dk^-}{2(2\pi)^4} \frac{m^6 \mathcal{P}_n^+(k;P,P')a_n}{\mathcal{D}_1\mathcal{D}_2\mathcal{D}_3\mathcal{D}_4\mathcal{D}_5\mathcal{D}_6\mathcal{D}_7\mathcal{D}_8\mathcal{D}_9}\ . \qquad (12)$$

here, are,

$$\begin{aligned}
\mathcal{D}_1 &= k^+\left(k_n^- - \frac{f_1 - i\epsilon}{k^+}\right);\ \mathcal{D}_2 = k^+\left(k_n^- - \frac{f_2 - i\epsilon}{k^+}\right); \\
\mathcal{D}_3 &= k^+\left(k_n^- - \frac{f_3 - i\epsilon}{k^+}\right);\ \mathcal{D}_4 = (P^+ - k^+)\left(P^- - k_n^- - \frac{f_4 - i\epsilon}{P^+ - k^+}\right); \\
\mathcal{D}_5 &= (P'^+ - k^+)\left(P'^- - k_n^- - \frac{f_5 - i\epsilon}{P'^+ - k^+}\right);\ \mathcal{D}_6 = (P^+ - k^+)\left(P^- - k_n^- - \frac{f_6 - i\epsilon}{P^+ - k^+}\right); \\
\mathcal{D}_7 &= (P'^+ - k^+)\left(P'^- - k_n^- - \frac{f_7 - i\epsilon}{P'^+ - k^+}\right);\mathcal{D}_8 = (P^+ - k^+)\left(P^- - k_n^- - \frac{f_8 - i\epsilon}{P^+ - k^+}\right); \\
\mathcal{D}_9 &= (P'^+ - k^+)\left(P'^- - k_n^- - \frac{f_9 - i\epsilon}{P'^+ - k^+}\right)\ .
\end{aligned}$$

We can identify nine propagators in electromagnetic current equation, J^+, with the following poles:

$$\begin{aligned}
k_1^- &= \frac{f_1}{k^+} - \frac{i\epsilon}{k^+},\ k_2^- = \frac{f_2}{k^+} - \frac{i\epsilon}{k^+}, k_3^- = \frac{f_3}{k^+} - \frac{i\epsilon}{k^+}, \\
k_4^- &= P^- - \frac{f_4}{P^+ - k^+} + \frac{i\epsilon}{P^+ - k^+},\ k_5^- = P'^- - \frac{f_5}{P'^+ - k^+} + \frac{i\epsilon}{P'^+ - k^+}, \\
k_6^- &= P^- - \frac{f_6}{P^+ - k^+} + \frac{i\epsilon}{P^+ - k^+}, k_7^- = P'^- - \frac{f_7}{P'^+ - k^+} + \frac{i\epsilon}{P'^+ - k^+}, \\
k_8^- &= P^- - \frac{f_8}{P^+ - k^+} + \frac{i\epsilon}{P^+ - k^+}, k_9^- = P'^- - \frac{f_9}{P'^+ - k^+} + \frac{i\epsilon}{P'^+ - k^+}.
\end{aligned} \tag{13}$$

To valence range of the electromagnetic current, we verified the poles contribution, Fig. (2).

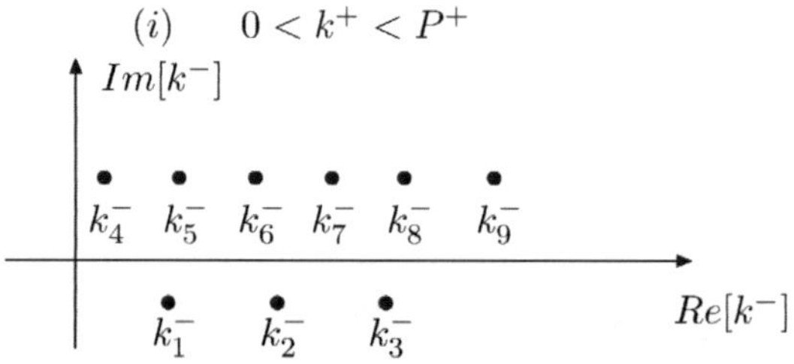

Figure 2: Poles position in Argand-Gauss plane for valence term of the electromagnetic current.

We have the valence contribuition term for the electromagnetic current, J^{+V}, which correspond the interval integration, $0 < k^+ < P^+$, in the light-front energy, k^-, (see the Fig.(2) above), is given below,

$$J^{+V} = \frac{N^2 N_c}{f_\pi^2} \sum_{n=1}^{3} \int \frac{d^2k_\perp dk^+}{2(2\pi)^3} \frac{m^6 \mathcal{P}_n^+ \mathcal{D}_n}{k^+ \mathcal{D}_1 \mathcal{D}_2 \mathcal{D}_3 \mathcal{D}_4 \mathcal{D}_5 \mathcal{D}_6 \mathcal{D}_7 \mathcal{D}_8 \mathcal{D}_9}, \tag{14}$$

From the electromagnetic current, we can obtain the pion space-like electromagnetic form factor, with the expression below:

$$\langle P'^+ | J^+ | P^+ \rangle = e(P'^+ + P^+) F_\pi(q^2) . \tag{15}$$

where e is the elementary charge, and $F_\pi(q^2)$ is the electromagnetic form factor; the constant normalization N is obtained from the condition of charge, $F_\pi(q^2 = 0) = 1$, [7, 8, 18, 29, 30].

4. Numerical results

For analysis of our model we present the results we obtained for the electromagnetic form factor of the pion. In the Fig. (3) on the left we can see the form factor as a function of the square transfer momentum. And on right we find the same function multiplied by q^2. Also we compare our results with two other models, they are Light-front symetric vertex [30] and non-symetric vertex [29]. Since the technique used in our calculations is also based on these models have drawn analytical valence wave function to separate the instant terms in LF. We compared our model for the form factor with the experimental data according to the references [31, 32, 33, 34]. These data are for describing the structure of the pion at low energies. In this case q^2 for values lower than 10 $[GeV/c]^2$.

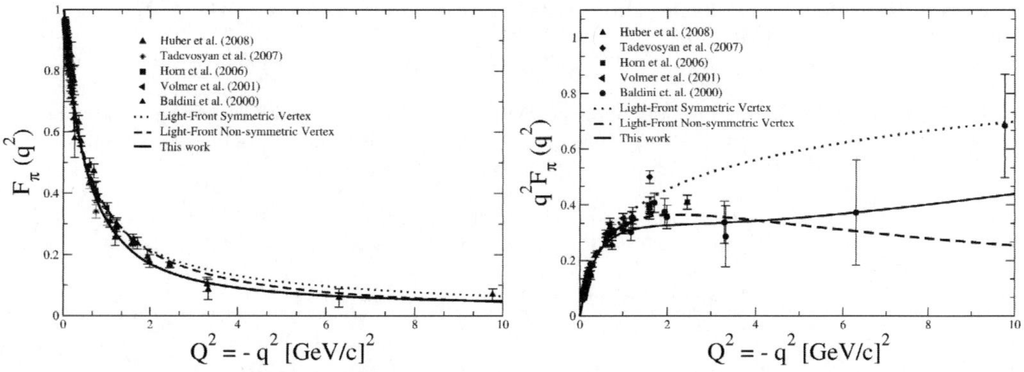

Figure 3: Pion electromagnetic form factor in terms of momentum transfer, q^2, with the light-front symmetric model, point line [30], dashed line, light-front non-symmetric vertex [29], solid line, present work, with the parameters from Lattice-QCD [6] and the experimental [35]. In this work, we use the following values parameters from the lattice-QCD, i.e, $\lambda^2 = 0.639\ GeV^2$, $m_0 = 0.014\ GeV$ and $m^3 = 0.189\ GeV^3$.

5. Conclusion

In this paper, we show how to describe the electromagnetic structure of a particle pseudoscalar, pion. Through a constituent quark model, we obtained some of the pion observable, but using the formalism in light-front we see across terms that are not invariant under Lorentz transformations. In our case we do not find non-valence terms to form factor when it is in Breit-frame and on Drell-Yan condition.

We study the electromagnetic current in respect to the valence term which we can extract the wave function of two fermions system, the quark-antiquark pair that composes the pion structure. From the wave function at the center of mass of the system, technique that has identified valence wave function in our model, we get the observable of pion, like the electromagnetic form factor. The electromagnetic current is decribed by constituent quarks model with a self-energy at pion vertex as expected. The form factor decreases with increasing of the transfer momentum. In relation to the form factor multiplied by square transfer momentum, it increases with increasing of q^2. The quark mass function describes the system in which particles dynamically gain mass, according to the constituent quark model. From the contribution of the photon we can study the internal structure of the pion as pion electromagnetic structure.

References

[1] James P. Vary, Nucl.Phys.Proc.Suppl. **251-252** (2014) 155.
[2] B. L. G. Bakker, Few-Body Syst **49** (2011) 177.
[3] P. Maris and C. D. Roberts , J. Mod. Phys. **E 12**, 297 (2003).
[4] A. Bashir, L. Chang, I. C. Cloët, B. El-Bennich, Y-X. Liu, C. D. Roberts and P. C. Tandy ,Commun. Theor. Phys. **58**, 79 (2012).
[5] E. Rojas, J. P. B. C. de Melo, B. El-Bennich, O. Oliveira and T. Frederico, JHEP **1310**, 193 (2013) [arXiv:1306.3022 [hep-ph]].
[6] D. Dudal, M. S. Guimarães, L. F. Palhares and S. P. Sorella, arXiv:1303.7134 [hep-ph].
[7] J.P.B.C. de Melo, T. Frederico, E. Pace, G. Salme, Phys.Lett. **B581** (2004) 75.
[8] J.P.B.C. de Melo, T. Frederico, E Pace and G. Salme, Phys.Rev. **D73** (2006) 074013.
[9] J. P. B. C. de Melo and T. Frederico, Phy. Rev. **C55** (1997) 2043.
[10] P. Maris; C. D. Roberts and P. C. Tandy, Physics Letters **B 420**, 267273 (1998).
[11] Clayton S. Mello, Anacé N. da Silva, J.P.B.C. de Melo and T. Frederico, Few Body Syst. **56** (2015) 6-9, 509.

[12] J.P.B.C. de Melo, Anac N. da Silva, Clayton S. Mello and T. Frederico, Few Body Syst. **56** (2015) 6-9, 503.
[13] Clayton S. Mello, Jose P.Cruz Filho, Edson O. da Silva, Bruno El-Bennich, J.P.B.C. de Melo and Victo S. Filho, AIP Conf.Proc. **1520** (2013) 333.
[14] Clayton S. Mello, Anacé Nunes da Silva, J.P.B.C. de Melo and T. Frederico, Nucl.Phys.Proc.Suppl. **251-252** (2014) 62.
[15] C. D. Roberts, M. S. Bhagwat, A. Holl and S. V. Wright Eur. Phys. J. Special Topics **140**,(2007) 53.
[16] Harindranath, A., Pramana **vol. 55**, n. 1 and 2, (2000), 141. And hep-ph 9612244 v 230, jun (1998).
[17] Y. Takahashi, "On the generalized Ward identity", Nuovo Cim. **6** (1957) 371.
[18] J. P. B. C. de Melo, Covariância na Frente de Luz, Tese de Doutorado, IFUSP, (1998).
[19] Itzykzon, C and Zuber, J B, Quantum Field Theory., New york, **vol. 1**, (1980).
[20] Drell, Sidney D and Bjorken, James D, Relativistic Qauntum Mechanics, Editor McGraw-Hill Book Company, **Vol.1**, (1964).
[21] Alan D Martin e Francis Halzen, Jhon Wiley and Sons, (1984), New York, USA.
[22] Lei Chang; I. C. Cloët; J. J. Çobos-Martinez; C. D. Roberts; S. M. Schmidt and P. C. Tandy, PRL **110**,132001 (2013).
[23] S.R. Beane, W. Detmold, K. Orginos, M.J. Savage, Prog.Part.Nucl.Phys. **66** (2011) 1.
[24] Quantum Fields on a Lattice, I. Montvay and G. Münster, Cambridge University Press, (1997).
[25] George F. Sterman, Paul Stoler, Ann.Rev.Nucl.Part.Sci. **47** (1997) 193.
[26] A. H. Mueller, Nucl. Phys. **A654** (1999) 37C.
[27] Edson O. da Silva, J.P.B.C. de Melo, Bruno El-Bennich, Victo S. Filho, Phys.Rev. **C86** (2012) 038202.
[28] George H. S. Yabusaki, Ishtiaq Ahmed, M. Ali Paracha, J.P.B.C. de Melo, Bruno El-Bennich, Phys.Rev. **D92** (2015) 3, 034017.
[29] J. P. B. C. de Melo, H. W. L. Naus and T. Frederico, Phy. Rev. C**59** (1999) 2278.
[30] J. P. B. C. de Melo, T. Frederico, E. Pace, and G. Salmé, Nucl. Phys. A**707** (2002) 399; Braz. Jour. Phys. **33** (2003) 301.
[31] J. Volmer *et al.*, Phy. Rev. Lett. **86** 1713 (2001).
[32] T. Horn *et al.*, Phys. Rev. Lett. **97**, 192001 (2006).
[33] V. Tadevosyan *et al.*, Phys Rev. C **75**, 055205 (2007).
[34] G. M. Huber *et al.*, Phys. Rev. C **78**, 045203 (2008).
[35] R. Baldini, E. Pasqualucci, S. Dubnicka, P. Gauzzi, S. Pacetti and Y. Srivastava, Nucl. Phys. A **666** (2000) 38.

Journal of Physics: Conference Series **706** (2016) 052009 doi:10.1088/1742-6596/706/5/052009

Comparative study between Glauber-Velasco and the Stochastic Vacuum models for high energy hadronic collision

D Szilard and T Kodama

Instituto de Física, Universidade Federal do Rio de Janeiro, Av. Athos da Silveira Ramos, 149 Centro de Tecnologia - bloco A - Cidade Universitária - Rio de Janeiro - RJ

E-mail: daniela@if.ufrj.br

Abstract.
The outlines of the Glauber-Velasco (GV) and the Stochastic Vacuum models (SVM) for describing pp elastic scattering are presented. The amplitudes and the cross-sections of both models are investigated and compared at $\sqrt{s} = 7\,\text{TeV}$. The asymptotic energy behaviour for the SVM is also studied.

1. Introduction

For decades, it has been a challenge to provide a suitable description of microscopic high-energy hadronic processes. The pp (and $p\bar{p}$) elastic collisions involve non-perturbative aspects of strong interactions and play a role in the understanding of QCD interactions and hadronic structure. Recently, the interest in this problem has increased due to the high-energy data avaiable from TOTEM [1, 2, 3] and ATLAS [4] experiments.

The understanding of the differential cross-section $d\sigma/dt$ depends directly on the interplay of the imaginary and real parts of the scattering amplitudes. For example, the existence of a dip in the $d\sigma/dt$ is related to the occurrence of a zero in the imaginary part and its height is influenced by the magnitude of the real part. The analytical representation of the amplitudes can only be achieved through models.

The literature provides a large number of phenomenological models. All of them should respect general quantum field theory principles, like analiticity and unitarity. Both properties may be assured in eikonalized models, which is an approach suitable to high energy scattering at small angles. An interesting way to probe these models is compare them with each other and, more importantly, with the data.

In the present work, the Stochastic Vacuum Model (SVM) and the Glauber-Velasco (GV) model were studied and compared at $\sqrt{s} = 7\,\text{TeV}$. The above two models reflect the two complementary aspects of QCD, particle and field. Our main goal is to clarify the relation between Glauber-Velasco model and the Stochastic Vacuum Model in the point of view of QCD. Some predictions for asymptotic energies were also investigated.

2. The Stochastic Vacum Model

More than two decades ago, E. Ferreira and H. G. Dosch [5, 6] developped the so-called Stochastic Vacuum Model (SVM), a model where the QCD vacuum structure is represented by a stochastic background.

Rather than calculating the quark-quark scattering amplitudes, the basic entities in this model are the Wilson loops. The Wilson loops are gauge-invariant field operators responsible for parallel transport in a closed path. A meson, for instance, can be treated as a quark and an antiquark, following an antiparallel trajectories (see Fig. 1). In this description, it is also necessary to transport color charges within the quark pair to obtain a colorless meson, and this role is played by a Wilson loop.

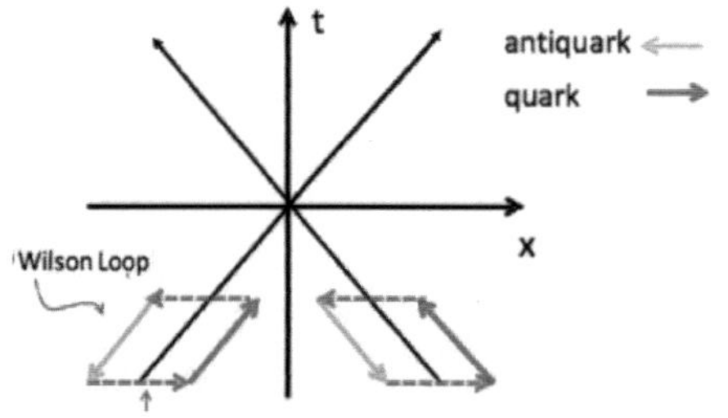

Figure 1. Schematic representation of a meson. Extracted from reference [7].

In order to evaluate the hadron-hadron scattering amplitude, some of the main assumptions of the model can be summarized as:

- the quark in study must be very energetic; thus, when it interacts with the background field it moves on a approximately straight lightlike line and the eikonal approximation is valid;
- the low-frequency contributions to the functional integral can be accounted as a stochastic process (the high-frequency contribution may be considered perturbatively);

The eikonal function is supposed to be linear with the quark-antiquark potential and this potential depends on Wilson expectation value $\langle W(C)\rangle$; only the lowest order on loop expansion $\langle W(C)\rangle$ is kept and it depends explicitly on gluon condensate, the correlation length of the vacuum field fluctuations and string tension.

A very specific feature of the model is the dependence of the cross section on the hadron size even if it is large compared to the correlation length. This effect spoils the quark additivity and is related to the same mechanism that leads to confinement.

Inspired by that formalism, A. K. Kohara, E. Ferreira and T. Kodama worked on a description of the available pp (and $p\bar{p}$) elastic scattering data from $\sqrt{s} \sim 20\,\text{GeV}$ to $7\,\text{TeV}$ [8, 9, 10, 11]. The impact parameter representation (b-space) is defined as the Fourier transform of the amplitudes in t-space. The impact parameter $\vec{b}$ is not an observable, but can give important information about geometrical aspects of the collision. In this sense, the amplitudes were built based of the profiles functions in the b-space and are given by

$$\tilde{T}_K(s,\vec{b}) = \frac{\alpha_K}{2\beta_K}\, e^{-b^2/4\beta_K} + \lambda_K\, \tilde{\psi}(s,b) \tag{1}$$

with the characteristic shape function

$$\tilde{\psi}_K(s,b) = \frac{2\, e^{\gamma_K - \sqrt{\gamma_K + b^2/a_0}}}{a_0\sqrt{\gamma_K + b^2/a_0}} \left[1 - e^{\gamma_K - \sqrt{\gamma_K + b^2/a_0}}\right], \tag{2}$$

where $K = R, I$ labels the real and the imaginary parts of the amplitude; $a_0 = 1.39\,\mathrm{GeV}^{-2}$ is related to the gluon correlation length of the correlation function.

The Fourier transform of Eq. 1 determines the real and imaginary scattering amplitudes in t-space

$$T_K^N(s,t) = \alpha_K(s)\, e^{-\beta_K|t|} + \lambda_K(s)\, \Psi_K(\gamma_K(s), t) + \delta_{K,R}\, R_{ggg}(t). \tag{3}$$

$R_{ggg}(t)$ represents the contribution from three-gluon exchange amplitude. It is responsible for the $|t|^{-8}$ tail behaviour in $d\sigma/dt$. The shape functions in t-space are

$$\tilde{\Psi}_K(\gamma_K(s), t) = 2e^{\gamma_K}\left[\frac{e^{-\gamma_K\sqrt{1+a_0|t|}}}{\sqrt{1+a_0|t|}} - e^{\gamma_K}\frac{e^{-\gamma_K\sqrt{4+a_0|t|}}}{\sqrt{4+a_0|t|}}\right]. \tag{4}$$

An interesting feature of this model is the fact that the amplitudes in both b and t-space have simple analytical forms. The amplitudes were built in order to be valid in a large $|t|$ range, accounting for the exponential decrease in the very foward region. For a complete analysis of elastic scattering, the Coulomb phase must be taken into account; thus, the amplitudes are modified as

$$\begin{aligned} T_R(s,t) &= T_R^N(s,t) + \sqrt{\pi}\, F^C(t)\cos(\alpha\Phi) \\ T_I(s,t) &= T_I^N(s,t) + \sqrt{\pi}\, F^C(t)\sin(\alpha\Phi), \end{aligned} \tag{5}$$

where α is the fine-structure constant, $\Phi(s,t)$ is the Coulomb phase and $F^C(t)$ is linked to the proton form factor. With the amplitude choice in Eq. 3, the differential cross-section is

$$\frac{d\sigma(s,t)}{dt} = (\hbar c)^2\left[T_I^2(s,t) + T_R^2(s,t)\right] \tag{6}$$

and the total cross-section,

$$\sigma_{tot}(s) = (\hbar c)^2 4\sqrt{\pi}\, T_I(s, t=0). \tag{7}$$

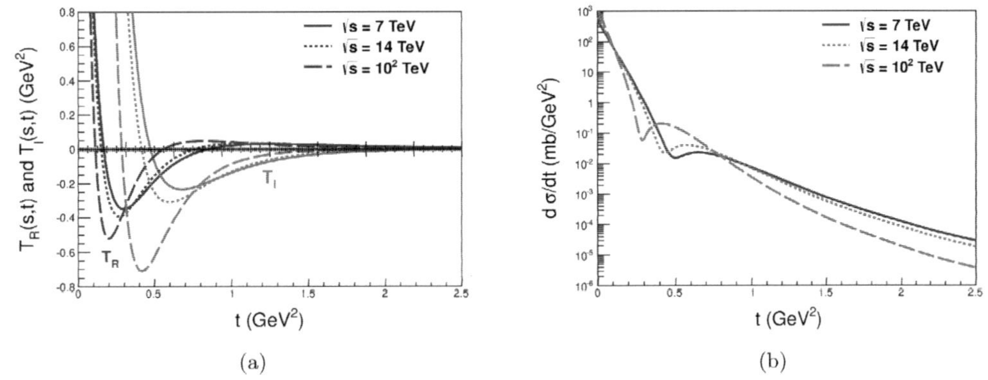

Figure 2. SVM predictions for three different energies. In Fig. (a), real and imaginary amplitudes $T_R(s,t)$ and $T_I(s,t)$ are shown and, in Fig. (b), the differential cross-section $d\sigma/dt$.

The energy-dependence of the amplitudes and the cross-section is given through eight energy-dependent parameters, four for each amplitude, $\alpha_K(s)$, $\beta_K(s)$, $\gamma_K(s)$ and $\lambda_K(s)$ [11] (see Fig. 2). Knowing the energy-dependence of the parameters allows us to estimate what happens in the asymptotic energy limit. The total cross-section increases with the energy and the position

Journal of Physics: Conference Series **706** (2016) 052009 doi:10.1088/1742-6596/706/5/052009

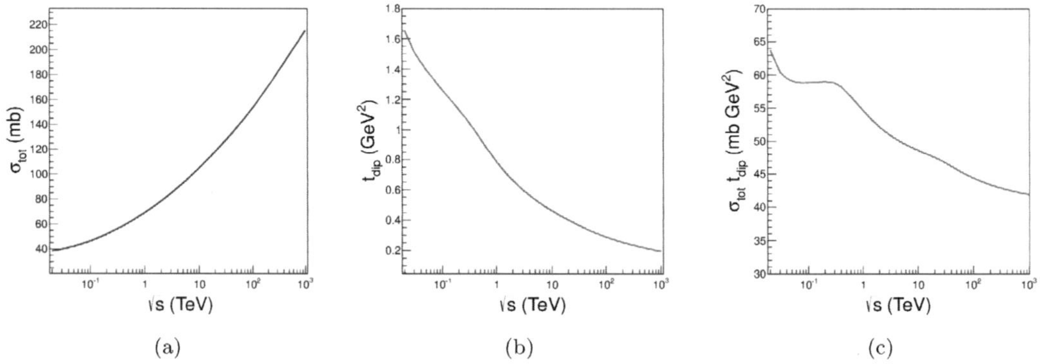

Figure 3. SVM prediction for the energy dependence of the total cross-section σ_{tot} (a), the dip position t_{dip} (b) and the product $\sigma_{tot}\, t_{dip}$ (c).

of the local minimum in the $d\sigma/dt$, $t = t_{dip}$, decreases. It was first noted by T. Csörgo [15] that there is a region where the product $\sigma_{tot}\, t_{dip}$ appears to be constant, but, in the SVM prediction, this seems to be accidental and does not extend for higher energies (see Fig. 3).

The probability of a inelastic interaction decreases smoothly as the impact parameter increases. This behaviour is shown in Fig. 4, where the probability of an inelastic collision $d^2\tilde{\sigma}_{in}/db^2$ is expressed in terms of the impact parameter b and with respect to the scaled variable x

$$x = \frac{b}{\sqrt{2\pi\, \sigma_{tot}(s)}}. \tag{8}$$

The differential cross-sections $d^2\tilde{\sigma}_{el}/db^2$, $d^2\tilde{\sigma}/db^2$ and $d^2\tilde{\sigma}_{in}/db^2$ are related to the amplitudes in Eq. 1 through

$$\begin{aligned} d^2\tilde{\sigma}_{el}(s,b)/db^2 &= \frac{(\hbar c)^2}{\pi}\, |\tilde{T}(s,b)|^2 \\ d^2\tilde{\sigma}(s,b)/db^2 &= \frac{2}{\sqrt{\pi}}\, (\hbar c)^2\, \tilde{T}_I(s,b) \\ d^2\tilde{\sigma}_{in}(s,b)/db^2 &= (\hbar c)^2 \left(\frac{2}{\sqrt{\pi}} \tilde{T}_I(s,b) - \frac{1}{\pi} |\tilde{T}(s,b)|^2 \right). \end{aligned} \tag{9}$$

It is clear that for asymptotic energies, the quantity $d^2\tilde{\sigma}_{in}/db^2$ as a function of x is getting steeper, but does not approach the Heaviside step function. This behaviour is related to the fact that the proton in the SVM does not approach the black disk limit for higher and higher energies.

3. Glauber-Velasco model

The multiple scattering theory was first developed in late 50's by Glauber, and then applied extensively in various areas to calculate high-energy scattering amplitude of composite particles [12].

The proton (or the antiproton) is pictured as a cluster of partons, which pass through each other, and interact through the partonic collisions. The pp scattering amplitude may be written as [13]

$$F(t) = i \int_0^\infty J_0\left(b\sqrt{-t}\right) \left[1 - e^{-\Omega(b)}\right] b\, db, \tag{10}$$

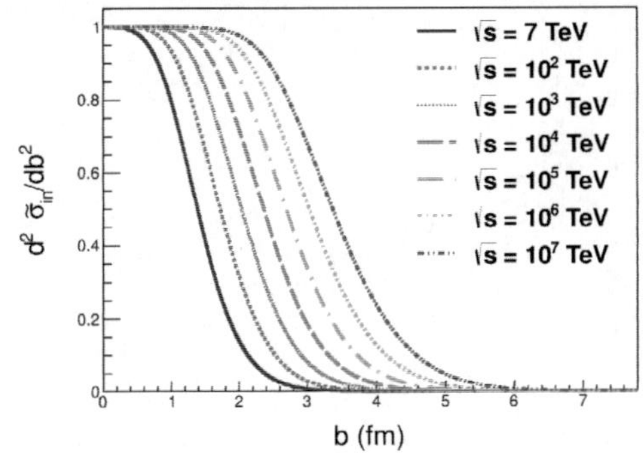

(a)

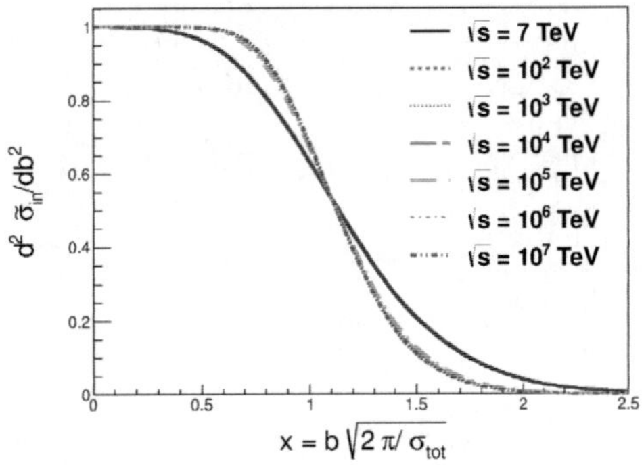

(b)

Figure 4. Energy evolution of the quantity $d^2\tilde{\sigma}_{in}/db^2$ as a function of the impact parameter b (a) and the variable $x = b/\sqrt{2\pi\,\sigma_{tot}(s)}$ (b).

where $\Omega(b)$ is the opacity function and b is the impact parameter. The elastic differential cross-section is given by $d\sigma_{el}/dt = \pi|F(t)|^2$. Any particular model depends on the choice for $\Omega(b)$, the opacity function (in general, it is complex valued). For two colliding nucleons, that scatter one into another with the parton-parton averaged scattering amplitude $f(t)$, $\Omega(b)$ is given by

$$\Omega(b) = \frac{\kappa}{4\pi}(1-\alpha)\int_0^\infty J_0(b\,q)\,G_{p,E}^2(-t)\frac{f(t)}{f(0)}\,q\,dq. \tag{11}$$

The parton-parton scattering amplitude depends on the parametrization. The parametrization investigated was [15]

$$\frac{f(t)}{f(0)} = \frac{e^{i(b_1|t|+b_2t^2)}}{\sqrt{1+a|t|}}, \tag{12}$$

where the parameters b_1, b_2 e a can be extrated from the data fit. In total, there are five free parameters in pp cross section data fit (a_1, a_2, b, κ and α). The $G_{p,E}(t)$ proton form adopted was a four pole parametrization [16]

$$G_{p,E}(q^2) = \sum_{i=1}^{n=4}\frac{a_i^E(m_i^E)^2}{(m_i^E)^2+q^2}, \quad \sum_{i=1}^{n=4}a_i^E = 1, \quad G_{p,E}(0) = 1 \tag{13}$$

with the parameters a_i e m_i kept fixed in the pp elastic cross section fit.

This model has been successfully applied to pp scattering data in TeV domain. Here, we have not performed a data fit, but only reproduced the model with the parameters of [14] for $\sqrt{s} = 7\,$TeV and we have not understood the energy dependence yet. In Fig. 5, the amplitudes and the cross section for $\sqrt{s} = 7\,$TeV for GV and SVM are plotted. Both models are in good agreement with the TOTEM 7 TeV data [3].

4. Final Comments

The GV and SVM models are in good agreement with the data. In order to better interpret the relation between them, it is fundamental to clarify the energy dependece of the Glauber-Velasco model. The SVM and the GV reveils complementary aspects of QCD, particle and field. As the

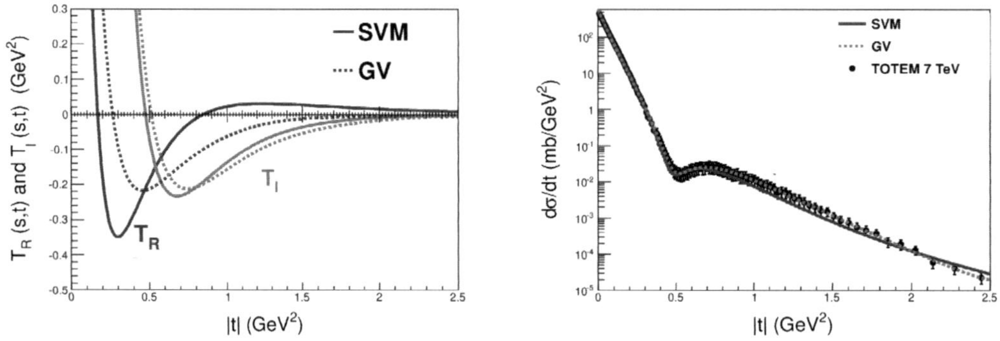

(a) (b)

Figure 5. In Fig. (a), real and imaginary amplitudes $T_R(s,t)$ and $T_I(s,t)$ for the SVM and GV for $\sqrt{s} = 7\,\mathrm{TeV}$. In Fig. (b), the differential cross-section $d\sigma/dt$ TOTEM 7 TeV data [3] compared to SVM and GV results.

collision energy increases, the parton-parton interaction range and rhe number of partons are expected to increase as well, which explains the enhacement in the elastic cross section at high energies. As showed in Fig. 3, this enhancement is expected in the SVM.

In the SVM, the proton does not behave as a black disk for asymptotic energies. This behaviour is related to the vacuum property of QCD and the mechanism of confinement and shoud be more profoundly studied. It will be interesting to compare these two models with other models used to describe *pp* elastic collisions.

Acknowledgments

The authors would like to thank E. Ferreira for careful reading and fruitful discussion. This work was financially supported by CNPq and CAPES.

References

[1] Antchev G et al (TOTEM Collaboration) 2011 *Europhys. Lett.* **95** 41001
[2] Antchev G et al (TOTEM Collaboration) 2011 *Europhys. Lett.* **96** 21002
[3] Antchev G et al (TOTEM Collaboration) 2013 *Europhys. Lett.* **101** 21002
[4] Aad G et al (ATLAS Collaboration) 2014 *Nucl. Phys.* B **889** 486
[5] Dosch H G 1987 *Phys. Lett.* B **190** 177
[6] Dosch H G, Ferreira E and Kramer A 1994 *Phys. Rev.* D **50** 1992
[7] Kohara A K 2012 *Espalhamento elstico pp e $p\bar{p}$* Master Thesis (Instituto de Física da Universidade Federal do Rio de Janeiro)
[8] Kohara A K, Ferreira E and Kodama T 2014 *J. Phys* G **41** 115003
[9] Kohara A K, Ferreira E and Kodama T 2013 *Phys. Rev.* D **87** 054024
[10] Kohara A K, Ferreira E and Kodama T 2013 *Eur. Phys. J.* C **73** 2326
[11] Kohara A K, Ferreira E and Kodama T 2014 *Eur. Phys. J.* C **74** 3175
[12] Glauber R J 1959 *Lectures in Theoretical Physics* vol 1, ed W E Brittin and L G Dunham (New York: Interscience Publishers) p 315
[13] Glauber R J and Velasco J 1984 *Phys. Lett.* B **147** 380
[14] Csörgö T 2014 *talk presented at International Workshop on Collectivity in Relativistic Heavy Ion Collisions* (Kolymbari, Crete) p 14
[15] Csörgö T, Glauber R J and Nemes F 2013 *arXiv:1311.2308v1*
[16] Borkowski F, Simon G G, Walther V H and Wendling R D 1975 *Nucl. Phys.* B **93** 461

Journal of Physics: Conference Series **706** (2016) 052010 doi:10.1088/1742-6596/706/5/052010

Relativistic BEC-BCS Crossover in a magnetized Nambu-Jona-Lasinio Model

Dyana C. Duarte[1], R.L.S. Farias[1,2], Pedro H.A. Manso[3,4], Rudnei O. Ramos[3]

[1] Departamento de Física, Universidade Federal de Santa Maria, 97105-900, Santa Maria, RS, Brazil

[2] Physics Department, Kent State University, Kent, OH 44242, United States

[3] Departamento de Física Teórica, Universidade do Estado do Rio de Janeiro, 20550-013 Rio de Janeiro, RJ, Brazil

[4] Centro Federal de Educação Tecnológica Celso Suckow da Fonseca, Campus Maria da Graça, Rua Miguel Ângelo 96, 20785-223 Rio de Janeiro, RJ, Brazil.

E-mail: dyana.c.duarte@gmail.com

Abstract. The BEC-BCS crossover in the NJL model is studied in the presence of an external magnetic field. Particular attention is given to two different regularization schemes used in the literature and we show how they compare to each other. The comparison is made for the case of a cold and magnetized two color NJL model. We also make a brief discussion about the $N_c = 3$ case without magnetic fields, as an extension of this work in the future.

1. Introduction

The phase structure of quantum chromodynamics (QCD) is of relevance in many different contexts, for example, in the study of relativistic heavy ion collisions, compact stars and in the early universe. Strong magnetic fields is expected to be produced in noncentral heavy-ion collisions, which motivates the study of their effects in the phase diagram of strongly interacting matter (see [1] and references). Perturbation theory of QCD is applicable in asymptotic regimes of temperatures and densities, while the study of the phase transitions at moderate temperatures and densities cannot be implemented on lattice QCD ($N_c = 3$) calculations due to the fermion sign problem. These facts have then motivated the study of many low energy effective models, such as chiral perturbation theory, linear sigma model, the Nambu-Jona-Lasinio (NJL) model among others models that are used to study the phase structure of strongly interacting matter [2]. It is generally expected that there should exist a crossover from Bose-Einstein condensation (BEC) to Bardeen-Cooper-Shriffer condensation (BCS) for diquarks at finite baryon density. This crossover can be observed in different ways, such as increasing the coupling constant of the attractive interations or changing the charge number through the variation of the chemical potential [3]. In this work we review the investigations of the BEC-BCS crossover in cold QCD at finite baryon chemical potential using the $SU(2)$ version of the NJL model for $N_c = 2$ and $N_c = 3$ in the mean field approximation. We study in details the effect that an external magnetic field on the crossover. We hope that our results will help to explain the role played by a magnetic field in the crossover BEC-BCS in real systems.

2. Thermodynamic potential for $N_c = 2$ QCD in the mean field approximation

2.1. Zero magnetic field case

We study the diquark condensation in a two color QCD. The starting point is from the Lagrangian density for the NJL model:

$$\begin{aligned}\mathcal{L} &= \bar{\psi}\left(i\gamma^\mu\partial_\mu - m_0\right)\psi + G_S\left[\left(\bar{\psi}\psi\right)^2 + \left(\bar{\psi}i\gamma_5\tau\psi\right)^2\right] \\ &+ G_D\left(\bar{\psi}i\gamma_5\tau_2 t_2 C\bar{\psi}^T\right)\left(\psi^T C i\gamma_5\tau_2 t_2\psi\right) \end{aligned} \quad (1)$$

where m_0 is the current mass, $C = i\gamma_0\gamma_2$ is the charge conjugation matrix, τ_i and t_i are the Pauli matrices in flavor and color spaces, G_S and G_D are the coupling constants, connected by a Fierz transformation in the color space, leading to $G_S = G_D = G$ in the two color case. The advantage of studying the two color QCD is that the confinement is less important than in three color QCD. The diquarks are colorless baryons and the diquark condensation breaks the baryon symmetry $U_B(1)$ rather than the color symmetry $SU_C(2)$ [3].

In this work our aim is to study the crossover BEC-BCS. It is well known that this phenomenon can happen at low temperatures. Thus, we start from the thermodynamic potential at zero temperature:

$$\Omega_0 = \frac{(m-m_0)^2 + \Delta^2}{4G} - 2N_cN_f\int\frac{d^3k}{(2\pi)^3}\left[\frac{E_\Delta^+}{2} + \frac{E_\Delta^-}{2}\right] \quad (2)$$

where the dispersion energies are $E_\Delta^\pm = \sqrt{(E_k \pm \mu)^2 + \Delta^2}$ and $E_k = \sqrt{k^2+m^2}$. Also, note that E_Δ^- corresponds to particles and E_Δ^+ corresponds to the antiparticles excitations. The NJL model is nonrenormalizable. Thus, some regularization scheme has to be used to account for the divergences. The most common procedure in the $B = 0$ regimes is to use a three dimensional momentum cutoff Λ. This is valid since it is well known that the crossover happens away from the asymptotic regions. In this way, all physical quantities become dependent on this parameter Λ. Besides the cutoff, we have the coupling constants G_S and G_D and the current mass m_0 to be specified. In three color QCD case these parameters are fixed such that the empirical values of the pion mass m_π, the pion decay constant f_π and chiral condensate in the vacuum $\langle\bar{\psi}\psi\rangle_0$ might be obtained. On the other hand, to determine the parameters used in numerical calculations for the $N_c = 2$ case we can follow the procedure proposed in reference [4], rescaling the physical quantities by N_c. It is well known that f_π is proportional to $N_c^{1/2}$ and $\langle\bar{\psi}\psi\rangle_0$ is proportional to N_c; therefore, we rescale the $N_c = 3$ values of these quantities by the factors $\sqrt{2/3}$ and and $2/3$, respectively. Therefore, for $N_c = 2$ the model parameters Λ, G and m_0 must be such that we would reproduce the rescaled values $f_\pi = 75.45$ MeV, $m_\pi = 140$ MeV and $\langle\bar{q}q\rangle_0^{1/3} = -218$ MeV.

In anticipation to the case of including a magnetic field, we call the three dimensional cutoff regularization Λ^{3d}, and also propose a smooth cutoff, with a Gaussian factor form $f_\Lambda(p) = e^{-p^2/\Lambda^2}$, implemented through the substitution

$$\int\frac{d^3k}{(2\pi)^3} \to \int\frac{d^3k}{(2\pi)^3}f_\Lambda(p) \quad (3)$$

where $p = \vec{k}$ in $B = 0$ case. In this work we refer to this regularization based in the Gaussian factor form as GR scheme. For each type of regularization used, we evaluate the corresponding model parameters, that are given in table 1.

It is important to observe that the values obtained for the Λ^{3d} scheme were obtained by using the N_c rescaling mentioned before (see reference [4]), but for the GR scheme, it is not possible

Table 1. Parameter sets for the NJL SU(2) model.

Parameter set	$m(0)$ MeV	m_0 MeV	G GeV^{-2}	Λ MeV
GR	305.385	3.557	4.77	777
Λ^{3d}	305.385	5.400	7.23	657

to obtain good values for the physical quantities in $N_c = 2$, using this rescaling. Therefore, we fix the vacuum mass and evaluate G, Λ and m_0 that reproduce $f_\pi = 75.45$ MeV, $m_\pi = 140$ MeV. The value obtained for the quiral condensate is $\langle \bar{q}q \rangle_0^{1/3} = -250.96$ MeV and the other parameters are shown in table 1.

Minimizing the potential with respect to Δ and m, we obtain the respective Gap equations. Namely

$$\frac{\partial \Omega_0}{\partial \Delta} = 0 \, , \qquad \frac{\partial \Omega_0}{\partial m} = 0 \tag{4}$$

To study the BEC-BCS crossover we decide to follow the same procedure of the reference [3], defining a reference chemical potential $\mu_N = \mu - m$, or equivalently, $\mu_N = \frac{\mu_B}{2} - m$ for the two-color QCD case. It is possible to note that for small values of μ_B we have $\mu_N < 0$, and the minimum of the dispersion is located at $|\vec{k}| = 0$ with particle gap energy $\sqrt{\mu_N^2 + \Delta^2}$. Otherwise, at larger values of μ_B we have $\mu_N > 0$, the minimum of the dispersion is shifted to $|\vec{k}| = \mu_B/2$, and the particle gap is Δ. This is a signal of the BEC-BCS crossover that happens due to the changes in the quark mass during the chiral symmetry restoration, when the chemical potential increases.

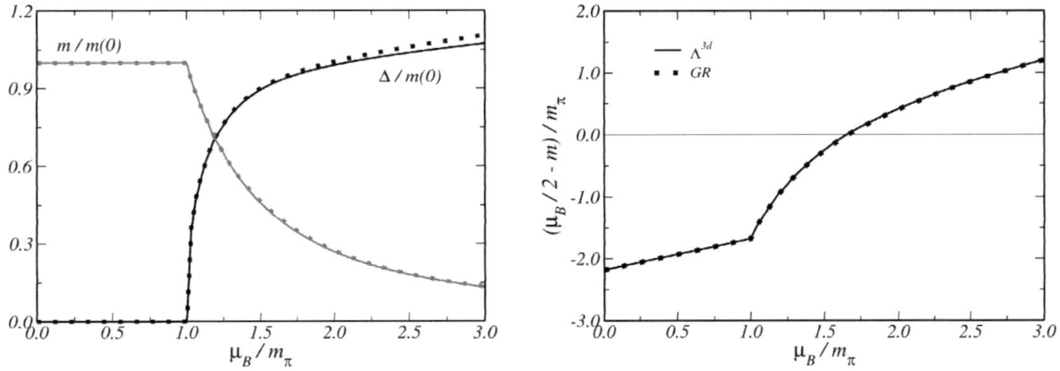

Figure 1. Left panel: (Color online) Effective quark mass m and diquark condensate Δ, rescaled by the vacuum quark mass $m(0)$, as functions of μ_B/m_π for the two regularization schemes used in this work. Continuum lines correspond to Λ^{3d}, and dotted lines correspond to the GR scheme. **Right panel:** Reference chemical potential $\mu_N = \mu - m$, for Λ^{3d} and GR schemes.

In the first panel of figure 1 we show the numerical solutions for the gap equations (4). Using both parametrizations, based on the N_c rescaling and for GR scheme, our qualitative results are similar to the ones obtained in the reference [3], where the authors work with parameters usually taken in the $N_c = 3$ case. It is possible to see from the behavior of the diquark condensates that a second order phase transition is happening at $\mu_B = m_\pi$, where the mass m begins to decrease from its vacuum value $m(0)$, while Δ becomes nonzero. On the other hand, in the right panel of figure 1 we see that for both regularization schemes the crossover happens at the same value, $\mu_{B_c}/m_\pi \sim 1.7$ (corresponding to a μ_{B_c} about 240 MeV). The critical baryon chemical potential μ_{B_c} is the value where μ_N changes its signal. When $\mu_N < 0$ the system is in the BEC state and with the increase of the chemical potential μ_N becomes positive and the system goes to BCS state.

2.2. Nonzero magnetic field case

In the presence of an external magnetic field, the corresponding Lagrangian density is given by

$$\begin{aligned}\mathcal{L} &= \bar{\psi}\left(i\not{D} - m_0\right)\psi + G_S\left[\left(\bar{\psi}\psi\right)^2 + \left(\bar{\psi}i\gamma_5\tau\psi\right)^2\right] \\ &\quad + G_D\left(\bar{\psi}i\gamma_5\tau_2 t_2 C\bar{\psi}^T\right)\left(\psi^T C i\gamma_5\tau_2 t_2\psi\right), \end{aligned} \tag{5}$$

with the same parameters defined in the $B = 0$ case; moreover, the coupling of the quarks to the electromagnetic field $\mathcal{A}_\mu$ is implemented through the covariant derivative $D_\mu = \partial_\mu - iQ\mathcal{A}_\mu$, where Q is the usual quark charge matrix. In this work we set $q_u = -q_d$ such that the Cooper pairs remain color neutral. In order of include the effects of an external magnetic field in the thermodynamic potential, it is usual to make use of the following prescription in all the d^3k integrals [5]:

$$N_f \int \frac{d^3k}{(2\pi)^3} \rightarrow \sum_{l=0}^{\infty} \sum_{f=u}^{d} \frac{|q_f|B}{4\pi}\alpha_l \int_{-\infty}^{+\infty} \frac{dk_z}{(2\pi)} \tag{6}$$

where the factor $\alpha_l = 2 - \delta_{l,0}$, takes into account the degeneracy of the Landau levels l that are all doubly degenerate (except for $l = 0$). This procedure is, after all, equivalent to take the $T = 0$ limit after perform a Matsubara sum in $k_0 \rightarrow i(\omega_n - i\mu))$ in the thermodynamic potential. The dispersion relation becomes $E_k \rightarrow E_{k,l} = \sqrt{k_z^2 + 2l|q_f|B + m^2}$ and $E_\Delta^\pm \rightarrow E_{\Delta,l}^\pm = \sqrt{(E_{k,l} \pm \mu)^2 + \Delta^2}$. After this prescription, Ω_0 becomes

$$\Omega_0(B) = \frac{\left(m - m_0\right)^2 + \Delta^2}{4G} - N_c \sum_{f=u}^{d} \frac{|q_f|\,B}{4\pi} \sum_{l=0}^{\infty} \alpha_l \int_{-\infty}^{+\infty} \frac{dk_z}{2\pi}\left[E_\Delta^+ + E_\Delta^-\right] \quad , \tag{7}$$

As mentioned before, it is well known that the NJL model is nonrenormalizable. Thus, it is necessary to regularize the momentum integrals. In the $B \neq 0$ case we use a similar procedure as in the $B = 0$ case studied above, but now we define a cutoff $\Lambda_l = \sqrt{\Lambda^2 - 2l|q_f|B}$, that is a function of the Landau levels, and so the superior limit for the sum over the Landau levels becomes the integer part of $\Lambda^2/(2|q_f|B)$ [6]. In the Gaussian factor form regularization scheme we now have

$$\sum_{l=0}^{\infty} \int_{-\infty}^{\infty} \frac{dk_z}{2\pi} \rightarrow \sum_{l=0}^{\infty} \int_{-\infty}^{\infty} \frac{dk_z}{2\pi} f_\Lambda(p) \tag{8}$$

where $p = k_z^2 + 2|q_f|Bl$ in $B \neq 0$ case [7, 8].

In the figure 2 we show the numerical results for the gap equations evaluated from the new thermodynamic potential. It is possible to see that there is a magnetic catalysis phenomena happening. As the magnetic field increases, the value of the mass, even at small values of μ_B, increases as well. We also note that the phase transition for the diquark condensate remains of second order, even in the presence of the external magnetic fields. The value of the critical chemical potential changes with qB. Note that at $qB = 0$ the phase transition happens at $\mu_B = m_\pi$, in agreement with the chiral effective theories [9], for both schemes presented here.

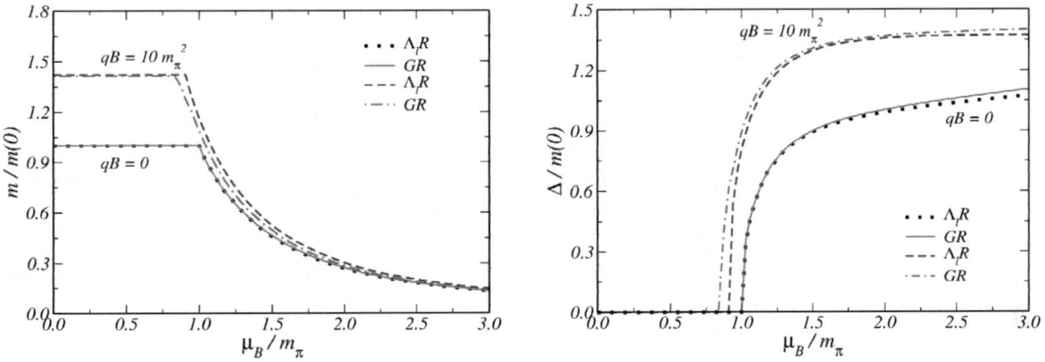

Figure 2. (Color online) Effective quark mass m (left panel) and diquark condensate Δ (right panel), both scaled by the vacuum quark mass, as functions of μ_B/m_π, for $B = 0$ and $B = 10m_\pi^2$.

In the figure 3 we have $\mu_N = \mu - m$ for $qB = 10m_\pi^2$ in comparison with $qB = 0$. Remember that negative values of μ_N corresponds to the BEC region, and positive ones corresponds to the BCS region. We see that the critical baryon chemical potential is also affected by the magnetic field. To determine how the crossover is affected through a large range of magnetic fields, we have the plot shown in the right panel of figure 3. Note that once this phenomenon happens when μ_N changes its signal, we set $m = \mu$ in the gap equations and solve it for μ_{B_c} and Δ. Here we observe that there are non-physical oscillations when working with the Λ_l scheme that are not present in the GR case. From the dashed curve shown in the right panel we may see that the critical baryon chemical potential increases with the magnetic field, i.e., the crossover happens for a higher value of μ_B, favoring the BEC region.

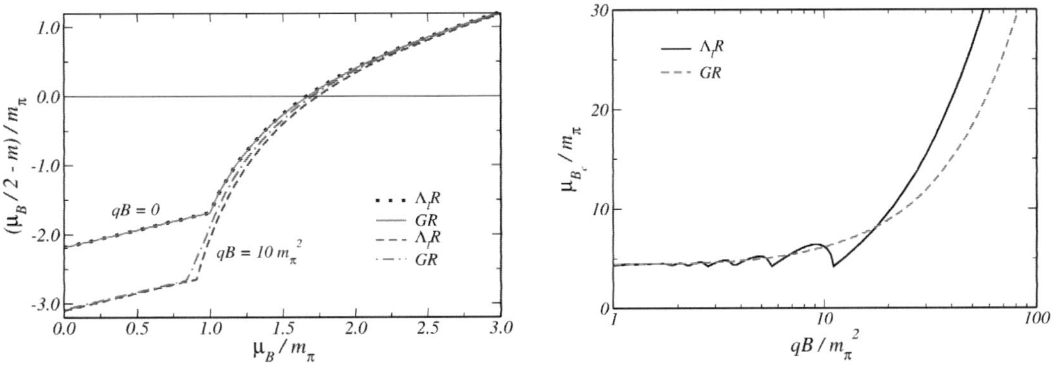

Figure 3. (Color online) **Left panel:** $\mu_N = \mu - m$ for $B = 0$ and $qB = 10m_\pi^2$, as a function of the baryon chemical potential (scaled by the pion mass). **Right panel:**, Critical chemical potential for the BEC-BCS crossover, as a function of the magnetic field.

3. NJL model with three colors at $B = 0$

In this section we consider a more realistic case, with three color degrees of freedom. The NJL model with $N_c = 3$ and including the scalar diquark channel is

$$\begin{aligned}\mathcal{L} &= \bar{\psi}\left(i\gamma^\mu\partial_\mu - m_0\right)\psi + G_S\left[(\bar{\psi}\psi)^2 + (\bar{\psi}i\gamma_5\tau\psi)^2\right] \\ &+ G_D\sum_{a=2,5,7}\left(\bar{\psi}i\gamma_5\tau_2\lambda_a C\bar{\psi}^T\right)\left(\psi^T C i\gamma_5\tau_2\lambda_a\psi\right)\end{aligned} \tag{9}$$

where λ_a are the Gell-Mann matrices in color space. Differently from the two color problem, now the confinement problem is important. Furthermore, the diquarks will no longer be colorless and the requirement of the color neutrality is not satisfied if we just include the diquarks.

The model parameters for the NJL model in the $N_c = 3$ case when $qB = 0$ are set to be $m_0 = 5$ MeV, $G_S = 4.93 \times 10^{-6}$ MeV^{-2}, $\Lambda = 653$ MeV and we choose to work with the ratio $\eta = G_D/G_S$ instead of G_D [3], to reproduce the physical parameters already discussed in previous section. The mean-field thermodynamic potential in the zero temperature limit becomes

$$\Omega_0 = \frac{(m-m_0)^2}{4G_S} + \frac{\Delta^2}{4G_D} - 4\int\frac{d^3k}{(2\pi)^3}\left[E_\Delta^+ + E_\Delta^- + \frac{1}{2}\left(E_k^+ + E_k^-\right)\right] \tag{10}$$

where $E_\Delta^\pm = \sqrt{(E_k \pm \frac{\mu_B}{3})^2 + \Delta^2}$ and $E_k^\pm = E_k \pm \frac{\mu_B}{3}$. Due to the SU(3) color symmetry we may set the gauge $\Delta_2 = \Delta$, and $\Delta_5 = \Delta_7 = 0$, so that only red and green quarks take place in the condensation [10].

If we do not take into account the effect of color neutrality, we may consider the same chemical potential for red, green and blue quarks, $\mu_r = \mu_g = \mu_b = \mu_B/3$. In the left panel of figure 4 we show some results for the mass m and diquark condensate Δ, without considering effects of color neutrality [3]. Those are the numerical solutions obtained when minimizing the thermodynamic potential with respect to Δ and m. In the right panel we have the reference chemical potential μ_N, as a function of μ_B, for different values of η. The BEC-BCS crossover in $N_c = 3$ QCD also happens when the quantity μ_N exchange its signal, and the increase of η favors the BCS region, once μ_N becomes positive in a smaller value of μ_B.

When considering the color neutrality, the gauge chosen for Δ may be kept, but it is necessary to include a color chemical potential μ_8, such that the quark chemical potentials are replaced by $\mu_r = \mu_g = \mu_B/3 + \mu_8/3$, and $\mu_b = \mu_B/3 - 2\mu_8/3$, and a new Gap equation, $-\frac{\partial\Omega_0}{\partial\mu_8} = 0$, has to be solved for μ_8. Our next step is study the influence of the magnetic field on the BEC-BCS crossover in $N_c = 3$ case, with and without the effects of color neutrality.

4. Final Remarks

In this work we have studied the effect of the application of an external magnetic field in the BEC-BCS crossover in the NJL model with $N_c = 2$. By using the GR scheme, we verify that when increasing the value of the magnetic field, the BEC region is strengthened, once the chemical potential μ_N changes its signal at higher values of μ_B. Interestingly, this result is the opposite of that found by the authors in the reference [8], where they study the effect of qB in the crossover in a different model, keeping constant the quark mass m. The result obtained in [8] was that at strong magnetic fields the system goes to the BCS regime.

It is important to note that the diquark condensate phase transition happens at $\mu_B = m_\pi$, in agreement with effective theories predictions, and when including effects of magnetic fields it remains of second order, for both regularization schemes that we have used.

To summarize, we have reached two important conclusions in this work. Firstly, the GR scheme does not reproduce the physical parameters in the vacuum. We have not found a way

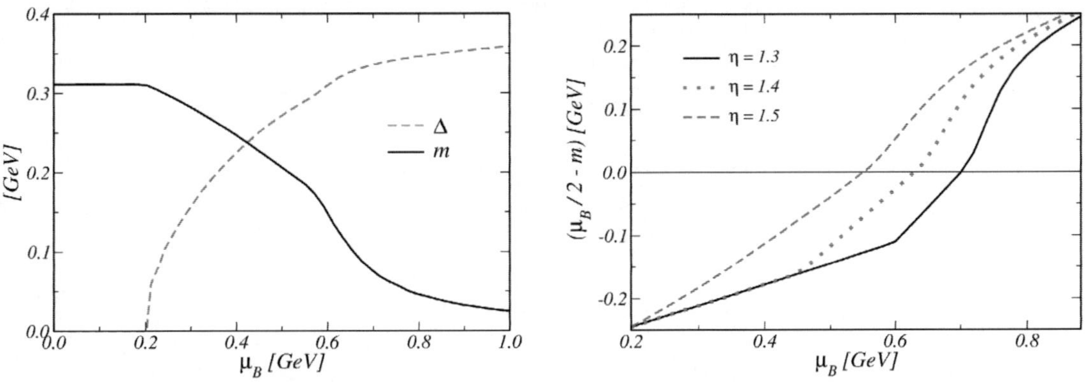

Figure 4. (Color online) **Left panel:** Effective quark mass m and diquark condensate Δ as functions of μ_B, ($\mu = \mu_B/3$ in $N_c = 3$ case, without considering the effects of color neutrality) for the ratio $\eta = G_D/G_S = 1.5$. **Right panel:** $\mu_N = \mu_B/3 - m$ as a function of the baryon chemical potential for different values of the ratio $\eta = G_D/G_S$.

to obtain the correct scaled values of the pion mass, chiral condensate and pion decay constant in the vaccum for $N_c = 2$ case. Secondly, in the Λ_l scheme there are non-physical oscillations that are not expected in this problem. To avoid this problems a new work is being developed using a different regularization procedure [11], with emphasis on the new scheme purposed in the reference [12]. In [12] it was found a way to extract completely the purely magnetic contributions, such that there are no divergences depending explicitly on the magnetic field. It seems that this method is well parametrized and does not present the non-physical oscillations that are very common to appear in other regularization schemes [6].

The discussion about $N_c = 3$ made in this work is just an introduction, where we reproduce the results obtained in reference [3]. It is a natural continuation of our original work to the more realistic three colors QCD case, with and without including the effects of color and charge neutralities. It will be interesting to see the competition between the magnetic field, that tends to favor the BEC region and the increase of the ratio G_S/G_D, that tends to favor the BCS region, and how this combination would affect the BEC-BCS crossover as a whole.

Acknowledgments
The authors thank N.N. Scoccola and P. Allen for discussions on related matters. Work partially supported by CAPES, CNPq and FAPERJ (Brazilian agencies).

References
[1] Liu H, Yu L, and Huang M 2015 *Phys. Rev.* D **91** 014017.
[2] Alford M G, Schmitt A, Rajagopal K, and Schäfer T 2008 *Rev. Mod. Phys.* **80** 1455.
[3] Sun G, He L and Zhuang P 2007 *Phys. Rev.* D **75** 096004.
[4] Brauner T, Fukushima K and Hidaka Y 2009 *Phys. Rev.* D **80** 074035, 2010 *Phys. Rev.* D **81** 119904.
[5] Menezes D P, Pinto M B, Avancini S S, Martínez A P and Providência C 2009 *Phys. Rev.* C **79** 035807, Fraga E S and Mizher A J 2008 *Phys. Rev.* D **78** 025016.
[6] Mandal T and Jaikumar P 2013 *Phys. Rev.* C **87** 045208.
[7] Noronha J L and Shovkovy I A 2007 *Phys. Rev.* D **76** 105030.
[8] Wang J-C, de la Incera V, Ferrer E J, and Wang Q 2011 *Phys. Rev.* D **84** 065014.
[9] Kogut J B, Sinclair D K, Hands S J and Morrison S E 2001 *Phys. Rev.* D **64** 09450.
[10] Huang M, Zhuang P, and Chao W 2002 *Phys. Rev.* D **65** 076012.

[11] Duarte D C, Farias R L S, Allen P G, Manso P H A, Ramos R O and Scoccola N N, in preparation.
[12] Allen P G, Grunfeld A G and Scoccola N N arXiv:150804724v1 [hep-ph] *Magnetized color superconducting cold quark matter within the $SU(2)_f$ NJL model: a novel regularization scheme.*

Cooling of neutron stars and emissivity of neutrinos by the direct Urca process under influence of a strong magnetic field

E L Coelho[1,2], **M Chiapparini**[1] **and R P Negreiros**[3]

[1]Instituto de Física, Universidade do Estado do Rio de Janeiro, 20559-900, Rio de Janeiro, Brazil

[2]Current address: Universidade Veiga de Almeida, 20271-020, Rio de Janeiro, Brazil

[3]Instituto de Física, Universidade Federal Fluminense, 24210-346, Rio de Janeiro, Brazil

E-mail: eduardo.coelho@uva.br

Abstract. Neutron stars are born with high temperatures and during a few seconds suffer rapid cooling by emission of neutrinos. The direct Urca process is the main mechanism to explain this loss of energy. In this work we study the influence of a strong magnetic field on the composition of nuclear matter at high densities and zero temperature. We describe the matter through a relativistic mean-field model with eight light baryons (baryon octet), electrons, muons magnetic field. As output of the numerical calculations, we obtain the relative population for a parametrized magnetic field. We calculate the cooling of neutron stars with different mass and magnetic fields due to direct Urca process

1. Introduction

Strong magnetic fields of magnitudes up to 10^{14} G are suppose to exists at the surface of pulsars. So, it is of interest to study the properties of nuclear matter in the presence of such strong magnetic fields. In this work we study the influence of a strong magnetic field on the composition of nuclear matter at $T = 0$, and investigate the cooling of neutron stars due to the direct Urca process using the resulting equation of state.

The matter at high densities is described using a relativistic mean field (MF) theory which describes correctly the nuclear ground state properties and elastic scattering of nucleons. The Lagrangian that describes this model, with a uniform magnetic field B along the z axis, is given by [1]

$$\begin{aligned}\mathcal{L} &= \sum_b \bar{\psi}_b[i\gamma_\mu D^\mu - m_b + g_{\sigma b}\sigma - g_{\omega b}\gamma_\mu\omega^\mu - \frac{1}{2}g_{\rho b}\gamma_\mu \boldsymbol{\tau}\cdot\boldsymbol{\rho}^\mu]\psi_b + \frac{1}{2}\partial_\mu\sigma\partial^\mu\sigma - \frac{1}{2}m_\sigma^2\sigma^2 \\ &-U(\sigma) - \frac{1}{4}\omega_{\mu\nu}\omega^{\mu\nu} + \frac{1}{2}m_\omega^2\omega_\mu\omega^\mu - \frac{1}{4}\boldsymbol{\rho}_{\mu\nu}\cdot\boldsymbol{\rho}^{\mu\nu} + \frac{1}{2}m_\rho^2\boldsymbol{\rho}_\mu\cdot\boldsymbol{\rho}^\mu \\ &+ \sum_{l=e^-,\mu^-}\bar{\psi}_l[i\gamma_\mu(\partial^\mu + iq_l A^\mu) - m_l]\psi_l - \frac{1}{4}F_{\mu\nu}F^{\mu\nu}, \qquad (1)\end{aligned}$$

where $D^\mu = \partial^\mu + iq_b A^\mu$, $A^0 = 0$, $\vec{A} = (0, xB, 0)$, q_b and q_l are the electric charge of baryons and leptons, ψ_b is the Dirac spinor for baryon b in the octet $\{n, p, \Lambda, \Sigma, \Xi\}$ with mass m_b; m_σ, m_ω,

m_ρ and $g_{\sigma b}$, $g_{\omega b}$, $g_{\rho b}$ are the masses and coupling constants of mesons σ, ω, ρ respectively. The summation in the first line represents the free Lagrangian of baryons together with the interaction between baryons and mesons; The mesonic and electromagnetic field strength tensors are

$$\omega_{\mu\nu} = \partial_\mu\omega_\nu - \partial_\nu\omega_\mu, \tag{2}$$

$$\rho_{\mu\nu} = \partial_\mu\rho_\nu - \partial_\nu\rho_\mu, \tag{3}$$

$$F_{\mu\nu} = \partial_\mu A_\nu - \partial_\nu A_\mu. \tag{4}$$

The baryon octet, showing their quantum numbers, charge q_b and isospin projection I_{3b} are shown in Table 1.

Table 1. The baryon octet, showing their quantum numbers, charge q_b, isospin projection I_{3b}.

Species	Mass (MeV)	q_b (e)	I_{3b}
p	939	1	1/2
n	939	0	−1/2
Λ	1115	0	0
Σ^+	1190	1	1
Σ^0	1190	0	0
Σ^-	1190	−1	−1
Ξ^0	1315	0	1/2
Ξ^-	1315	−1	−1/2

The Lagrangian of leptons (electrons and muons) is written in the third line while the scalar self-interactions term is given by

$$U(\sigma) = \frac{1}{3}b\,m_n(g_{\sigma n}\sigma)^3 + \frac{1}{4}c(g_{\sigma n}\sigma)^4, \tag{5}$$

where b and c are constants.

The dynamic equations of nucleon and mesons (Dirac and Klein-Gordon equations respectively) are obtained from the Euler-Lagrange equations

$$\frac{\partial\mathcal{L}}{\partial\phi(x)} - \partial_\mu\frac{\partial\mathcal{L}}{\partial(\partial_\mu\phi)} = 0, \tag{6}$$

where $\phi(x)$ is the corresponding field. The resulting equations of motion in the mean field approximation are given by

$$g_{\omega n}\omega_0 = \left(\frac{g_{\omega n}}{m_\omega}\right)^2 \sum_b \chi_{\omega b}\rho_b, \tag{7}$$

$$g_{\rho n}\rho_{03} = \left(\frac{g_{\rho n}}{m_\rho}\right)^2 \sum_b \chi_{\rho b} I_{3b}\rho_b, \tag{8}$$

$$m_n^* = m_n + \left(\frac{g_{\sigma n}}{m_{\sigma n}}\right)^2 \left[b\,m_n\,(m_n - m_n^*)^2 + c(m_n - m_n^*)^3 - \sum_b \chi_{\sigma b} n_s\right], \tag{9}$$

where I_{3b} is the 3-component of the isospin of the baryon b, $m_n^* = m_n - \chi_{\sigma b} g_{\sigma n}\sigma$, $\chi_{\sigma b} = g_{\sigma b}/g_{\sigma n}$, $\chi_{\omega b} = g_{\omega b}/g_{\omega n}$, $\chi_{\rho b} = g_{\rho b}/g_{\rho n}$; The scalar density is given by

$$n_s = n_s^{q=0} + n_s^{q\neq 0}, \tag{10}$$

Journal of Physics: Conference Series **706** (2016) 052011 doi:10.1088/1742-6596/706/5/052011

where

$$n_s^{q=0} = \frac{m_b^*}{2\pi^2}\left[\mu_b^* k_b - m_b^{*2}\ln\left(\frac{\mu_b^* + k_b}{m_b^*}\right)\right], \tag{11}$$

$$n_s^{q\neq 0} = \frac{m_b^*|q_b|B}{2\pi^2}\sum_{\nu=0}^{\nu_{\max(b)}} g_\nu \ln\left[\frac{\mu_b^* + k_{b,\nu_b}}{m_{b,\nu_b}^*}\right], \tag{12}$$

with

$$m_{b,\nu_b}^{*2} = m_b^{*2} + 2\nu_b|q_b|B, \tag{13}$$

$$k_{b,\nu_b}^2 = \mu_b^{*2} - m_b^{*2} - 2\nu_b|q_b|B, \tag{14}$$

where q_b and μ_b^* are the electric charge and effective chemical potential of baryon b, k_b and k_{b,ν_b} are the Fermi momentum of neutral and charge baryons respectively. ν_b is the Landau principal quantum number, which can take all possible positive integer values including zero. The upper limit $\nu_{\max(b)}$ is defined by the condition $k_{b,\nu_b}^2 \geq 0$, then

$$\nu_{\max(b)} = \text{int}\left[\frac{\mu_b^{*2} - m_b^{*2}}{2|q_b|B}\right]. \tag{15}$$

The effective chemical potential of baryons are given by

$$\mu_b^* = \mu_b - \chi_{\omega b} g_{\omega n}\omega_0 - \chi_{\rho b} g_{\rho n} I_{3b}\rho_{03}. \tag{16}$$

They are constrained due to the β-equilibrium condition, which reads

$$\mu_b = \mu_n - q_b\mu_e, \tag{17}$$

$$\mu_{e^-} = \mu_{\mu^-}, \tag{18}$$

where μ_n and μ_e are the chemical potentials of neutron and electron respectively. The baryon densities are

$$\rho_b^{q=0} = \frac{k_b^3}{3\pi^2}, \tag{19}$$

$$\rho_b^{q\neq 0} = \frac{|q_b|B}{2\pi^2}\sum_{\nu=0}^{\nu_{\max(b)}} g_\nu k_{b,\,\nu_b} \tag{20}$$

while the lepton (electrons and muons) densities are

$$\rho_l = \frac{|q_l|B}{2\pi^2}\sum_{\nu=0}^{\nu_{\max(l)}} g_\nu k_{l,\nu_l}, \tag{21}$$

with

$$k_{l,\nu_l}^2 = u_l^2 - m_l^2 - 2\nu_l|q_l|B \tag{22}$$

and

$$\nu_{\max(l)} = \text{int}\left[\frac{\mu_l^2 - m_l^2}{2|q_l|B}\right], \tag{23}$$

where $m^{*2}_{b,\nu_b} = m^{*2}_b + 2\nu_b|q_b|B$, ν is the Landau principal quantum number and the Landau level degeneracy g_ν is 1 for $\nu = 0$ and 2 for $\nu > 0$.

Neutron star matter satisfies the constraints of conservation of baryon number and neutrality of electric charge, which reads

$$\rho = \sum_b \rho_b, \tag{24}$$

$$0 = \sum_b q_b \rho_b + \sum_{l=e^-,\mu^-} q_l \rho_l. \tag{25}$$

The energy density due to the matter is given by [2]

$$\begin{aligned}\varepsilon_m = &\frac{1}{3} b\, m_n (g_{\sigma n}\sigma)^3 + \frac{1}{4} c\, (g_{\sigma n}\sigma)^4 + \frac{1}{2}\left(\frac{m_{\sigma n}}{g_{\sigma n}}\right)^2 (g_{\sigma n}\sigma)^2 + \frac{1}{2}\left(\frac{m_{\omega n}}{g_{\omega n}}\right)^2 (g_{\omega n}\omega_0)^2 \\ &+\frac{1}{2}\left(\frac{m_{\rho n}}{g_{\rho n}}\right)^2 (g_{\rho n}\rho_{03})^2 + \sum_{b(q=0)} \frac{1}{8\pi^2}\left[2\mu_b^{*3} k_b - m_b^{*2}\mu_b^* k_b - m_b^{*4} \ln\left\{\frac{\mu_b^* + k_b}{m_b^*}\right\}\right] \\ &+\frac{|q|B}{4\pi^2} \sum_{b(q\neq 0)} \sum_{\nu_b=0}^{\nu_{\max(b)}} g_\nu \left[\mu_b^* k_{b,\nu_b} + m^{*2}_{b,\nu_b} \times \ln\left\{\frac{\mu_b^* + k_{b,\nu_b}}{m^*_{b,\nu_b}}\right\}\right] \\ &+\frac{|q|B}{4\pi^2} \sum_{l=e^-,\mu^-} \sum_{\nu_l=0}^{\nu_{\max(l)}} g_\nu \left[\mu_l\, k_{l,\nu_l} + m^2_{l,\nu_l} \ln\left\{\frac{\mu_l + k_{l,\nu_l}}{m_{l,\nu_l}}\right\}\right],\end{aligned} \tag{26}$$

where

$$m^2_{l,\nu_l} = m_l^2 + 2\nu_l |q_l| B. \tag{27}$$

The matter pressure is given by

$$P_m = \mu_n \rho_b - \varepsilon_m. \tag{28}$$

Next, we add to these the contribution from electromagnetic field tensor, obtaining the total energy density and pressure as

$$\varepsilon = \varepsilon_m + \frac{B^2}{2}, \tag{29}$$

$$P = P_m + \frac{B^2}{2}. \tag{30}$$

The magnetic field is parametrized by

$$B(\rho/\rho_0) = B^{surf} + B_0[1 - \exp\{-\beta(\rho/\rho_0)^\gamma\}], \tag{31}$$

where $B^{surf} = 10^8$G and $B_0 = 10^{19}$G are the magnetic fields at the surface and the center of the star respectively, with the parameters $\beta = 10^{-4}$ and $\gamma = 17$, and ρ_0 is the saturation density.

The system of coupled nonlinear equations (7-9) with constraints (24-25) is solved numerically by iteration. We use the Newton-Raphson method with global search of the solution. We adopt natural units. As output, we obtain the relative population of each specie of particles as a function of the baryon density, and the energy density and pressure with and without magnetic field. The coupling constants are given in Tables 2, 3 e 4.

Table 2. Nucleon-meson coupling constants to compression $K = 300$ MeV and $m^*/m = 0.70$ [1].

$(\frac{g_{\sigma n}}{m_\sigma})^2$ (fm^2)	$(\frac{g_{\omega n}}{m_\omega})^2$ (fm^2)	$(\frac{g_{\rho n}}{m_\rho})^2$ (fm^2)	b	c
11.79	7.148	4.410	0.002947	-0.001070

Table 3. Parametrizations used for the hyperon coupling constants [3].

$\chi_{\sigma\Lambda}$	$\chi_{\sigma\Sigma}$	$\chi_{\sigma\Xi}$	$\chi_{\omega\Lambda} = \chi_{\omega\Sigma}$	$\chi_{\omega\Xi}$
0.6106	0.4046	0.3195	2/3	1/3

Table 4. Parametrizations used for the hyperon coupling constants [3].

$\chi_{\rho\Lambda}$	$\chi_{\rho\Sigma}$	$\chi_{\rho\Xi}$
1	1	1

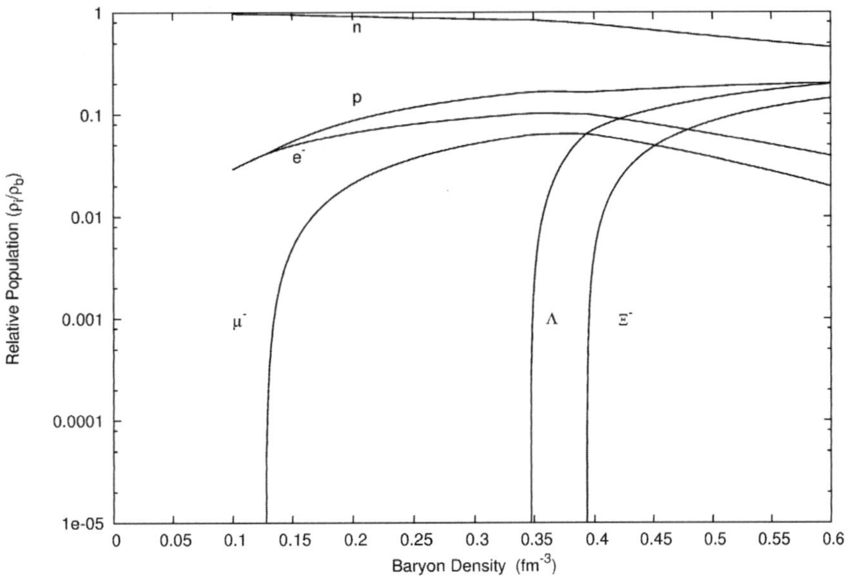

Figure 1. The particle fractions in cold β-equilibrated neutron star without magnetic field.

The Fig 1 and Fig 2 show the results without and with an parametrized magnetic field, respectively. It can be seen that the incorporation of strong magnetic field increases the proton and electron fraction. This effect is important for the neutrino emissivity due to the direct Urca process, having an impact on the cooling of neutron stars.

The direct Urca process is the most powerful mechanism of neutrino emission in the core of neutron stars. Neutron star cooling by Urca process may provide important informations about the interior composition of the star. The reaction is given by

$$n \;\rightarrow\; p + e^- + \bar{\nu}_e,$$

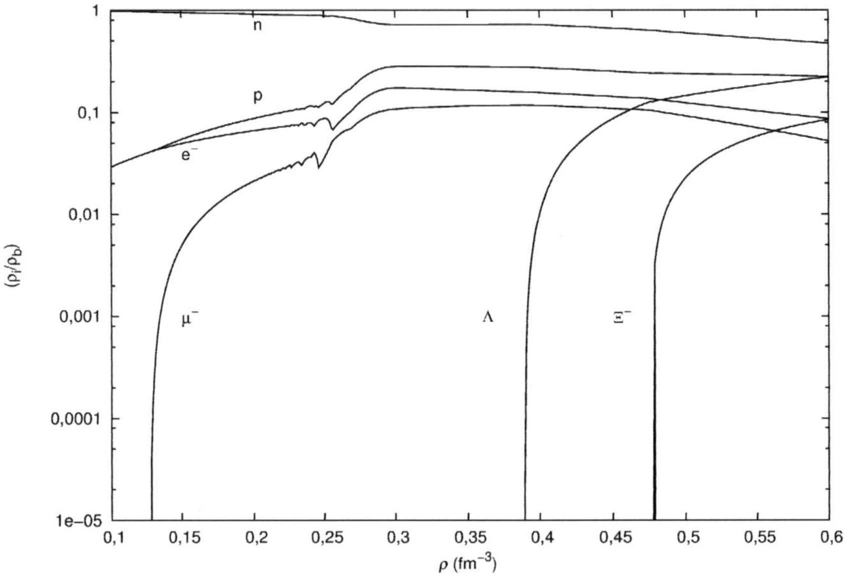

Figure 2. The particle fractions in cold β-equilibrated neutron star for the parametrized magnetic field.

$$p + e^- \;\;\rightarrow\;\; n + \nu_e.$$

This process may occur if the proton fraction is large enough

$$k_{Fn} \leq k_{Fp} + k_{Fe}, \tag{32}$$

where

$$k_{F\alpha} = (3\pi^2 \rho_\alpha)^{1/3}, \tag{33}$$

in order to conserve momentum in the reaction. As we have showed above, strong magnetic fields lead to an increase of the proton fraction and the cooling of neutron stars is more efficient.

In the Weinberg-Salam theory for weak interactions, the interaction Lagrangian is given by [4]

$$\mathcal{L}_{\text{weak}} = \frac{G_F}{\sqrt{2}} \cos\theta_c l_\mu j^\mu,$$

where G_F is the Fermi weak coupling constant and θ_c is the Cabibbo angle. The Lepton and nucleon charged weak currents are

$$l_\mu \;=\; \bar{\psi}_4 \gamma_3 \mu (1 - \gamma_5) \psi_2, \tag{34}$$

$$j^\mu \;=\; \bar{\psi}_3 \gamma^\mu (g_V - g_A \gamma_5) \psi_1, \tag{35}$$

g_V and g_A are vector and axial-vector coupling constants and the indices $i = 1 - 4$ refer to the $n, \bar{\nu}_e$, p and e^-, respectively. The wave functions for neutron and antineutrino are plane wave functions. The wave functions for both protons and electron in the presence of a magnetic field strong enough that only the ground Landau level is occupied are given by

$$\psi_3(X) = \frac{1}{\sqrt{L_y L_z}} \exp(-iE_3^* t + ik_{3y}y + ik_{3z}z) f_{k_{3y},k_{3z}}^{\nu_3=0}, \tag{36}$$

$$\psi_4(X) = \frac{1}{\sqrt{L_y L_z}} \exp(-iE_4 t + ik_{4y}y + ik_{4z}z) f_{k_{4y},k_{4z}}^{\nu_4=0}, \tag{37}$$

where $f_{k_{3y},k_{3z}}^{\nu_3=0}$ and $f_{k_{4y},k_{4z}}^{\nu_4=0}$ are the 4-component spinor solutions of the corresponding Dirac equation. The only positive energy spinor for protons in the chiral representation is [5]

$$f_{k_{3y};k_{3z}}^{\nu_3=0}(x) = N_{\nu_3=0} \begin{pmatrix} E_3^* + k_{3z} \\ 0 \\ -m_3^* \\ 0 \end{pmatrix} I_{\nu_3=0;k_{3y}}, \tag{38}$$

where

$$N_{\nu_3=0} = \frac{1}{[2E_3^*(E_3^* + k_{3z})]^{1/2}}, \quad E_3^* = (k_{3z}^2 + m_3^{*2})^{1/2}, \tag{39}$$

$$I_{\nu_3=0;k_{3y}} = \left(\frac{eB}{\pi}\right)^{1/4} \exp\left[-\frac{1}{2}eB\left(x - \frac{k_{3y}}{eB}\right)^2\right] \frac{1}{\sqrt{\nu_3!}} H_{\nu_3}\left[\sqrt{2eB}\left(x - \frac{k_{3y}}{eB}\right)\right],$$

and are the Hermite polynomial and e is the electron charge.

The emissivity due to the antineutrino emission process in presence of a uniform magnetic field B along z-axis is [?]

$$\epsilon_\nu = 2\int \frac{Vd^3k_1}{(2\pi)^3} \int \frac{Vd^3k_2}{(2\pi)^3} \int_{-qBL_x/2}^{qBL_x/2} \frac{L_y dk_{3y}}{2\pi} \int \frac{L_z dk_{3z}}{2\pi} \int_{-qBL_x/2}^{qBL_x/2} \frac{L_y dk_{4y}}{2\pi}$$
$$\times \int \frac{L_z dk_{4z}}{2\pi} E_2 W_{fi}\, f_1[1 - f_3][1 - f_4], \tag{40}$$

where the pre-factor 2 takes into account the neutron spin degeneracy and f_i is the Fermi-Dirac distribution functions. By the Fermi's golden rule, the transition rate per unity volume W_{fi} is

$$W_{fi} = \frac{\langle |M_{fi}|^2 \rangle}{tV}, \tag{41}$$

where t is the time, $V = V_x V_y V_z$ is the normalization volume and the matrix element for the V-A interaction is given by

$$M_{fi} = \frac{G_F}{\sqrt{2}} \int d^4X \bar{\psi}_1(X)\gamma^\mu(g_V - g_A\gamma_5)\psi_3(X)\bar{\psi}_2(X)\gamma_\mu(1 - \gamma_5)\psi_4(X), \tag{42}$$

where $\langle . \rangle$ denotes an averaging over the initial spin of n and a sum over spins of final particles (p, e). Then, the transition rate per unit volume is

$$W_{fi} = \frac{G_F^2}{E_1^* E_2 E_3^* E_4} \frac{1}{V^3 L_y L_z} \exp\left(-\frac{(k_{1x} - k_{2x})^2 + (k_{3y} + k_{4y})^2}{2eB}\right)$$
$$\times [(g_V + g_A)^2 (k_1.k_2)(k_3.k_4) + (g_V - g_A)^2$$
$$\times (k_1.k_4)(k_3.k_2) - (g_V^2 - g_A^2) m^{*2} (k_4.k_2)](2\pi)^3 \delta(E_1^* - E_2 - E_3^* - E_4)$$
$$\times \delta(k_{1y} - k_{2y} - k_{3y} - k_{4y})\delta(k_{1z} - k_{2z} - k_{3z} - k_{4z}).$$

Then, the emissivity is

$$\epsilon_\nu = \frac{457\pi}{5040} G_F^2 \cos^2\theta_c (eB)[(g_V + g_A)^2 \left(1 - \frac{k_{F3}}{\mu_3^*}\right) + (g_V - g_A)^2 \left(1 - \frac{k_{F_1}}{\mu_n^*}\cos\theta_{14}\right) - (g_V^2 - g_A^2)\frac{m^{*2}}{\mu_3^*\mu_1^*}\Bigg] \exp\left[\frac{(k_{F_3} + k_{F_4})^2 - k_{F_1}^2}{2eB}\right] \frac{\mu_1^*\mu_3^*\mu_4}{k_{F_3}k_{F_4}} T^6 \Theta,$$

$\cos\theta_{14} = (k_{F_1}^2 + k_{F_4}^2 - k_{F_3}^2)/2k_{F_1}k_{F_4}$, T is the temperature, k_{Fi} is the Fermi momentum and the threshold factor is $\Theta = \theta(k_{F_3} + k_{F_4} - k_{F_1})$, with $\theta(x) = 1\,(x > 0)$, $\theta(x) = 0$ (otherwise).

2. Results

Figure 3 shows the cooling due to the direct Urca process of neutron stars with 1.4 and 1.6 solar masses (continuous and dashed lines, respectively) for the cases $B = 0$ (blue line) and $B(\rho/\rho_0)$ (red line). We can see that the cooling is more intense with the increase in the mass of the star and for the case $B(\rho/\rho_0)$. Theses differences may be attributed to the increase of proton and electron fractions with the mass and magnetic field of the star and to the phase space modifications.

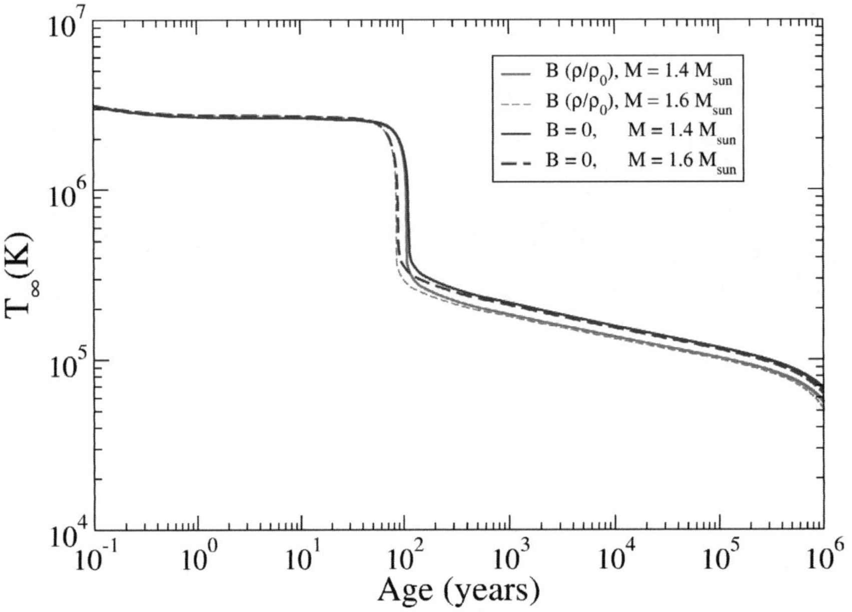

Figure 3. Cooling of a neutron star mass with 1.4 and 1.6 solar masses (continuous and dashed lines, respectively) for the cases $B = 0$ (blue line) and $B(\rho/\rho_0) = B^{surf} + B_0[1 - \exp\{-\beta(\rho/\rho_0)^\gamma\}]$ (red line).

[1] Glendenning N K 2000 *Compact Stars* (New York: Springer).
[2] Chakrabarty S, Bandyopadhyay D and Pal S 1997 *Phys. Rev. Letter* **78** 2898.
[3] Chiapparini M, Bracco M E, Delfino A, Malheiro M, Menezes D P and Providência C 2009 *Nuclear. Phys.* A **826** 178.

[4] Bandyopadhyay D, Chakrabarty S, Dey P and Pal S 1998 *Phys. Rev.* D **58** 121301.
[5] Kobayashi M and Sakamoto M 1983 *Progress of Theoretical Phys.* **70** 5.

Journal of Physics: Conference Series **706** (2016) 052012 doi:10.1088/1742-6596/706/5/052012

Ward identities and the analogous Goldberger-Treiman relation in a three-flavor Spectral Quark Model

E A Reis, A L Mota and E W Dias

Departamento de Ciências Naturais, Universidade Federal de São João del Rei, C.P. 110, CEP 36301-160, São João del Rei, Brazil

E-mail: eduardoantonioreis@gmail.com, motaal@ufsj.edu.br, edsondias@ufsj.edu.br

Abstract. This work presents the first results of an extension of the spectral quark model which includes different flavors. The spectral quark model is an approach based on a generalization of the Lehmann representation for the quark propagator. Gauge and chiral invariance are ensured with the help of gauge technique which provides particular solutions to the Ward-Takahashi identities. General conditions on the quark spectral function follow from natural physical requirements. In particular, the function is normalized, its positive momenta must vanish, while the physical observables depend on negative moments and the so-called log moments. As a consequence, the model is made finite. To allow the description of mesons constituted by different flavors of quarks we introduce different spectral functions and obtain vertex functions constructed from Ward-Takahashi identities that includes two different spectral (constituent) quark masses, allowing the physical description of strange mesons, for example. We obtain some observables based on the current approach and, in particular, the spectral version of the Kaon analogous Goldberger-Treiman relation.

1. Introduction

The non-perturbative behavior of Quantum-Chromodynamics (QCD) at low energies is an ubiquitous feature of this theory. The running coupling constant of the strong interaction at this level assumes large values, not allowing the perturbative treatment that is successfully employed in the high energy limit or in other theories (Quantum Electrodynamics, for example). Thus, the study of the hadronic states that lie on this range of energy has to be carried out by using non-perturbative approaches such as lattice QCD or effective models. Belonging to this second class, chiral quark models, i.e., effective models that present the quarks as the relevant degrees of freedom and incorporate the relevant symmetries of QCD, are very successful on the description of the QCD low energy phenomenology. In this case, the non-perturbative behavior of QCD manifests itself by the presence of a constituent mass for the quarks, the consequence of a non-perturbative vacuum. The constituent mass is the effective mass that includes the effects of quantum corrections due the interaction of quarks with gluons and the self interaction of the gluons.

As a price to be paid, effective models are not fundamental, and need the introduction of parameters external to the theory. Some of these models include other degrees of freedom, such as mesonic [1, 2] or gluonic fields [3] that, even ensuring their renormalizability, introduce extra couplings and masses not present on the underlying theory. Other purely fermionic models, as the well established Nambu-Jona-Lasinio model [4, 5] are non renormalizable, and results are dependent of the regularization schemes and regularization parameters employed on their treatment.

The Spectral Quark Model (SQM) is a recently formulated effective chiral quark model that naturally incorporate some of the essential features discussed above. The non perturbative behavior is introduced by using the Lehmann representation for the quark propagators. These propagators include the constituent mass as a spectral running mass, reproducing the effect of the dependence of the constituent mass with the energy scale of the processes involved. SQM is also constructed in such a way to preserve important symmetries of QCD, by means of the gauge technique [6, 7].

A surprisingly and relevant feature of SQM is its finiteness, constructed via the correct selection of the so called spectral conditions and implying in the independence of the regularization scheme and regularization parameters employed on the intermediary steps of the calculations. Although the introduction of the Lehmann spectral distributions is *per se* an introduction of extra parameters, the SQM provides, by imposing physical conditions (as unitarity, finiteness and symmetries expressed by Ward identities [8, 9]) the necessary relations to determine the relevant parameters of the spectral distribution - the spectral conditions. The model was successfully applied to describe the low energy hadrons phenomenology [10, 11, 12] and other effects such as the Polyakov loop [13], Chiral anomaly [14] and so on.

As originally formulated, the SQM includes only one quark flavor (expressed by an unique spectral distribution), being adequate to describe the observables related to one flavored states, as the mesons Pion, Eta, Rho and so on. As a consequence, SQM still could not be used to describe the phenomenology involving two different quarks, like strange mesons for example.

In this contribution we take the first steps towards the construction of a three flavored version of the Spectral Quark Model by postulating two different spectral distributions for the up/down and for the strange quarks. We will show here that it is possible to construct vector and axial vectors vertex functions with one current that satisfy the Ward Identities and involving only the two spectral distributions mentioned above. Besides, by using the axial vertex function, we obtain the spectral version of the analogous Goldberger-Treiman relation for the Kaon.

2. The Model

In order to develop the three flavored version of SQM, we closely follow the results presented in [10], employing the Lehmann representation for the up/down and strange quarks propagators:

$$S_s(p) = \int_C d\omega \frac{\rho_s(\omega)}{\not{p} - \omega} \qquad S_u(p) = \int_C d\omega \frac{\rho_u(\omega)}{\not{p} - \omega}, \tag{1}$$

where ω is the spectral constituent mass, $\rho_u(\omega)$ e $\rho_s(\omega)$ are the spectral distributions for quarks up/down and strange respectively and C denotes a contour in the complex ω plane. As in the one flavored version, physical requirements imposed some relations (the spectral conditions) for the moments of the spectral distributions, such as the normalization condition

$$\rho_{q,0} \equiv \int d\omega \rho_q(\omega) = 1. \tag{2}$$

We also define the positive, negative and log spectral momenta respectively as

$$\rho_{q,n} \equiv \int d\omega \omega^n \rho_q(\omega) \qquad n = 1, 2, ..., \tag{3}$$

$$\rho_{q,-n} \equiv \int d\omega \omega^{-n} \rho_q(\omega), \tag{4}$$

and

$$\rho'_{q,n} \equiv \int d\omega \log(\omega^2)\omega^n \rho_q(\omega). \tag{5}$$

In equations (3)-(5) the distribution $\rho_q(\omega)$ denotes generically any of the distributions $\rho_u(\omega)$ and $\rho_s(\omega)$. Other relevant quantities to be considered on the current approach are the one current vector and axial unamputated vertex functions, defined as

$$\Lambda_V^{\mu,a}(p,p') = iS_u(p')\Gamma_V^{\mu,a}(p,p')iS_s(p) = \int d^4x d^4x' \langle 0|T\{J_V^{\mu,a}(0)q(x')\overline{q}(x)\}|0\rangle e^{i(p'x'-px)}, \tag{6}$$

and

$$\Lambda_A^{\mu,a}(p,p') = iS_u(p')\Gamma_A^{\mu,a}(p,p')iS_s(p) = \int d^4x d^4x' \langle 0|T\{J_A^{\mu,a}(0)q(x')\overline{q}(x)\}|0\rangle e^{i(p'x'-px)}, \tag{7}$$

respectively. Here

$$J_V^{\mu,a}(x) = \overline{q}(x)\gamma^\mu \frac{\lambda_a}{2} q(x), \tag{8}$$

and

$$J_A^{\mu,a}(x) = \overline{q}(x)\gamma^\mu \gamma_5 \frac{\lambda_a}{2} q(x). \tag{9}$$

are the vector and axial currents and the Γ's represent the corresponding amputated vertex functions. Also, λ_a are the Gell-Mann matrices and γ^μ are the Dirac matrices. The Ward-Takahashi (WT) identity for the full vector vertex reads

$$(p'-p)_\mu \Lambda_V^{\mu,a}(p,p') = S_u(p')\frac{\lambda_a}{2} - \frac{\lambda_a}{2} S_s(p), \tag{10}$$

and, for the axial vertex,

$$(p'-p)_\mu \Lambda_A^{\mu,a}(p,p') = S_u(p')\frac{\lambda_a}{2}\gamma_5 + \gamma_5 \frac{\lambda_a}{2} S_s(p). \tag{11}$$

The SQM also makes use of the gauge technique [15, 16] in order to determine particular representations of the vertex functions Eq.(6) and Eq.(7) that satisfy the WT identities, Eq.(10) and (11). We will determine these quantities, in the context of the three flavor version of SQM, on section 5. In the next sections, we will obtain some single expressions for the quark condensates and the vacuum energy density.

3. $\langle \bar{u}u \rangle$ and $\langle \bar{s}s \rangle$ quark condensates

The quark condensate is the vacuum expected value of a chiral fermion/antifermion pair, taken as the order parameter for the chiral phase transition from the chiral symmetric phase($\langle \bar{q}q \rangle = 0$) to the phase where this symmetry is broken ($\langle \bar{q}q \rangle \neq 0$). The quark condensate for a single flavor is given by

$$\langle \bar{q}q \rangle = -iN_c \int d\omega \rho_q(\omega) \int \frac{d^4p}{(2\pi)^4} Tr \frac{1}{\not{p} - \omega}, \tag{12}$$

where the trace is done on the Dirac matrices and $N_c = 3$ is the number of colors. As we need to perform the integration on p before to proceed the spectral integration, we need to use an auxiliary regularization method - which will be removed at the end of calculation, once that the integral over the momentum p is quadratically divergent. Here, we employ the sharp-cutoff regularization, obtaining

$$\langle \bar{q}q \rangle = -\frac{N_c}{4\pi^2} \int d\omega \rho_q(\omega)\omega \left[\Lambda^2 + \omega^2 \log\left(\frac{\omega^2}{\Lambda^2 + \omega^2} \right) \right]. \tag{13}$$

As before, the subscript q denotes both u or s flavors of the quarks. As we can see from Eq. 13, finiteness of the result at $\Lambda \to \infty$ requires that

$$\rho_{q,1} = \int d\omega \omega \rho_q(\omega) = 0, \tag{14}$$

and

$$\rho_{q,3} = \int d\omega \omega^3 \rho_q(\omega) = 0. \tag{15}$$

and thus

$$\langle \bar{q}q \rangle = -\frac{N_c}{4\pi^2} \int d\omega \log(\omega^2)\omega^3 \rho_q(\omega) = -\frac{N_c}{4\pi^2} \rho'_{q,3}, \tag{16}$$

or

$$\rho'_{q,3} = -\frac{4\pi^2}{N_c} \langle \bar{q}q \rangle. \tag{17}$$

allowing the determination of the spectral momenta $\rho'_{q,3}$ from the experimental estimates of the quark condensates. Except for the possibility of employing different spectral conditions, this result is the same previously obtained on [10].

4. Vacuum energy density

The vacuum energy density (denoted as B) is the vacuum energy related to the spontaneous creation of virtual particle/antiparticle pairs. Within SQM, it can be evaluated from the vacuum expected value of the energy-momentum tensor for a purely quark model, given by, after explicitly taken the trace on flavors

$$\begin{aligned} g^{\mu\nu} B &= \langle \theta^{\mu\nu}(x) \rangle - \langle \theta^{\mu\nu}(x) \rangle_0 \\ &= -iN_c \int d\omega (2\rho_u(\omega) + \rho_s(\omega)) \int \frac{d^4p}{(2\pi)^4} Tr \left\{ \frac{1}{\not{p} - \omega} \times \left[\frac{1}{2}(\gamma^\mu p^\nu + \gamma^\nu p^\mu) - g^{\mu\nu}(\not{p} - \omega) \right] \right\} \\ &= -4iN_c \int d\omega (2\rho_u(\omega) + \rho_s(\omega)) \int \frac{d^4p}{(2\pi)^4} \frac{p^\mu p^\nu - g^{\mu\nu}(p^2 - \omega^2)}{p^2 - \omega^2}, \end{aligned} \tag{18}$$

where $\langle \theta^{\mu\nu} \rangle_0$ is the expected value of the energy-momentum tensor evaluated for the free theory (with $\rho(\omega) = \rho(\omega') = \delta(\omega)$).

Thus, we obtain, for the vacuum energy density,

$$B = -iN_c \int d\omega (2\rho_u(\omega) + \rho_s(\omega)) \int \frac{d^4p}{(2\pi)^4} \frac{\omega^2}{p^2 - \omega^2}. \tag{19}$$

Employing again a sharp cut-off intermediate regularization procedure, we obtain

$$B = -\frac{N_c}{16\pi^2} \int d\omega (2\rho_u(\omega) + \rho_s(\omega)) \omega^2 \left[\Lambda^2 + \omega^2 \log\left(\frac{\omega^2}{\Lambda^2 + \omega^2} \right) \right]. \tag{20}$$

Hence, the conditions that must be fulfilled for finiteness of B are

$$\rho_{q,2} = \int d\omega \omega^2 \rho_q(\omega) = 0, \tag{21}$$

and

$$\rho_{q,4} = \int d\omega \omega^4 \rho_q(\omega) = 0. \tag{22}$$

Finally,

$$\begin{aligned} B &= -\frac{N_c}{16\pi^2}\Big\{2\int d\omega\rho_u(\omega)\omega^4\log(\omega^2) + \int d\omega\rho_s(\omega)\omega^4\log(\omega^2)\Big\} \\ &= -\frac{N_c}{16\pi^2}(2\rho'_{u,4} + \rho'_{s,4}). \end{aligned} \tag{23}$$

Written in terms of the expected value for the gluon condensate, $\langle \frac{\alpha_s}{\pi} G^2 \rangle$, we have [7]

$$B = -\frac{N_c}{16\pi^2}(2\rho'_{u,4} + \rho'_{s,4}) = -\frac{9}{32}\langle\frac{\alpha}{\pi}G^2\rangle, \tag{24}$$

or

$$(2\rho'_{u,4} + \rho'_{s,4}) = -\frac{3\pi^2}{2}\langle\frac{\alpha}{\pi}G^2\rangle. \tag{25}$$

where α_s is the perturbative strong interaction constant, $G^2 = G^a_{\mu\nu}G^{a\mu\nu}$ and $G^{\mu\nu}$ is the gluon field strength tensor.

By collecting the results for the momenta of the spectral distributions, such as Eq.(17) and (25), one can find a set of equations relating the unknown spectral momenta to physical observables that allow the determination of the relevant momenta of ρ_u and ρ_s and giving predictive power to the SQM. In this contribution, however, we will concentrate on the obtaining of vertex functions that satisfy the Ward identities in the context of quarks with different flavors.

5. Vertices with one current from the gauge technique

The gauge technique consists on finding solutions for the chiral and electromagnetic WT identities, Eqs. (10) and (11), determining a particular representation for the vector and axial-vector vertices. For two different flavors, we write a first attempt solution for the vector WT identity as

$$\Lambda^{\mu,a}_V(p,p') = iS_u(p')\Gamma^{\mu,a}_V(p,p')iS_s(p) = \int d\omega' d\omega \rho_{us}(\omega,\omega')\frac{i}{\not p' - \omega'}\gamma^\mu\frac{\lambda_a}{2}\frac{i}{\not p - \omega}. \tag{26}$$

One should expect that the spectral distribution for two different flavors, $\rho_{us}(\omega,\omega')$, could be written as a separable function of the two spectral masses expressed in terms of the quark distribution functions ρ_u and ρ_s. In fact, this is really the case we will treat here, and is a remarkable feature that this possibility is assured by the normalization spectral condition, Eq. (2), as we shall see. So, we will assume

$$\rho_{us}(\omega,\omega') = \rho_u(\omega)\rho_s(\omega'). \tag{27}$$

We should stress, however, that the separable solution is not the only possibility. For example, the results for the one flavor SQM can be recovered from our approach if we write ρ_{us} (in fact, ρ_{uu}, in this case) as

$$\rho_{uu}(\omega,\omega') = \rho_u(\omega)\delta(\omega - \omega'), \tag{28}$$

and this solution is non-separable.

Journal of Physics: Conference Series **706** (2016) 052012 doi:10.1088/1742-6596/706/5/052012

Inserting the attempt solution (26) on the vector WT identity, Eq.(10), and using Eq.(27), we obtain

$$
\begin{aligned}
(p'-p)_\mu iS_u(p')\Gamma_V^{\mu;a}(p,p')iS_s(p) = &-\frac{\lambda_a}{2}\int d\omega'\rho_u(\omega')\int d\omega\frac{\rho_s(\omega)}{\not p-\omega} \\
&+\frac{\lambda_a}{2}\int d\omega\rho_s(\omega)\int d\omega'\frac{\rho_u(\omega')}{\not p'-\omega'} \\
&+\int d\omega' d\omega\rho_u(\omega')\rho_s(\omega)\frac{i}{\not p'-\omega'}(\omega'-\omega)\frac{\lambda_a}{2}\frac{i}{\not p-\omega},
\end{aligned}
\tag{29}
$$

where we can explicitly see that the normalization spectral condition, Eq.(2), ensures that

$$
\begin{aligned}
(p'-p)_\mu iS_u(p')\Gamma_V^{\mu,a}(p,p')iS_s(p) = &\, S_u(p')\frac{\lambda_a}{2}-\frac{\lambda_a}{2}S_s(p) \\
&+\int d\omega' d\omega\rho_u(\omega')\rho_s(\omega)\frac{i}{\not p'-\omega'}(\omega'-\omega)\frac{\lambda_a}{2}\frac{i}{\not p-\omega}.
\end{aligned}
\tag{30}
$$

The first attempt solution does not satisfy the WT identity, but can be used as a guide for a second *ansatz* for the vector vertex function, written as

$$
\Lambda_V^{\mu,a}(p,p') = \int d\omega' d\omega\rho_u(\omega')\rho_s(\omega)\frac{i}{\not p'-\omega'}\left(\gamma^\mu-\frac{(\omega'-\omega)q^\mu}{q^2}\right)\frac{\lambda_a}{2}\frac{i}{\not p-\omega},
\tag{31}
$$

where $q^\mu = (p'-p)^\mu$. One can easily show that Eq.(31) is a solution of the vector WT identity, Eq.(10). Also, if we replace the spectral two quarks distribution ρ_{us} by the one flavor non separable solution, Eq.(28), we recover the result of the original SQM.

Following the same steps, we obtain, for the solution of the axial WT identity, Eq. (11), the axial unamputated vertex,

$$
\Lambda_A^{\mu,a}(p,p') = \int d\omega' d\omega\rho_u(\omega')\rho_s(\omega)\frac{i}{\not p'-\omega'}\left(\gamma^\mu-\frac{(\omega'+\omega)q^\mu}{q^2}\right)\gamma^5\frac{\lambda_a}{2}\frac{i}{\not p-\omega}.
\tag{32}
$$

The results showed here are similar to those obtained in ref. [10], but present two different features that are significant to the development of a three flavor version of SQM: (a) the vector current term inside the spectral integral of the vertex function presents a dependence on the spectral masses ω and ω', implying on the rising of extra terms on the expressions of observables involving the vector function, in contrast with the one flavored version, where the vector current contains only the γ^μ term; (b) Both vertex functions involve two spectral distributions integrated over two different spectral masses. This fact allows the emergence of new spectral moments involving non separable integrals in ω and ω' (convolution integrals of the distribution functions ρ_u and ρ_s). Nevertheless, none of the results presented in this contribution depends on these new spectral moments.

6. Analogous Goldberger-Treiman Relation for the Kaon

In this section, we present a version of the Goldberger-Treiman Relation (GT) for the kaon, in the context of the three flavor version of SQM. As it is well known, any chiral model which intends to be a prototype of QCD in low energies should satisfy the GT Relation, once that such relation is not dependent of any model, but, instead of, it is a consequence of the chiral symmetry breaking and the partial conservation of axial current hypothesis (PCAC), two of the most important elements of QCD. GT relation establishes a connection between quark and mesonic properties, like masses, meson weak constant decays and quark-meson coupling constants in low energies.

We can obtain the analogous GT Relation in the SQM from the axial unamputaded vertex expression. In such vertex, we take the limit $q^2 \to 0$, where q^μ is the meson momentum. In this limit, we mean that there is a coupling between the Goldstone bosons (kaons or pions) and the axial current, and such coupling is proportional to the meson momentum. As we can see, the axial unamputaded vertex has a pole in $(q^2 \to 0)$ and, the kaon decay, characterized by the axial vertex (32), is dominated by the pseudoscalar coupling and the axial vertex is associated with the wave function of the kaon (corresponding to the vertex $k \to q\bar{q}$) by the relation

$$\Lambda_A^{\mu,a}(p,p')|_{q\longrightarrow 0} \to -2f_k \frac{q^\mu}{q^2} \Lambda_k^a(p,p'), \tag{33}$$

where f_k is the kaon constant decay. From (33), we get

$$\Lambda_k^a(p,p') = \int d\omega' d\omega \rho_u(\omega') \rho_{\bar{s}}(\omega) \frac{i}{\not{p}' - \omega'} \frac{(\omega' + \omega)}{2f_k} \gamma^5 \frac{\lambda_a}{2} \frac{i}{\not{p} - \omega}, \tag{34}$$

and we can identify the coupling between quarks and the kaon by

$$g_k(\omega', \omega) = \frac{\omega' + \omega}{2f_k}. \tag{35}$$

which is the analogous Goldberger-Treiman relation for the kaon [17], connecting the coupling constant between the kaon meson and the up and strange quarks, the spectral quark masses and the kaon constant decay. It is immediate to see that the analogous GT relation reproduces the expected result in the one flavor SQM [10], where the two spectral masses are equal.

7. Conclusion

In this work we obtained the vector and axial vector vertex functions in the context of the Spectral Quark Model by using an approach that includes different flavors of quarks. The approach is based on employing two different spectral functions and, by using the gauge technique, obtaining vertex functions that are related to these spectral propagators and satisfy the corresponding vector and axial vector Ward-Takahashi identities. From the propagators and vertex functions we obtained the expressions for the quark condensate, vacuum energy density and the analogous Goldberger-Treiman relation for the kaon. We discussed the relevant differences between the current approach and the one flavor version, and the spectral two quarks distribution that leads the current version to the original one. We also obtained the spectral version of the kaon analogous Goldbeger-Treiman relation. The approach opens the possibility to obtain other observables of mesons with different content of quarks, such as the weak decay and the electromagnetic form factor of the strange mesons (work in progress).

8. Acknowledgments

The authors would like to thank financial support from CNPq and FAPEMIG (Brazilian funding agencies).

References

[1] Gell-Mann M and Levy M 1960 *The axial vector current in beta decay* Nuovo Cim. vol 16 p 705
[2] Mota A L, Nemes M C, Hiller B and Walliser H 1999 *Meson properties in a renormalizable version of the NJL model* Nucl. Phys. vol A 652 p 73
[3] Langfeld K, Kettner C and Reinhardt H 1996 *A Renormalizable extension of the NJL model* Nucl. Phys. vol A 608 p 331
[4] Vogl U and Weise W 1991 *The Nambu and Jona Lasinio model: Its implications for hadrons and nuclei* Prog. Part. Nucl. Phys vol 27 pp 195-272

[5] Klevansky S P 1992 *The Nambu-Jona-Lasinio model of quantum chromodynamics* Rev. Mod. Phys. vol 64 pp 649-708
[6] Ioffe B L 2003 *Condensates in quantum chromodynamics* Phys. Atom. Nucl. vol 66 pp 30-43
[7] Ioffe B L and Zyablyuk K N 2003 *Gluon condensate in charmonium sum rules with 3-loop corrections* Eur. Phys. J. vol C 27 pp 229-241
[8] Itzykson C and Zuber J B 1980 *Quantum Field Theory* (New York: Mcgraw-hill) p 705
[9] Peskin M E and Schroeder D V 1995 *An Introduction to Quantum Field Theory* ed Westview Press p 238
[10] Ruiz E A and Broniowski W 2003 *Spectral quark model and low energy hadron phenomenology* Phys. Rev. vol D 67 p 074021
[11] Broniowski W and Ruiz E A 2007 *Pion-photon Transition Distribution Amplitudes in the Spectral Quark Model* Phys. Lett. vol B 649 p 49
[12] Ruiz E A, Broniowski W and Golli B 2007 *Confined Chiral Solitons in the Spectral Quark Model* Phys. Rev. vol D 76 p 014008
[13] Megias E, Ruiz E A and Salcedo L L 2006 *Chiral Lagrangian at finite temperature from the Polyakov-Chiral Quark Model* Phys. Rev. vol D 74 p 114014
[14] Ferreira L C and Mota A L 2009 *Axial-vector coupling and chiral anomaly in the spectral quark model* Phys. Rev. vol D 79 p 085002
[15] Delbourgo R and West P C 1997 *A Gauge Covariant Approximation to Quantum Electrodynamics* J. Phys. vol A 10 p 1049
[16] Delbourgo R 1979 *The Gauge Technique* Nuovo Cim. vol A 49 pp 484-496
[17] Okabayashi T 1969 *The Goldberger-Treiman Relation and Its Kaon Analogue as a Non-Group-Theoretical Kinematics Generating an SU(3) Symmetry* Prog. Theor. Phys. vol 41 p 1566

SU(2)-flavor NJL couplings from polarization effect

Fábio L. Braghin

Instituto de Física, Univ. Fed. de Goiás, P.B.131, Campus II, CEP 74001-970, Goiânia, GO, Brazil

E-mail: braghin@ufg.br

Abstract. In this work corrections to the quark Nambu-Jona-Lasinio (NJL) coupling are presented in terms of the QCD parameters by considering polarization effects in terms of a generic shape of the gluon propagator. For that, the quark field is splitted into two parts and one of them, assumed to be a weak field quark, is integrated out. The resulting determinant is expanded to the 2-nd order in quark bilinears and some of the resulting couplings are exhibited.

1. Introduction

The Nambu-Jona-Lasinio model [1] is one of the most celebrated effective models for the low energy regime of Quantum Chromodynamics [2] because it provides the basic mechanism for dynamical chiral symmetry breaking whose consequences in hadron and nuclear physics are broad and quite well known. Several observables are calculable from it with good quantitative agreement [2]. However, it is important that the fundamental mechanisms, in the quark and gluon dynamics, that give rise to the effective interactions be very clearly understood. Simple gluon exchange is a fundamental and basic mechanism that is not enough to provide physical understanding of low energy observable hadron properties although it can be dressed by non linear strong gluon (virtual) interactions that are absolutely important at low energies. From this part of the QCD effective action the NJL appears from a Fierz transformation [2, 3]. Besides instantons effects [4] polarization effects have also been show to provide corrections to low energy quark effective interactions [5, 6, 7]. This contribution shows a computation of polarization corrections for the NJL coupling by following Ref. [6]. In that last paper a gluon propagator with not realistic shape was chosen and in the present work it is shown that the same analytical results are obtained without fixing the gluon propagators. Nevertheless to provide numerical estimations the gluon propagator would had to be fixed. As explained in Refs. [6, 7], by departing from a basic (dressed) gluon exchange interaction, the quark field is separated into two components:(ψ_1) remains as interacting quarks and (ψ_2) gives rise to light quark-antiquark meson states. In the present work, because we are interested in the quark-quark effective interaction sector, the self interactions of second component, that might give rise to quark-antiquark states, will be neglected. This corresponds to considering weak fourth order terms compared to the same terms of the component (ψ_1). This might be implemented by considering the second component to be weakly interacting, although in this case this component would be associated to higher energy modes and not to constituent quarks of light quark-antiquark mesons, in particular pions. The second component is integrated out approximatedly and the determinant is expanded up to the second order in the quark bilinears. The resulting NJL coupling constants are presented.

2. Effective interaction and quark field splitting

The quark-quark effective term in QCD effective action is given by [8]:

$$S_{eff}[\bar{\psi}, \psi] = \int_x \left[\bar{\psi} \left(i\partial\!\!\!/ - m \right) \psi - \frac{g^2}{2} \int_y j^b_\mu(x) (\tilde{R}^{\mu\nu}_{bc})(x-y) j^c_\nu(y) \right] + S_A, \quad (1)$$

Where $j^\mu_a = \bar{\psi}\lambda_a\gamma^\mu\psi$ is the color quark current, a, b and i, h stand respectively for color and flavor indices, the sums in color, flavor and Dirac indices are implicit, $\int_x = \int d^4x$, the gluon propagator is 1 $\tilde{R}^{\mu\nu}_{bc}$, and S_A corresponds to other terms due the gluon integration and sources. 3 and 4 body quark interactions are neglected. The kernel $\tilde{R}^{\mu\nu}_{bc}$ can be written in terms of transversal and longitudinal components such that:

$$\tilde{R}^{\mu\nu}_{ab} = \delta_{ab} \left[R_T \left(g^{\mu\nu} - \frac{\partial^\mu \partial^\nu}{\partial^2} \right) + R_L \frac{\partial^\mu \partial^\nu}{\partial^2} \right]. \quad (2)$$

With a Fierz transformation [2, 3, 8], by considering only the color singlet sector, the effective quark interaction of the type (color) current-current can be expressed in terms of bilocal quark bilinears, $j^q_i(x, y) = \bar{\psi}(x)\Gamma^q\psi(y)$ where $q = s, p, v, a$ where Γ_q stands for Dirac and SU(2) operators $\Gamma_s = I$ (for the 2x2 flavor and 4x4 identities), $\Gamma_p = \sigma_i i\gamma_5$, $\Gamma^\mu_v = \gamma^\mu\sigma_i$ and $\Gamma^\mu_a = \gamma^\mu i\gamma_5\sigma_i$, where σ_i are the flavor SU(2) Pauli matrices.

The quark field is separated into two parts, one will be considered to be a weak field component or the component that form bound states of quark-antiquark pairs such as light mesons and the chiral condensate, and which generate virtual lines in polarization calculation. The other component will be dressed generating (constituent) interacting quarks. This can implemented by means of the following substitution for each of the bilinears: $\bar{\psi}\Gamma^q\psi \to (\bar{\psi}\Gamma^q\psi)_2 + (\bar{\psi}\Gamma^q\psi)_1$, The interaction terms can then be rewritten with terms denoted by Ω_{12} that corresponds to the mixed terms. The component ψ_2 will be neglected since it gives rise to light quark-antiquark mesons [7] which will not be considered here. It might be a weak field and, in this case, its interactions neglected as well. By integrating out the component $(\bar{\psi}\psi)_2$, and rewriting the determinant as: $\det(A) = e^{Tr \ln A}$, the following non linear (non local) effective action for quarks $(\bar{\psi}\psi)_1$ is obtained:

$$\begin{aligned} S_{eff} &= -i\, Tr \log \Big\{ (S^c_0)^{-1} - \alpha t_1 g^2 \bar{R}^{\mu\nu} \gamma_\mu \sigma_i \left[(\bar{\psi}\gamma_\nu\sigma_i\psi)_1 + i\gamma_5(\bar{\psi} i\gamma_5\gamma_\nu\sigma_i\psi)_1 \right] \\ &+ 2\alpha t_1 g^2 R \left[(\bar{\psi}\psi)_1 + i\gamma_5\sigma_i(\bar{\psi} i\gamma_5\sigma_i\psi)_1 \right] \Big\} - \int \mathcal{L}_1 + S_A, \end{aligned} \quad (3)$$

where Tr stands for traces of discrete internal indices and integration of spacetime coordinates for the quark component ψ_2. The following kernel has been defined: $(S^c_0)^{-1} \equiv (S^c_0(x-y))^{-1} = (i\gamma_\mu \cdot \partial^\mu - m^*)\delta^4(x-y)$. In this expression m^* is an effective mass defined from the dynamical chiral symmetry breaking, therefore it depends on the scalar quark-antiquark condensate that, in the present approach, is found from the quark component ψ_2 [6]. Note that the Lagrangian mass of the component ψ_1 did not received any dynamical correction so far due to the dynamical chiral symmetry breaking. The last term can be written as:

$$\mathcal{L}_1 = t_1 \bar{\psi}_1 \left(i\gamma \cdot \partial - m \right) \psi_1 - \frac{g^2 t_1^2}{2} \int_y j^{a,(1)}_\mu R^{\mu\nu}_{ab} j^{b,(1)}_\nu. \quad (4)$$

As discussed above the last term in this expression already provides a NJL coupling for the component ψ_1 after a Fierz transformation.

3. Polarization correction to the NJL coupling

By considering the 't Hooft large N_c limit [9], an expansion of the determinant above for weak coupling, $g = \tilde{g}/\sqrt{N_c}$, is done next, and it provides fourth order quark effective interactions in the second order term. The expansion will be performed by writing the kernel S_0^c as: $(i\partial\!\!\!/+m^*)S_0(k)$, where the part diagonal in momentum space is $S_0(k) = 1/(k^2 - (m^*)^2)$. The first and some of the second order terms, in the zero order derivative expansion and in the scalar chanel (by neglecting the vector quark bilinears), can be written as [6, 7]:

$$S_d^{(2)} = \int_x \left(g_2\, \bar{\psi}\psi + g_4 \left[(\bar{\psi}\psi)^2 - (\bar{\psi}\sigma_i i\gamma_5\psi)^2 \right] + g_{4d} \left[[\partial_\mu(\bar{\psi}\psi)]^2 - [\partial_\mu(\bar{\psi}\sigma_i i\gamma_5\psi)]^2 \right] \right). \quad (5)$$

The first order term (g_2) is correction to the quark effective mass, whose origin is different from the usual effective mass from the gap equation [6, 7]. The second order terms are: g_4 which is a SU(2) Nambu-Jona-Lasinio coupling and g_{4d} which is a derivative NJL coupling. Both are $\mathcal{O}(N_c^{-1})$ as expected [10], since $g \to \tilde{g}/\sqrt{N_c}$, being of the same order as the NJL coupling obtained from a Fierz transformation of the color quark current gluon exchange effective interaction. All the higher order n-quark effective interactions scales as N_c^{n-1} as expected. These effective parameters can be written as [7]:

$$g_2 = -i\, g^2 N_c \alpha\, Tr'\, S_0 R, \quad (6)$$
$$g_4 = -2ig^4 N_c \alpha^2\, Tr'\, m^{*2} S_0^2 R^2, \quad (7)$$
$$g_{4d} = -2ig^4 N_c \alpha^2\, Tr'\, S_0^2 R^2. \quad (8)$$

In these expressions Tr' stands for traces in flavor and Dirac indices and also spacetime or momentum integration, $\alpha = 8/9$ is a factor from the Fierz transformation of SU(2) flavor and the following kernel was used: $R = 3R_T + R_L$. By carrying out the traces in momentum space g_2 becomes logarithmically ultraviolet divergent and g_4 and g_{4d} are finite. Note that these couplings g_4 and g_{4d} have different dimensions however. The shape of the gluon propagator was unimportant so far although it is necessary to provide numerical estimations. In Refs. [5, 6] a too much simplified gluon propagator was used with the single aim of providing numerical estimations for the resulting effective parameters, although it does not contribute for the form of the resulting interactions. An extensive analysis of the influence of the gluon propagator with other resulting couplings from this expansion will be shown elsewhere.

Acknowledgments

This work was partially supported by CNPq-Brazil.

References

[1] Y. Nambu, G. Jona-Lasinio, Phys. Rev. 122, 345 (1961); Phys. Rev. 124, 246 (1961).
[2] S.P. Klevansky, Rev. Mod. Phys. **64**, 649 (1992). U. Vogl, W. Weise, Progr. in Part. and Nucl. Phys. **27**, 195 (1991).
[3] C. Itzykson, J.B. Zuber, *Quantum Field Theory*, McGraw Hill (1985).
[4] Yu. A. Simonov, Phys. Rev. **B 65** , 094018 (2002). E. Shuryak, Phys. Rep. **264**, 357 (1996); Phys. Rep. **391**, 381 (2004). Th. Schaffer, E. Shuryak, arXiv:hep-lat/0005025, and references therein.
[5] E. Barros Jr., F.L. Braghin, Phys. Rev. **D 88**, 034011 (2013).
[6] A. Paulo Jr., F.L. Braghin, Phys. Rev. **D 90**, 014049 (2014).
[7] F.L.Braghin, Submitted to publication (2015).
[8] C.D. Roberts, R.T. Cahill, J. Praschifka, Ann. of Phys. **188**, 20 (1988). D. Ebert, H. Reinhardt, M.K. Volkov, Progr. Part. Nucl. Phys. **33**, 1 (1994). H. Kleinert, *in Erice Summer Institute 1976, Understanding the Fundamental Constituents of Matter*, 289, Plenum Press, New York, ed. by A. Zichichi (1978).
[9] G. 't Hooft, Nucl. Phys. **B 72**, 461 (1974). B. Lucini, M. Panero, Phys. Rept. **526**, 93 (2013). A.V. Manohar, arXiv:[hep-ph] 9802419.
[10] E. Witten, Nucl. Phys. **B160**, 57 (1979).

XIII International Workshop on Hadron Physics IOP Publishing
Journal of Physics: Conference Series **706** (2016) 052014 doi:10.1088/1742-6596/706/5/052014

On some Aspects of Gravitomagnetism and Correction for Perihelion Advance

Rocha, F.[1†]; Malheiro, M.[1]; Marinho Jr, R.[1]

[1]Departamento de Física, Instituto Tecnológico de Aeronáutica , Vila das Acácias, São José dos Campos, SP, 12228-900, Brazil

E-mail: [†]rocha.pereira.flavia@gmail.com

Abstract. In 1918 Joseph Lense and Hans Thirring, discovered the gravitomagnetic effect when studied solutions to the Einstein field equations using the weak field and slow motion approximation of rotating systems. They noted that when a body falls towards a massive object in rotation it feels a force perpendicular to its movement. The equations that they obtained were similar to Maxwell's equations of electromagnetism, now known as Maxwell's equations for gravitomagnetism. Some authors affirm that the gravitomagnetic effect can cause precession then in this paper we calculate the precession that gravitomagnetic effect cause in Mercury's perihelion advance. To make this we calculate the field between dipoles to measure the influence that the Sun has on Mercury, taking into account the gravitomagnetic field that the Sun and Mercury produces when they rotate around themselves. In addition, we calculate the ratio of the dipole force (of all solar system planet's) and the Newton's gravitational force to see how much is smaller.

1. Introduction

In 1918, Joseph Lense and Hans Thirring found gravitomagnetism (GM), while studying solutions to the Einstein field equations, using the weak field approximation for systems with rotation. They noted that when a body falls, towards a rotating massive object, it feels a perpendicular force to its movement. The equations that they obtained were similar to Maxwell's equations of electromagnetism, now known as Maxwell's equations for GM presented in (1), (2), (3) and (4)

$$\nabla \cdot \vec{g} = -4\pi G \rho_G \tag{1}$$

$$\nabla \cdot \vec{b} = 0 \tag{2}$$

$$\nabla \times \vec{g} = -\frac{1}{c}\frac{\partial \vec{b}}{\partial t} \tag{3}$$

$$\nabla \times \vec{b} = \frac{1}{c}\left(-4\pi G \vec{J}_G + \frac{\partial \vec{g}}{\partial t}\right), \tag{4}$$

where c is the speed of light in vacuum, G is the gravitational constant, $\vec{g}$ refers to gravito-electric field and $\vec{b}$ gravito-magnetic field.

The GM is expected by Einstein's theory of General Relativity. When a planet, star, black hole or something massive rotates, it drags space-time around it, this action is call "frame dragging".

Journal of Physics: Conference Series **706** (2016) 052014 doi:10.1088/1742-6596/706/5/052014

Einstein theory shows that all gravitational forces correspond to the curvature of space-time, the twist here is GM.
However, when the GM can cause some effects? Some researchers[4],[5] and [8] say that it can cause precession orbiting satellites, and this would make a gyroscope (for example, placed in Earth orbit) swing. Some NASA researchers led by physicist Ignazio Ciufolini tried to detect equatorial bulge pulls satellites as well, which causes a precession billions of times greater than GM. Ciufolini and their researchers then had to subtract that effect, which then left the doubt whether the precision was enough to detect the GM. Many scientists accept these results, however others are skeptical.
In fact, there are some satellites in Earth's orbit to study about the gravitomagnetic precession[1] and [2], for this study they used the Laser Geodynamic Satellites, the LAGEOS and LAGEOS II. The researchers found a small precess (about 5 per cent) consistent with GM. Gravitomagnetic effects as can be seen in[1],[2] and [5] can be theoretically measured for some systems, such as black holes, super massive stars and even rotation curve of galaxies.
Bearing in mind the gravitomagnetic concepts and formalism, we calculate the field between two dipoles to measure the influence that the Sun has on Mercury, taking into account the gravitomagnetic field that the Sun and Mercury produces to rotate around themselves, and whether the addition of this force influence the perihelion advance of Mercury. Moreover, we calculate the ratio of the dipole force (of all solar system planet's) and the Newton's gravitational force to see how much is smaller.

2. Dipole moment of a Planet

In this section, we will calculate the dipole moment equation of a planet, to do this we will consider a sphere spinning in your own axis and we used the expression for dipole moment taken from[7] and doing appropriate modifications to the gravitomagnetic formalism, we have:

$$\vec{\mu} = \frac{1}{2c}\int \vec{r} \times \vec{J} d^3x,$$

And

$$J = \rho\omega r \sin\theta, \tag{5}$$

But,

$$v = \omega r \sin\theta,$$

This expression becomes:

$$\vec{\mu} = \int (xJ\cos\phi\hat{e}_z + yJ\sin\phi\hat{e}z - zJ\sin\phi\hat{e}_y - zJ\cos\phi\hat{e}_x)d^3x, \tag{6}$$

Applying coordinate transformation and, replacing (5) in (6), we have:

$$\vec{\mu} = \frac{\rho\omega}{2c}\left(\frac{r^5}{5}\frac{4\pi}{3}\hat{e}_z + \frac{r^5}{5}\frac{4\pi}{3}\hat{e}_z\right) = \frac{\rho\omega r^5 4\pi}{15c}\hat{e}_z.$$

But, into the case of homogeneous sphere

$$\rho = \frac{M}{\frac{4}{3}\pi r^3},$$

Replacing ρ and simplifying, the expression becomes:

$$\vec{\mu} = \frac{\sqrt{G}Mwr^2}{5c}\hat{e}_z. \tag{7}$$

Where G is the gravitational constant, c is the speed of light, M is the planet mass, r is the planet radius and w is the angular velocity of the planet.

3. Sun Gravitomagnetic field

The expression below is the Sun's gravitomagnetic dipole field produced by its mass in rotation, this expression was taken from Jackson[7] and gives the value of gravitomagnetic field outside of the Sun:

$$\vec{B}_\odot = \frac{3(\vec{\mu}_\odot \cdot \vec{R})\vec{R} - R^2 m}{R^5}.$$

In spherical coordinates, with $\vec{\mu}_\odot = \mu_\odot \vec{e}_z$, this equation can be rewritten as

$$\vec{B}_\odot = \frac{\mu_\odot}{R^3}(2\cos\theta \hat{e}_r + \sin\theta \hat{e}_\theta). \tag{8}$$

Where $\vec{\mu}_\odot$ is the Sun dipole moment and $\vec{R}$ is the position vector from the Sun to the planet.

4. Potential and interaction force between dipoles

Now in this section, we can calculate the gravitomagnetic potential to finally find the expression for the interaction force between the dipoles. For this we will use the expression for the potential energy of a dipole in a gravitomagnetic field $\vec{B}$

$$U = -(\vec{\mu}_p \cdot \vec{B}_\odot),$$

where $\vec{\mu}_p$ is the planet's dipole moment. Now working with

$$\vec{\mu}_p = \frac{\sqrt{G} M_p w_p r_p^2}{5c} \hat{e}_z = K_p \hat{e}_z,$$

and

$$\vec{\mu}_\odot = \frac{\sqrt{G} M_\odot W_\odot r_\odot^2}{5c} \hat{e}_z = K_\odot \hat{e}_z,$$

the potential reads

$$U = -\frac{K_p K_\odot}{R^3}(2\cos^2\theta - \sin^2\theta).$$

The force between dipoles is given by:

$$F_d = -\nabla U = \frac{3K_p K_\odot}{R^4}[(3\sin^2\theta - 6\cos^2\theta)\hat{e}_r + (2\sin\theta\cos\theta + 4\cos\theta\sin\theta)\hat{e}_\theta].$$

Considering that $\theta = \frac{\pi}{2}$, we have that the force of the interaction between the dipoles is:

$$F_d = \frac{3K_p K_\odot}{R^4}\hat{e}_r.$$

Therefore, the force have dependence in the radial component. Now, we are in position to calculate the ratio between the gravitomagnetic gravitational forces.

Table 1. Ratio between the dipole force and the gravitational for the solar system planets.

Planet	F_d/F_{grav}
Mercury	$3,7931 \times 10^{-10}$
Venus	$2,6149 \times 10^{-10}$
Earth	$9,3934 \times 10^{-11}$
Mars	$6,0768 \times 10^{-12}$
Jupiter	$3,6582 \times 10^{-11}$
Saturn	$3,1088 \times 10^{-12}$
Uranus	$4,8549 \times 10^{-14}$
Neptune	$9,4391 \times 10^{-15}$

As we can see in Tab.(1), the gravitomagnetic force is much smaller than gravitational force.

5. The orbit equation

In this section, we will calculate the orbit equation starting from the Lagrangian:

$$L = T - V. \tag{9}$$

In our case the potential energy will be an effective potential with one term coming from the gravitational potential and the other from the dipole's potential energy

$$V_{eff} = -\frac{k}{r} + \frac{d}{r^3}.$$

Here r is the distance from the Sun to the planet, $k = GM_{\odot}M_p$ and d is strength of the potential energy of the dipoles and m the mass of the planet. Replacing this potential in Eq.(9) results

$$L = \frac{1}{2}mv_r^2 + \frac{l^2}{2mr^2} - \left(\frac{k}{r} + \frac{d}{r^3}\right),$$

where l is the angular momentum of the planet. Doing some manipulations, using $r = 1/u$ we reach:

$$\frac{d^2u}{d\theta^2} + u = \frac{mk}{l^2} + \frac{3dmu^2}{l^2}. \tag{10}$$

Notice that unless the term $3dmu^2/l^2$ the equation corresponds to a Newtonian equation for the orbit of a particle test in a gravitational field produced by a point mass body. So, this term corresponds to a perturbation when compared to mk/l^2.

In the Tab.(1) we have the comparison between the last two terms of the Eq.(10), for all solar system planets. As we can see the Mercury is the planet subjected to the strongest field, according to the data

$$r_{\text{orbital}} = 5{,}79 \times 10^{10} \text{ m},$$

$$\text{period} = 7{,}60 \times 10^{6} \text{ s}.$$

Thus,

$$\frac{3dmu^2}{mk} = 3{,}79 \times 10^{-10}.$$

This shows clearly that $3dmu^2/l^2$ is a correction term to mk/l^2. The Eq.(10) can be put in the form:

$$\frac{d^2u}{d\theta^2} + u = \frac{mk}{l^2} + \epsilon u^2, \tag{11}$$

where $\epsilon = 3dm/l^2$ and solved perturbatively [12], choose:

$$u = \sum_{k=0}^{\infty} \epsilon^k u_k = \epsilon^k u_k. \qquad \text{Einstein summation convention implied,}$$

substitute this expansion in the Eq.(11), given

$$\epsilon^k \frac{d^2 u_k}{d\theta^2} + \epsilon^k u_k = \frac{mk}{l^2} + \epsilon(\epsilon^k u_k)^2.$$

and collect terms with the same power of ϵ. The zero order of perturbation is the equation:

$$\frac{d^2 u_0}{d\theta^2} + u_0 = \frac{mk}{l^2},$$

whose solution with the appropriate initial conditions and with the semi-major axis a pointing in the x direction is

$$u_0 = \frac{mk}{l^2}(1 + e\cos\theta),$$

where e is the eccentricity given by the expression $l^2 = mka(1 - e^2)$. The next order of perturbation is obtained taking the terms in ϵ^1 and using value of u_0 obtained above, it reads

$$\frac{d^2 u_1}{d\theta^2} + u_1 = u_0^2.$$

The solution of the Eq.(10) up to the first order is

$$u = \frac{mk}{l^2}(1 + e\cos(\theta - \delta\theta)),$$

where

$$\delta\theta = \frac{3dm^2k}{l^4}\theta.$$

After one revolution ($\theta = \delta\theta + 2\pi \approx 2\pi$), so that the precession is

$$\delta\theta = \frac{6\pi dm^2k}{l^4}.$$

Using the experimental data for Mercury we have: $\delta\theta \approx 1.27 \times 10^{-12}$ arcsec/Cy.

6. Conclusions

In this work we have focused on the gravitomagnetic field produced by two rotating spheres, in particular we considered that the field produced by the Sun and its influence on Mercury. Our work is a first step to analyse the gravitomagnetic field influence in solar system's planets. We concluded that as was expected the value for the advance of perihelion obtained by GM is very small comparing with the one obtained by GR that is $\delta\theta \approx 43$ arcsec/Cy. However we still search if it's possible that the GM have a significant effect in some system.

References

[1] Biemond, J. Which gravitomagnetic precession rate will be measured by Gravity Probe B? *ArXiv Physics e-prints* (Nov. 2004).

[2] Chashchina, O., Iorio, L., and Silagadze, Z. Elementary Derivation of the Lense-Thirring Precession. *Acta Physica Polonica B 40* (July 2009), 2363.

[3] Ciufolini, I. Frame-dragging, gravitomagnetism and lunar laser ranging. Tech. Rep. TEAT-7150, 2009.

[4] Ciufolini, P. A confirmation of the general relativistic prediction of the Lense-Thirring effect. *nature* (2004).

[5] Iorio, L., Lichtenegger, H. I., Ruggiero, M. L., and Corda, C. Phenomenology of the lense-thirring effect in the solar system. *Astrophysics and Space Science 331*, 2 (2011), 351–395.

[6] J. Lense, H. Thirring, E. t.-t. b. B. M. F. W. H. D. S. T. J. lense, h. thirring, phys. z. 19 (1918) 156 (english transla-tion by b. mashhoon, f. w. hehl, d. s. theisis, gen. rel. grav. Tech. Rep. TEAT-711, 1984.

[7] Jackson, J. D. *Classical electrodynamics*, 3rd ed. ed. Wiley, New York, NY, 1999.

[8] Karlsson, A. A gravitomagnetic thought experiment for undergraduates. Tech. Rep. TEAT-7150, 2006.

[9] NASA. Mercury, 2015.

[10] Ruggiero, M. L. Gravitomagnetic Field of Rotating Rings. *ArXiv e-prints* (July 2015).

[11] S.O. Kepler e Saraiva, M. F. O. *Astronomia e Astrofísica.* Departamento de Astronomia, Instituto de Física, UFRGS, Porto Alegre, Rio grande do Sul, 2012.

[12] Thornton, S., and Marion, J. *Classical Dynamics of Particles and Systems.* Brooks/Cole, 2004.

Forward particle production in the CGC formalism: average transverse momentum and τ scaling

F O Durães[1,2], A V Giannini[2], F S Navarra[2] and V P B Gonçalves[3]

[1] Curso de Física, Escola de Engenharia, Universidade Presbiteriana Mackenzie, CEP 01302-907, São Paulo, SP, Brazil
[2] Instituto de Física, Universidade de São Paulo, CEP 05315-970 São Paulo, SP, Brazil
[3] Instituto de Física e Matemática, Universidade Federal de Pelotas, CEP 96010-900, Pelotas, RS, Brazil

E-mail:
[1,2]duraes@mackenzie.br;fduraes@if.usp.br;[2]avgiannini@usp.br;[2]navarra@if.usp.br;
[3]barros@ufpel.edu.br

Abstract. In this work we have used the Color Glass Condensate (CGC) formalism of forward particle production to describe the transverse momentum spectra of charged particles in proton-lead collisions at the LHC. We have also investigated the rapidity dependence of the average transverse momentum, $< p_T >$, and the so called scaling variable, $< \tau = p_T^2/Q_s^2 >$, where Q_s represents the saturation scale of these processes. We have computed the ratios of the first quantity at a given rapidity y to the value at $y = 0$ (going from $y = 0$ towards the proton fragmentation region) at the LHC energies. The ratios of the second quantity have been computed to the normalized τ distributions. Our analysis, based on gluon saturation and geometrical scaling, shows that these ratios decrease strongly with y above midrapidities and increase slightly with the energy. We briefly discuss the implications of our results and present predictions for the forthcoming LHC data.

Recent experiments at the LHC have measured charged hadron rapidity and transverse momentum distributions in proton-proton (pp), proton-lead (pPb) and lead-lead ($PbPb$) collisions in a new domain of collision energy. The quantities measured in these collisions are very sensitive to the underlying dynamics, to the initial conditions of the created system and to the hadronic final state interactions.

Many questions about the dynamics of interaction between partons in these processes remain still open. It is expected that a system of gluons at high energy (or small Bjorken-x) forms a new state of matter, the Color Glass Condensate (CGC), where the gluon density increases inside the hadron wave function and becomes very large in comparison to all other parton species (valence and sea quarks). This growth eventually saturates and reaches the regime of "parton saturation", characterized by a momentum scale (Q_s), the so called saturation scale, which marks the onset of nonlinear (gluon recombination) effects and grows with the reaction energy. In fact, the CGC has now become an effective theory which describes the gluon content of a high energy hadron or nucleus in the saturation regime and many saturation models have been proposed (see [1] and references therein).

It is important to emphasize that the data from the LHC will allow to probe the nuclear gluon distribution at very small Bjorken-x ($\approx 10^{-6}$) and, therefore, to test the nature and the properties of the CGC. Of course, many models will also be put to the test.

In this work we adopt the *hybrid formalism* of the CGC [2, 3, 4] to compute the cross section of inclusive charged hadron production at the LHC energies. In this formalism the large-x degrees of freedom of the proton are described in terms of usual parton distribution functions (PDF) of collinear factorization with a scale dependence given by the DGLAP evolution equations. The small-x glue of the nucleus is described in terms of its unintegrated gluon distribution (UGD).

At leading order, the cross section for single-inclusive forward hadron production (with rapidity y and transverse momentum p_T) in high-energy nucleon-nucleus collisions is described in terms of the dipole scattering amplitude in the following way:

$$\begin{aligned}\frac{dN^{pA\to hX}}{dyd^2p_T} &= \frac{K}{(2\pi)^2}\int_{x_F}^{1}\frac{dz}{z^2}\sum_q x_1 f_q(x_1,Q^2)\tilde{N}_F(x_2,\frac{p_T}{z})D_q(z,Q^2)\\ &+ x_1 f_g(x_1,Q^2)\tilde{N}_A(x_2,\frac{p_T}{z})D_g(z,Q^2)\,, \qquad (1)\end{aligned}$$

where the K-factor effectively accounts for NLO corrections, $x_{1,2} = e^{\pm y}\, p_T/z/\sqrt{s}$ and $z \equiv x_F/x_1$.

In order to compare our results to the final state distributions of measured charged particles one has to translate the rapidity distributions into pseudo-rapidity distributions through the corresponding Jacobian transformation, $y \to \eta$, (, $x_F = e^{\eta}\sqrt{p_T^2 + m^2}/\sqrt{s}$, where m represents an average mass of the produced particles). The parton distribution functions ($f_{q(g)}$) and the fragmentation functions ($D_{q(g)}$) are considered at the scale Q^2, and $\tilde{N}_{F(A)}$ describes a quark (gluon) scattering off the nucleus, in the fundamental (adjoint) representation and encodes the information about the hadronic scattering and thus about the non-linear and quantum effects in the hadron wave function.

In the hybrid formalism, the forward scattering amplitude of a color dipole of radius r_T off a nucleus target, $\tilde{N}_A$, is given by

$$\begin{aligned}\tilde{N}_A(x_2,q_T) &= \int d^2r_T\, e^{i\vec{q}_T\cdot\vec{r}_T} N_A(r_T,x_2,q_T)\,,\\ N_A(r_T,x_2,q_T) &= 1-\exp\{-\frac{1}{4}[r_T^2Q_s^2(x_2)]^{\gamma(x_2,q_T)}\} \qquad (2)\end{aligned}$$

The corresponding expression for quarks, $\tilde{N}_F$, is easily obtained from N_A by the replacement $(r_T^2Q_s^2)^{\gamma} \to ((C_F/C_A)r_T^2Q_s^2)^{\gamma}$, with $C_F/C_A = 4/9$.

The main difference among the distinct phenomenological models comes from the behavior of the "anomalous dimension" of the target gluon distribution, $\gamma(x_2, q_T)$, which determines the transition from the nonlinear to the extended "geometric scaling" regime (it is well known that the small-x Deep Inelastic Scattering data show geometric scaling [5]), as well as from the extended geometric scaling to the DGLAP regime. This means that the cross sections depend on $Q^2/Q_s^2(x)$ only, instead of Q^2 and x independently. In such a regime, of course, the geometric scaling arises as a feature of saturation from nonlinear evolution equations.

In this work we restrict our analysis to the so called BUW model [4], which is able to describe the ep HERA data for the proton structure function and the hadron spectra measured in pp and dAu collisions at RHIC energies. Another feature of the BUW model which motivates this analysis is that it explicitly satisfies the property of geometric scaling, which is predicted for the solutions of the BK equation in the asymptotic regime of large energies.

In the BUW model, the anomalous dimension is given by $\gamma(x, q_T) = \gamma_s + \Delta\gamma_{BUW}(x, q_T)$, with

$$\Delta\gamma_{BUW}(x,q_T) = (1-\gamma_s)(\omega^a-1)\,[\,(\omega^a-1)+b]^{-1}\,, \qquad (3)$$

where $\omega \equiv q_T/Q_s(x)$, γ_s, a and b are free parameters of the model.

Moreover in the BUW model the behavior expected for the unintegrated gluon distribution in the large p_T limit (linear regime) is recovered: $\varphi(x_2, q_T) \propto 1/q_T^4$ at large q_T (or, equivalently, for small r_T). For larger ω the exponent in Eq. (2) can be expanded and the dipole scattering amplitude simplifies to:

$$N_A(q_T) \approx \frac{2\pi\, 2^{2\gamma(w)-1}\, Q_s^{2\gamma(w)}}{q_T^{2\gamma(w)+2}} \frac{\Gamma(1+\gamma(w))}{-\Gamma(-\gamma(w))} \tag{4}$$

$$\overset{\gamma(w)\to 1}{\approx} \frac{4\pi\, Q_s^2}{q_T^4}\,(1-\gamma(w)) \propto \frac{Q_s^{2+a}}{q_T^{4+a}}\,, \tag{5}$$

if possible y dependencies are suppressed.

In what follows we assume that the saturation scales in adjoint and fundamental representations are given, respectively, by $Q_{s,A}^2 = A^{1/3}\, Q_0^2\,(x_0/x_2)^\lambda$ and $Q_{s,F}^2 = 4/9\, Q_{s,A}^2$, with $Q_0^2 = 1\,GeV^2$, $x_0 = 3\times 10^{-4}$, $\lambda = 0.288$ and $A \equiv A_{Pb} = 208$. We also assume the CTEQ5-LO [6] and KKP [7] parameterizations for the parton distribution and fragmentation functions, respectively, at the scale $Q^2 \equiv m^2 + p_T^2$ and a y-dependent K-factor for NLO corrections.

In Figure 1 we present our results (in the large ω limit) for charged particle distributions in pPb collisions, at LHC with energy $\sqrt{s} = 5.02\,TeV$ [8], for three different windows of pseudo-rapidity: a) $\mid \eta \mid < 0.3$, b) $0.3 < \eta < 0.8$ and c) $0.8 < \eta < 1.3$. In these calculations we have adopted, respectively, $<\eta>= 0$ ($K = 2.0$), $<\eta>= 0.525$ ($K = 2.1$) and $<\eta>= 1.05$ ($K = 2.2$). In all cases, the (free) parameters of the model are $\gamma_s = 0.80$, $a = 2.73$, $b = 470$ and $m = m_\pi$. As can be seen our results exhibit a reasonable agreement with data.

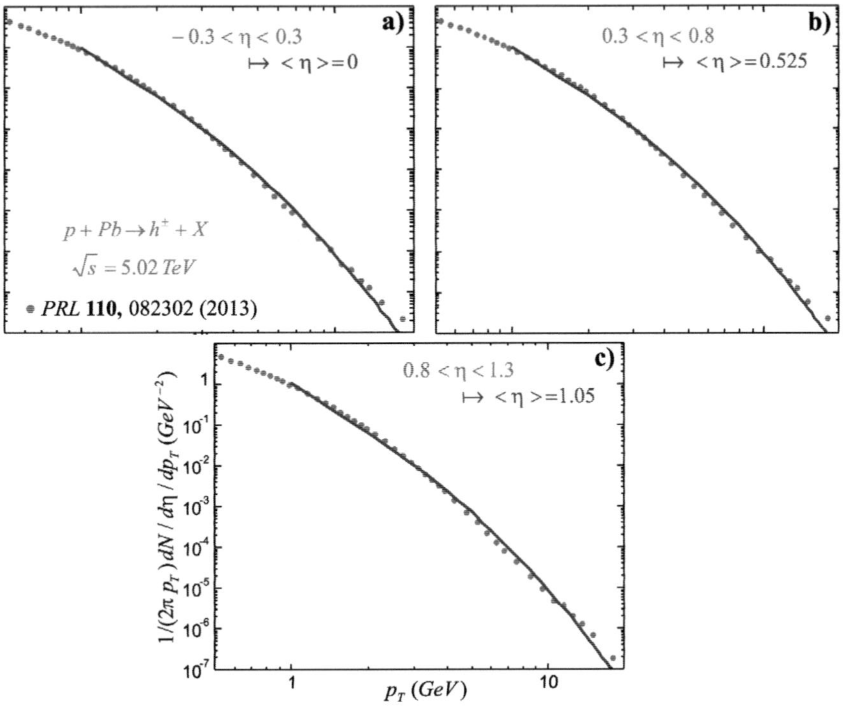

Figure 1. Charged particle distributions in pPb collisions at LHC.

Inspired by Ref. [9], we shall investigate the rapidity dependence of the average transverse momentum ($< p_T >$) and the scaling variable ($< \tau = p_T^2/Q_{s,A}^2 >$). To this end we have computed the following ratios:

$$R_1(y,s) = \frac{< p_T > (y,s)}{< p_T > (y,s)\,|_{y=0}} \;\; ; \;\; < p_T > (y,s) = \frac{\int d^2p_T\, p_T\, \frac{dN}{dy\, d^2_{p_T}}}{\int d^2p_T\, \frac{dN}{dy\, d^2p_T}} \tag{6}$$

$$R_2(y,s) = \frac{< \tau > (y,s)}{< \tau > (y,s)\,|_{y=0}} \;\; ; \;\; < \tau > (y,s) = \frac{\int d\tau\, \tau\, \frac{dN}{dy\, d\tau}}{\int d\tau\, \frac{dN}{dy\, d\tau}} \tag{7}$$

With the same parameters fixed before, in Figure 2 we present the ratios above at the LHC energies. As can be seen our analysis, based on gluon saturation and geometrical scaling, shows that these ratios decrease strongly with y (above midrapidities) and increase slightly with the energy. It is important to note that our first result is compatible with that predicted by the collective expansion picture in the hydrodynamic framework. Our second result tells us that the geometrical scaling violation increase with y but is slightly smaller for higher energies.

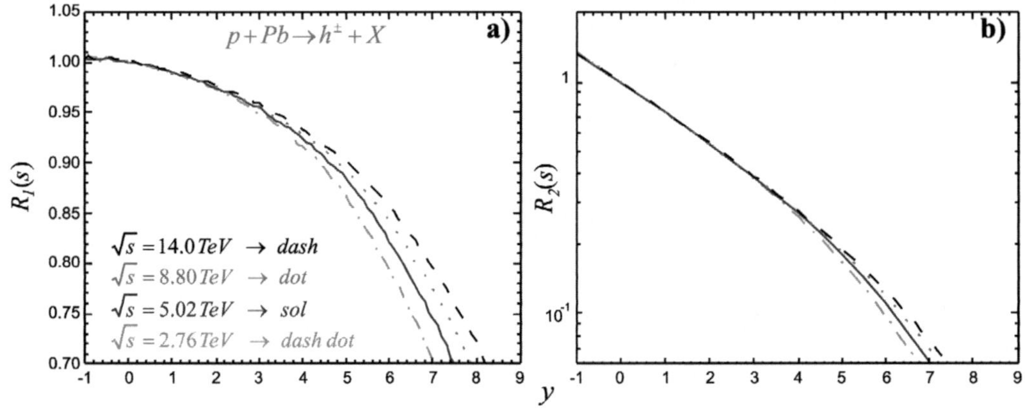

Figure 2. Rapidity dependence of the average transverse momentum (a) and the geometrical scaling variable (b).

Acknowledgments

This work was partially financed by the Brazilian funding agencies CAPES, CNPq, FAPESP and FAPERGS.

References

[1] Durães F O, Giannini A V, Navarra F S and Gonçalves V P, 2014 *Phys. Rev.* C **89** 035205
[2] Dumitru A, Hayashigaki A and Jalilian-Marian J, 2006 *Nucl. Phys.* A **765** 464
[3] Dumitru A, Hayashigaki A and Jalilian-Marian J, 2006 *Nucl. Phys.* A **770** 57
[4] Boer D, Utermann A and Wessels E, 2008 *Phys. Rev.* D **77** 054014
[5] Staśto A M, Golec-Biernat K J and Kwieciński J 2001, *Phys. Rev. Lett.* **86** 596
[6] Lai H L *et al.* (CTEQ Coll.) 2000 *Eur. Phys. J.* C **12** 375
[7] Kniehl B A, Kramer G and Pötter B (KKP Coll.) 2000 *Nucl. Phys.* B **582** 514
[8] Abelev B *et al.* (ALICE Coll.) 2013 *Phys. Rev. Lett.* **110** 082302
[9] Bożek P, Bzdak A and Skokov V, 2014 *Phys. Lett.* B **728** 662

Mass-Radius Relation for White Dwarfs Models at Zero Temperature

G. A. Carvalho[1†]; R. M. Marinho Jr[1] and M. Malheiro[1]

[1]Departamento de Física, Instituto Tecnológico de Aeronáutica , Vila das Acácias, São José dos Campos, SP, 12228-900, Brazil

E-mail: †geanderson.araujo.carvalho@gmail.com

Abstract. In this work we investigate the structure of WD stars using the Tolman-Oppenheimer-Volkoff equations and compare with the Newtonian equations of gravitation in order to put in evidence the importance of the General Relativity in the study of these stars. We solved the equations using the exact relativistic energy equation for the model of completely degenerate electron gas and we also use the politropic EoS for ultra and non-relativistic limit. We find a good fit of the TOV solution with the general EoS for the WD mass-radius diagram. We propose that our fit has to be used as relation between mass and radius for general relativistic WD instead of that Newtonian $M \sim 1/R^3$, this fit is given by $M = R/(a+bR+cR^2+dR^3+kR^4)$, where a, b, c and d are parameters and $1/k$ is the constant of the Newtonian mass-radius relation and it can be used in simulation study of binary systems that occurs accretion.

1. Introduction

In this work we did a continuation of the previous work[3] then a lot of this paper can be found in the old one.

Since their discovery WD constitute a laboratory of tests in general relativity[15][17][10]. We decide to study the WD mass-radius relation taking into account general relativity[1]. The WD are formed of matter in high densities, on average 10^6 g/cm^3. In case of the white dwarfs they are constituted essentially of degenerate electron gas[15]. The typical mass is 0.5 1 $M_\odot$ ($M_\odot$ represents the solar mass) and they have radii of the order of 10000 km.

2. Differential Equations

Basically we analyze three cases. In general in our accounts the configuration of the mass of the star is static and spherically symmetric, in this case the pressure and the density of the star are functions of only the radial coordinate r[1].

In the first case we use the Newtonian equations, then we have[15]

$$\frac{dp}{dr} = -\frac{Gm(r)\rho}{r^2} \tag{1a}$$

$$\frac{dm}{dr} = 4\pi r^2 \rho, \tag{1b}$$

where p denotes pressure, G is the gravitational constant, ρ is the mass density, and m represents the mass inside of the radius r. We can improve the Newtonian case if we replace mass density by energy density and add the contribution of the electron energy density to the energy density of the star and write

$$\frac{dp}{dr} = -\frac{Gm(\rho c^2 + e_e)}{r^2 c^2} \tag{2a}$$

$$\frac{dm}{dr} = \frac{4\pi r^2(\rho c^2 + e_e)}{c^2}, \tag{2b}$$

where e_e is electron energy density and c is the speed of light. We call this as the improved Newtonian case. If taking into account the general relativistic effects, the structure of the star in hydrostatic equilibrium for a static and isotropic ideal fluid sphere is described by the TOV[18][13]

$$\frac{dp}{dr} = -\frac{Gm\epsilon}{c^2 r^2}\left[1 + \frac{p}{\epsilon}\right]\left[1 + \frac{4\pi r^3 p}{mc^2}\right]\left[1 - \frac{2Gm}{c^2 r}\right]^{-1} \tag{3a}$$

$$\frac{dm}{dr} = \frac{4\pi r^2 \epsilon}{c^2}, \tag{3b}$$

where ϵ represents the energy density. The new three terms in the equilibrium Eq.(3a) are corrections terms.

3. Equation of State EoS

To solve numerically the coupled differential equations in both cases, general relativistic and Newtonian, we must have an EoS. We use the well known politropic EoS and we use the EoS for completely degenerate electron gas given by[4]

$$p = \frac{m_e c^2}{\lambda_e^3}\frac{1}{24\pi^2}\left[(2x^3 - 3x)(1 + x^2)^{1/2} + 3\sinh^{-1}(x)\right], \tag{4}$$

where x is a dimensionless variable $x = P_f/m_e c$ (P_f is the Fermi moment and m_e is the electron mass) and λ_e is the Compton wavelength of the electron. In terms of the parameter x the density is given by

$$\rho = 9.738 \times 10^5 \mu_e x^3 \text{ g/cm}^3, \tag{5}$$

μ_e is the ratio between the mass number and atomic number. If we assume that the white dwarf is made of ^4He, ^{12}C or ^{16}O, hence $\mu_e = 2$.

4. Mass-radius relation

By solving the Eq.(1) and Eq.(3) with the ultra-relativistic we see that in both solutions the stars are unstable due to secular instability $\partial M/\partial\rho_c < 0$ and the non-relativistic limit for EoS is good only below $0.25M_\odot$ (fig.(1)). We also solved the set of Eq.(1), (2) and (3) with the EoS (4).

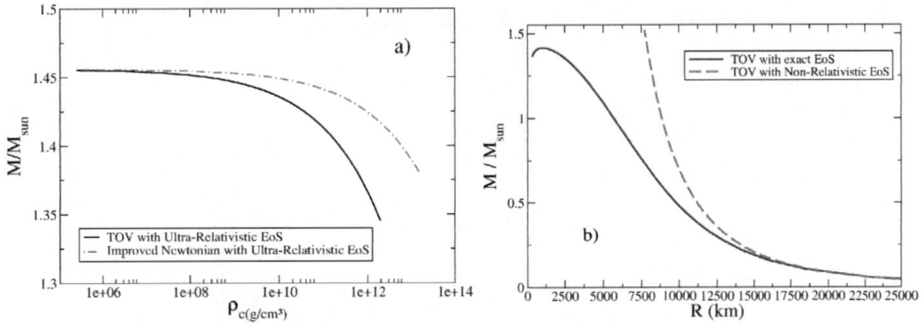

Figure 1. a) TOV and Improved Newtonian case using politropic equation with index $n = 3$ b) TOV with EoS (4) and politropic with index $n = 1.5$.

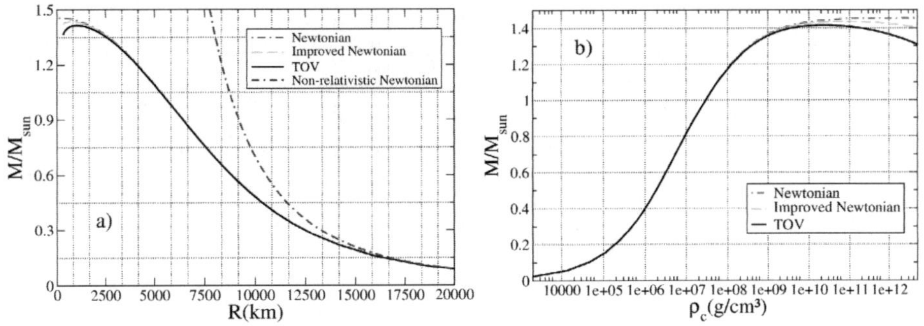

Figure 2. a) Mass-radius relation for general relativistic and Newtonian ODE and b) Mass-Central density relation for the same models.

In Fig.(2) we show the numerical results for mass-radius relation and mass-density relation. The purely Newtonian Eq.(1), i.e., without special relativity corrections don't have the secular instability $\partial M/\partial R > 0$. While TOV and even "Improved Newtonian" case presents instability, when pass from stability to instability we compute the mass and the correspondent radius (tab.(1)).
The differences between the maximum masses are small, however the minimum radii are different.

Table 1. Maximum mass for the models of white dwarfs stars.

Model	Mass	Radius [km]
Newtonian $P = P(\rho)$	1.4546 $M_\odot$	329
Improved Newtonian $P = P(\epsilon)$	1.4358 $M_\odot$	739
TOV $P = P(\epsilon)$	1.4154 $M_\odot$	1021
Non-relativistic Newtonian $P = K\rho^{5/3}$	1.4564 $M_\odot$	7833

As we can see in tab.(1) the "Newtonian" minimum radius is about three times larger than "TOV" one.

We use the expression below to fit the TOV curve in fig.(2).

$$\frac{M}{M_\odot} = \frac{R}{a + bR + cR^2 + dR^3 + kR^4}. \tag{6}$$

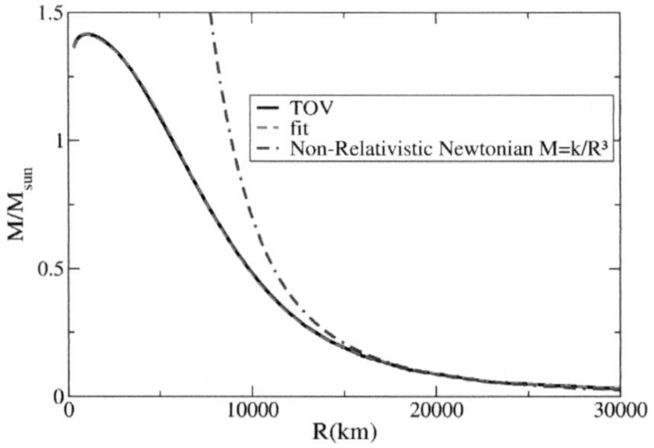

Figure 3. General relativistic mass-radius diagram fit.

The constants a, b, c and d in TOV case using $\mu_e = 2$ (Fig. (3)) are

$$\begin{aligned} a &= 20.86 \text{ km} \\ b &= 0.66 \\ c &= 2.48 \times 10^{-5} \text{ km}^{-1} \\ d &= -2.43 \times 10^{-5} \text{ km}^{-2}. \end{aligned} \tag{7}$$

the Eq.(6) with above constants is a better expression that we found in[3] to calculate the mass in function of the radius considering GR effects in static case.

5. Concluding Remarks

We show that the general relativity can be relevant to mass-radius relation of WD. Eq.(6) can be applied in another formulas like magnetic dipole field, gravitational potential[4][9] and surface gravity. We shown also that is better to use the exact relativistic EoS.

6. Acknowledgments

I would like to thank CAPES (Coordenação de Aperfeiçoamento de Pessoal de Nível Superior) for the financial support.

References

[1] Adler, R., Bazin, M., and Schiffer, M. *Introduction to general relativity.* McGraw-Hill, 1965.
[2] Althaus, L., Crsico, A., Isern, J., and Garca-Berro, E. Evolutionary and pulsational properties of white dwarf stars. *The Astronomy and Astrophysics Review 18* (2010), 471–566.
[3] Carvalho, G. A., Marinho Jr, R. M., and Malheiro, M. Mass-radius diagram for compact stars. *Journal of Physics: Conference Series 630* (2015), 012058.
[4] Chandrasekhar, S. *An Introduction to the Study of Stellar Structure.* Dover Publications, 1957.

[5] Hamada, T., and Salpeter, E. E. Models for zero-temperature stars. *The Astrophysical Journal 134* (1961), 683.

[6] Howell, D. A., Sullivan, M., Nugent, P. E., Ellis, R. S., Conley, A. J., Le Borgne, D., Carlberg, R. G., Guy, J., Balam, D., Basa, S., Fouchez, D., Hook, I. M., Hsiao, E. Y., Neill, J. D., Pain, R., Perrett, K. M., and Pritchet, C. J. The type ia supernova snls-03d3bb from a super-chandrasekhar-mass white dwarf star. *Nature 443* (2006), 308–311.

[7] Jackson, C. B., Taruna, J., Pouliot, S. L., Ellison, B. W., Lee, D. D., and Piekarewicz, J. Compact objects for everyone: I. white dwarf stars. *European Journal of Physics 26*, 5 (2005), 695.

[8] Kepler, S. O., Kleinman, S. J., Nitta, A., Koester, D., Castanheira, B. G., Giovannini, O., Costa, A. F. M., and Althaus, L. White dwarf mass distribution in the sdss. *Monthly Notices of the Royal Astronomical Society 375* (2007).

[9] Kepler, S. O., and Saraiva, M. F. O. *Astronomia e Astrofísica.* Departamento de Astronomia, Instituto de Física, Universidade Federal do Rio Grande do Sul, 2012.

[10] Lobato, R. V., Coelho, J., and Malheiro, M. Particle acceleration and radio emission for sgrs/axps as white dwarf pulsars. *Journal of Physics: Conference Series 630* (2015), 012015.

[11] Marsh, T. R., Nelemans, G., and Steeghs, D. Mass transfer between double white dwarfs. *Monthly Notices of the Royal Astronomical Society 350* (2004), 113–128.

[12] Mathew, A., and Nandy, M. K. General relativistic calculations for white dwarf stars. *ArXiv e-prints* (2014).

[13] Oppenheimer, J. R., and Volkoff, G. M. On massive neutron cores. *Physical Review 55* (1939), 374–381.

[14] Postnov, K. A., and Yungelson, L. R. The evolution of compact binary star systems. *Living Reviews in Relativity 17* (2014), 1–166.

[15] Sagert, I., Hempel, M., Greiner, C., and Schaffner-Bielich, J. Compact stars for undergraduates. *European Journal of Physics 27* (2006), 577.

[16] Shapiro, S., and Teukolsky, S. *Black Holes, White Dwarfs and Neutron Stars: The Physics of Compact Objects.* 2008.

[17] Silbar, R. R., and Reddy, S. Neutron stars for undergraduates. *American Journal of Physics 72*, 7 (2004), 892–905.

[18] Tolman, R. C. Static solutions of einstein's field equations for spheres of fluid. *Physical Review 55* (1939), 364–373.

Journal of Physics: Conference Series **706** (2016) 052017 doi:10.1088/1742-6596/706/5/052017

Exclusive photoproduction of light vector meson in coherent collisions at the LHC energies

G Sampaio dos Santos and M V T Machado

High Energy Physics Phenomenology Group, GFPAE IF-UFRGS
Caixa Postal 15051, CEP 91501-970, Porto Alegre, RS, Brazil

E-mail: magno.machado@ufrgs.br

Abstract. In this work we present our predictions for the rapidity distribution for ρ^0 and ϕ photoproduction in coherent pp and AA collisions at the LHC energies using the color dipole approach. In particular, we perform an analysis on the uncertainties associated to the choice of vector meson wavefunction and the phenomenological models for the dipole cross section. Comparison is done with the recent ALICE analysis on coherent production of ρ^0 at 2.76 TeV in PbPb collisions.

1. Exclusive vector meson photoproduction in ultraperipheral collisions

The exclusive photoproduction of vector mesons has been investigated recently both experimentally and theoretically [1, 2, 3, 4, 5] as it allows to test the interface between the Quantum Chromodynamics scenarios. In particular, the light vector mesons as ρ and ϕ have not a perturbative scale associated to the process in photoproduction limit and so they test the non-perturbative regime of QCD. At high energies, i.e. small-x region, it is expected a transition between the regime described by the linear dynamics of emissions chain and a new regime where the physical process of recombination of partons turn out to be crucial, which can be addressed by the so-called saturation approaches [6, 7, 8] within the color dipole formalism [9]. In such theoretical framework, the $q\bar{q}$ fluctuation (color dipole) of the incoming quasi-real photon (from the protons or nuclei) interacts with the target via the dipole cross section and the result is projected in the wavefunction of the observed hadron. The typical scale driven the dynamics of light meson production is the saturation scale, which characterizes the limitation on the maximum phase-space parton density that can be reached in the hadron wavefunction. For this energy domain hadrons and photons can be considered as color dipoles in the mixed light cone representation, where their transverse size can be considered frozen during the interaction [10]. Accordingly, the scattering process is characterized by the color dipole cross section describing the interaction of those color dipoles with the nucleon or nucleus target. For interaction at large impact parameter and at ultrarelativistic energies, e.g. ultraperipheral collisions, where the impact parameter should be larger than the sum of the hadron/nuclei radius it is expected that the electromagnetic interaction to be dominant. Therefore, in this case the photoproduction is highly favored. The main advantage of using colliding hadrons and nuclear beams for studying photon induced interactions is the high equivalent photon energies and luminosities achieved at the LHC. Consequently, studies of γp or γA interactions at the LHC could provide valuable information on the QCD dynamics at high energies. The basic idea in coherent hadronic collisions is that the cross section for a given process can be factorized in terms of the equivalent flux of photons into the hadron projectile and the photon-photon or photon-target production cross section [11, 12, 13]. Namely, the cross section of vector meson production can be evaluated within the Weizsäcker-Williams approximation as a product of the photon flux emitted by one of the colliding participants ad the cross section of vector meson photoproduction on the remaining hadron or nucleus. The photon energy spectrum for protons and nuclei, $dN_\gamma^p/d\omega$ and $dN_\gamma^A/d\omega$, which depends on the photon energy ω, are well known [11, 12, 13]. The rapidity distribution y for vector meson photoproduction in pp collisions can be written down as,

$$\frac{d\sigma}{dy}(pp \to p \otimes V \otimes p) = \left[\omega \frac{dN_\gamma^p}{d\omega}\,\sigma(\gamma p \to Vp) + (y \to -y)\right], \tag{1}$$

with $y \simeq \ln(2\omega/m_V)$ being the rapidity of the produced state with mass m_V and the square of the γp center-of-mass energy is given by $W_{\gamma p}^2 \simeq 2\omega\sqrt{s}$. In the present analysis, we consider the photon-hadron/nucleus scattering in the color dipole formalism, where the probing projectile fluctuates into a quark-antiquark pair with transverse separation r (and momentum fraction z) long after the interaction, which then scatters off the target (proton or nucleus). Similarly, the rapidity distribution y in nucleus-nucleus collisions has the same factorized form,

$$\frac{d\sigma}{dy}(AA \to A \otimes V \otimes Y) = \left[\omega \frac{dN_\gamma^A}{d\omega}\,\sigma(\gamma A \to V + Y) + (y \to -y)\right], \tag{2}$$

where $Y = A$ represents the coherent case. The cross section for exclusive photoproduction of vector meson off a nucleon target is given by [10]

$$\sigma(\gamma p \to Vp) = \frac{1}{16\pi B_V}\left|\int dz\, d^2r\, \Phi_T^{\gamma^* V}(z, r, m_q)\,\sigma_{dip}(x, r)\right|^2, \tag{3}$$

where the dipole cross section is denoted by $\sigma_{dip}(x,r)$ and the diffractive slope parameter by B_V. Moreover, the nuclear version of the cross section can be written as [14]

$$\sigma(\gamma A \to VA) = \int d^2b \left|\sum_{h,\bar{h}} \int dz\, d^2r\, \Phi_T^{\gamma^* V}\left\{1 - \exp\left[-\frac{1}{2}T_A(b)\,\sigma_{dip}(x,r)\right]\right\}\right|^2, \tag{4}$$

being $T_A(b)$ the nuclear profile function. The expressions for the overlap functions we have used appropriately summed over the helicity and flavor indices,

$$\Phi_T^{\gamma^* V} = \hat{e}_q \frac{\sqrt{4\pi\,\alpha_{em}}}{(2\pi)^2} N_c \left\{m_q^2 K_0(rm_q)\phi_T(r,z) - [z^2 + (1-z)^2]\, m_q K_1(rm_q)\partial_r \phi_T(r,z)\right\}, \tag{5}$$

where the constant $\hat{e}_q$ stands for an effective charge. In the numerical evaluations, we have considered for the vector meson wavefunctions the Boosted Gaussian [15] (BG) and the Light-Cone Gaussian [16] (LCG) wavefunctions

$$\phi_T^{BG} = N_T\, 4\sqrt{2\pi R^2}\, \exp\left[-\frac{m_q^2 R^2}{8z(1-z)} + \frac{m_q^2 R^2}{2} - \frac{2z(1-z)r^2}{R^2}\right], \tag{6}$$

$$\phi_T^{LCG} = N_T\, z(1-z)\, \exp\left[-r^2/(2R^2)\right]. \tag{7}$$

The parameters R and N_T are constrained by unitarity of the wavefuntion as well as by the electronic decay widths. Besides, for the phenomenological models for the dipole-proton cross section we have adopted the GBW model [17]

$$\sigma_{dip}^{GBW}(x,r) = \sigma_0 \left[1 - \exp\left(-\frac{r^2 Q_{\text{sat}}^2}{4}\right)^{\gamma_{\text{eff}}}\right], \tag{8}$$

where the the effective anomalous dimension is taken as $\gamma_{\rm eff} = 1$ and the IIM model [18]

$$\sigma_{dip}^{IIM}(x, r) = \sigma_0 \begin{cases} 0.7 \left(\frac{\bar{\tau}^2}{4}\right)^{\gamma_{\rm eff}(x, r)}, & \text{for } \bar{\tau} \leq 2, \\ 1 - \exp\left[-a \ln^2(b\,\bar{\tau})\right], & \text{for } \bar{\tau} > 2, \end{cases}$$

where $\bar{\tau} = rQ_{\rm sat}(x)$ and the expression for $\bar{\tau} > 2$ (saturation region) has the correct functional form, as obtained from the theory of the Color Glass Condensate (CGC). For the color transparency region near saturation border ($\bar{\tau} \leq 2$), the behavior is driven by the effective anomalous dimension, $\gamma_{\rm eff}(x, r) = \gamma_{\rm sat} + \frac{\ln(2/\bar{\tau})}{\kappa \lambda y}$ with $\kappa = 9.9$. The saturation scale is defined as $Q^2_{sat}(x) = (x_0/x)^{\lambda}$ and $\sigma_0 = 2\pi R_p^2$. We have considered two set of parameters for the IIM parameterization. The first set (labeled by IIM-old [19]) considers the previous DESY-HERA data and the values for parameters are $\gamma_{\rm sat} = 0.7376$, $\lambda = 0.2197$, $x_0 = 0.1632 \times 10^{-4}$ and $R_p = 3.344$ GeV^{-1} ($\sigma_0 = 27.33$ mb). The second set (labeled IIM-new [20]) considered the extremely small error bars on the recent ZEUS and H1 combined results for inclusive DIS. In this case, the parameters are $\gamma_{\rm sat} = 0.762$, $\lambda = 0.2319$, $x_0 = 0.6266 \times 10^{-4}$ and $\sigma_0 = 21.85$ mb.

2. Results

Let us start by calculating the rapidity distribution for ρ^0 and ϕ production in proton-proton collisions at the energy of 7 TeV. In Fig.1-a we present the predictions for ρ^0 taking into account the BG wavefunction and the GBW and IIM models for the dipole cross section. The dot-dashed curve stands for the IIM-old model, the solid line represents the IIM-new model, whereas the dashed curve stands for the GBW parametrisation. At large rapidities the behavior is similar for the distinct models. However, at mid-rapidities there is an evident model dependence. It is obtained the values $d\sigma/dy(y = 0) = 0.9, 0.83, 0.95$ μb for IIM-old, IIM-new and GBW, respectively. The deviation is order 14% in that case. Now, in Fig.1-b the results are presented for the LCG wavefunction, where we adopt the same set for phenomenological models for the dipole cross section. In the mid-rapidity region it is found $d\sigma/dy(y = 0) = 0.80, 0.45, 0.85$ μb for IIM-old, IIM-new and GBW, respectively. The predictions taking into account the LCG wavefunction are smaller than the results for BG wavefunction. We notice that there is an intense suppression when IIM-new model is considered (a reduction by a factor 1.8). In Fig.2-a we show the results for the rapidity distribution of ϕ using BG wavefunction, while the corresponding values for LCG wavefunction are presented in Fig.2-b. In both plots the notation is that same as for the ρ^0 case. For the ϕ meson production, at mid-rapidity we get $d\sigma/dy(y = 0) = 108.8$ nb (IIM-old), 101 nb (IIM-new) and 121.3 nb (GBW) considering the BG wavefunction and $d\sigma/dy(y = 0) = 132.4$ nb (IIM-old), 80.4 nb (IIM-new) and 147.7 nb (GBW) using the LCG wavefunction. The mid-rapidity value of cross sections are higher using LCG instead of BG wavefunction by a factor 20% at least for IIM-old and GBW models, in contrast with the ρ^0 predictions. Once again, when we choose the IIM-new model a reduction is observed.

Finally, in Fig.3 the investigation for the photonuclear production of ρ meson in PbPb collisions at the energy of 2.76 TeV is done. We present the results for the rapidity distributions of ρ^0 photoproduction, where the preliminary ALICE data [21] for coherent ρ^0 production, $\frac{d\sigma}{dy}(y = 0) = 420 \pm 10\,(\text{stat.})\,^{+30}_{-55}\,(\text{syst.})$ mb, is also presented and the notation follows the same as for the proton-proton case. In Fig.3-a the predictions are obtained considering the BG wavefunction (without nuclear break up). It was found the following estimates $d\sigma/dy(y = 0) =$ 661.5 mb (IIM-old), 747 mb (IIM-new) and 685.6 mb (GBW), respectively. In any case, the results are in average 50% larger than the experimental result. In Fig.3-b we presented the values for $d\sigma/dy$ taking into account the LCG wavefunction, including the previous BG prediction (dotted line) that also considered the color dipole approach. We checked the results $d\sigma/dy(y = 0) = 469.5$ mb (GM), 585.1 mb (IIM-old), 409 mb (IIM-new) and 603.3 mb (GBW),

Journal of Physics: Conference Series **706** (2016) 052017 doi:10.1088/1742-6596/706/5/052017

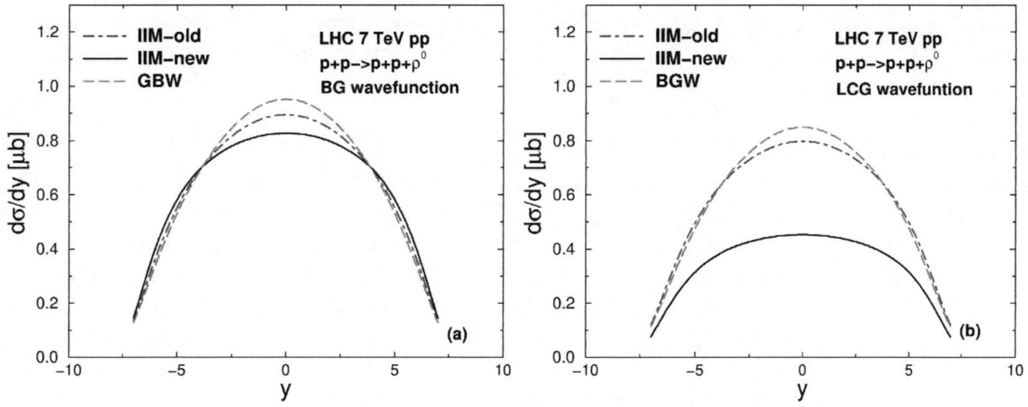

Figure 1. Rapidity distribution for ρ^0 production in pp collisions at the LHC.

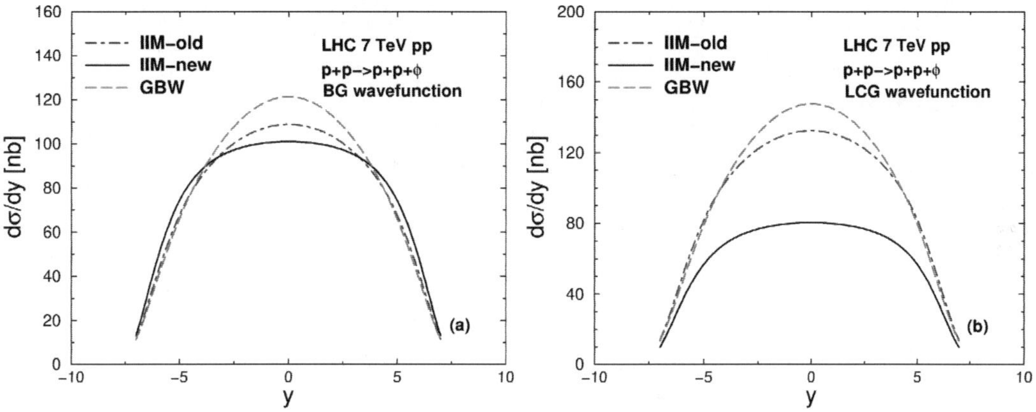

Figure 2. Rapidity distribution for ϕ production in pp collisions at the LHC.

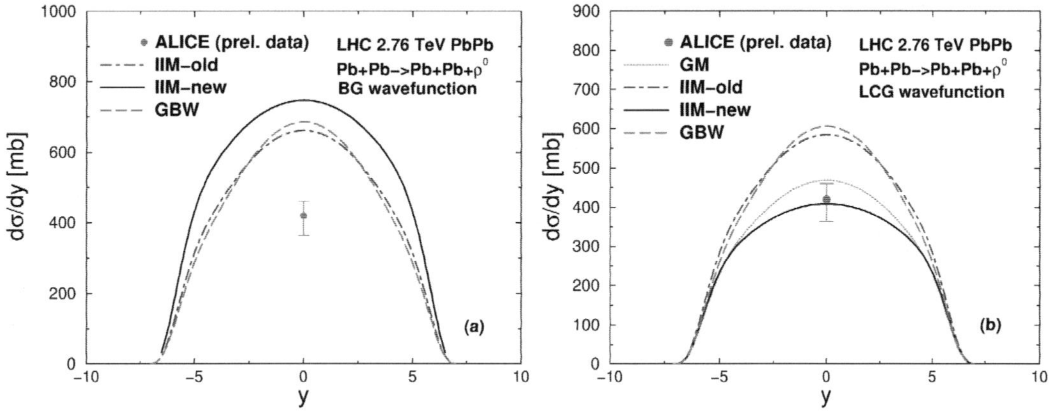

Figure 3. Rapidity distribution for ρ^0 production in $PbPb$ collisions at the LHC.

respectively. The values are smaller that for the BG wavefunction and the IIM-new result is consistent with data within the error bars. For the recent investigation of heavy vector meson photoproduction using also the color dipole approach see [22].

3. Conclusions

An investigation was done on the coherent photoproduction of the light vector as ρ and ϕ in interactions at the LHC energies for pp and $PbPb$ collisions. Predictions for the rapidity distributions are presented using the color dipole formalism and including saturation effects that are expected to be relevant at high energies. We show that the theoretical uncertainty is considerably large and the main sources are the models for the meson wavefunction and the phenomenological models for the dipole cross section. At mid-rapidity, we verified that BG wavefunction provides larger cross sections compared to the LCG wavefunction. Moreover, the dependence of the overall normalization with the color dipole model is important. In particular, the set LCG wavefunction and IIM-new model seems to be mode consistent in describing the recent ALICE preliminary data. Our results demonstrate that the production rates in LHC are fairly described by the color dipole approach.

Acknowledgments

This work was partially financed by the Brazilian funding agency CNPq and by the French-Brazilian scientific cooperation project CAPES-COFECUB 744/12.

References

[1] Aaij R *et al* 2013 *J. Phys.* G **40** 045001
[2] Aaij R *et al* 2014 *J. Phys.* G **41** 055002
[3] Abelev B *et al* 2013 *Phys. Lett.* B **718** 1273
[4] Abbas E *et al* 2013 *Eur. Phys. J.* C **73** 2617
[5] Abelev B *et al* 2014 *Phys. Rev. Lett.* **113** 232504
[6] Weigert H 2005 *Prog. Part. Nucl. Phys.* **55** 461
[7] Jalilian-Marian J and Kovchegov Y V 2006 *Prog. Part. Nucl. Phys.* **56** 104
[8] Gelis F, Iancu E, Jalilian-Marian J and Venugopalan R 2010 *Ann. Rev. Nucl. Part. Sci.* **60** 463
[9] Nikolaev N N and Zakharov B G 1994 *Phys. Lett.* B **332** 184
[10] Nemchik J, Nikolaev N N, Predazzi E and Zakharov B G 1996 *Phys. Lett.* B **374** 199
[11] Baur G, Hencken K, Trautmann D, Sadovsky S and Kharlov Y 2002 *Phys. Rep.* **364** 359
[12] Bertulani C A, Klein S R and Nystrand J 2005 *Ann. Rev. Nucl. Part. Sci.* **55** 271
[13] Hencken K *et al* 2008 *Phys. Rept.* **458** 1
[14] Kopeliovich B Z and Zakharov B G 1991 *Phys. Rev.* D **44** 3466
[15] Nemchik J, Nikolaev N N, Predazzi E and Zakharov B G 1997 *Z. Phys.* C **75** 71
[16] Dosch H G, Gousset T, Kulzinger G and Pirner H J 1997 *Phys. Rev.* D **55** 2602
[17] Golec-Biernat K and Wüsthoff M 1999 *Phys. Rev.* D **59** 014017
[18] Iancu E, Itakura K and Munier S 2004 *Phys. Lett.* B **590** 199
[19] Soyez G 2007 *Phys. Lett.* B **655** 32
[20] Rezaeain A H and Schmidt I 2013 *Phys. Rev.* D **88** 074016
[21] Nystrand J 2014 Photonuclear production of vector mesons in ultra-peripheral Pb-Pb collisions at the LHC *Preprint* hep-ex/14080811
[22] Sampaio dos Santos G and Machado M V T 2014 *Phys. Rev.* C **89** 025201

Dynamical chiral symmetry with an infrared finite gluon propagator

J. C. Cardona and A. C. Aguilar

University of Campinas - UNICAMP, Institute of Physics "Gleb Wataghin", 13083-859 Campinas, SP, Brazil

E-mail: jeinerc@ifi.unicamp.br

Abstract.
In this work we study dynamical quark mass generation using an infrared finite gluon propagator obtained from quenched lattice simulations. The quark gap equation is solved using a purely non-Abelian Ansatz for the quark-gluon vertex, which displays a dependence on the ghost dressing function and the scalar component of quark-ghost scattering kernel. For the former quantity we use quenched lattice results, while for the latter we derive its own integral equation at the one-loop-dressed approximation. This latter quantity is then coupled to the system of equations governing the two Dirac structures of the quark propagator. It turns out that when a current quark mass of 5 MeV is introduced, the constituent quark mass generated from the gap equation is of the order of 310 MeV. In addition, the pion decay constant computed from the resulting quark propagator is in good agreement with the physical value.

1. Introduction

One of the major challenges of the strong interactions is to understand the underlying mechanism that generates masses for the quarks and triggers the subsequent breaking of the chiral symmetry (CS) [1, 2]. It is well known that QCD must be endowed with a very efficient mass generation mechanism, given that the Higgs mechanism accounts for the generation of only 2% of the total mass of the light quarks; thus, the remaining 98% must be accounted for by the internal QCD dynamics [2, 3, 4].

The CS breaking is an inherently nonperturbative phenomenon, whose study in the continuum leads almost invariably to a treatment based on the Schwinger-Dyson equation (SDE) for the quark propagator (gap equation) [1, 5, 6, 7].

Besides the quark propagator, the kernel of this nonlinear integral equation is expressed in terms of three others Green's functions, namely (i) the full gluon propagator, (ii) the full ghost propagator and, (iii) the complete quark-gluon vertex; evidently, each one of these Green's functions satisfies its own SDE. One of the characteristic features of the gap equation is that, in order to give rise to nontrivial solutions, the support of its kernel in the region of few hundred MeV must overcome a critical amount. Within this context, the nonperturbative quark-gluon vertex plays a crucial role in providing the required enhancement at the correct momentum scale [6].

Despite its physical importance, the nonperturbative behavior of this special vertex is still only partially known, mainly due to a variety of serious technical difficulties. In particular, its rich tensorial structure (composed by 4 longitudinal and 8 transverse structures) leads to the determination of twelve independent form factors, which depend on three kinematic variables [8, 9]. In addition, this vertex controls the way that the ghost sector enters into the gap equation, and introduces a pivotal dependence on the ghost dressing function and the quark-ghost scattering amplitude. This important quantity satisfies its own dynamical equation, which was studied in Ref. [6] under certain simplifying approximations.

In Ref. [6] an approximate version of the gap equation was studied in the chiral limit, using the gluon and the ghost propagators obtained from the quenched lattice simulations [10, 11, 12]. Here we will follow the same steps presented in the Ref. [6] to study the case where the CS is already explicitly broken by the presence of a current quark mass. In addition, we present an integral equation that furnishes an improved one-loop dressed approximation for the scalar component of the quark-ghost scattering kernel.

These proceedings are organized as follows. In Section 2 we introduce the main ingredients and the crucial relations necessary for the study of the CS breaking. In Section 3, we present the structure of the quark-gluon vertex entering into the definition of the gap equation, and derive the dynamical equation for the scalar component of the quark-ghost scattering kernel. In Section 4 the numerical results for the quark dynamical mass are presented and the corresponding value for the pion decay constant is computed from them. Finally, in Section 5 we summarize our conclusions.

2. Basics equations and conventions

In this section we introduce the basic definitions and ingredients used throughout this work. Let us start with the definition of the full gluon propagator, that in Landau gauge, takes the form

$$i\Delta_{\mu\nu}(q) = -i\left[g_{\mu\nu} - \frac{q_\mu q_\nu}{q^2}\right]\Delta(q^2)\,. \tag{1}$$

In addition, it is usual to decompose the quark propagator, $S(p)$, in terms of a Dirac vector component $A(p^2)$, and a scalar component $B(p^2)$ in the following way [1, 6]

$$S^{-1}(p) = A(p^2)\not{p} - B(p^2) - m_0 = A(p^2)[\not{p} - \mathcal{M}(p^2)] - m_0\,. \tag{2}$$

where we have defined the dynamical quark mass function as $\mathcal{M}(p^2) = B(p^2)/A(p^2)$. Notice that when $m_0 \neq 0$ the chiral symmetry is explicitly broken by the presence of a current quark mass. Moreover, the dynamical breaking only occurs when the scalar function $B(p^2)$ acquires a nonvanishing value.

In Fig. 1 we show the diagrammatic representation of the quark SDE. The gray circles represent the complete quark and gluon propagators and the full quark-gluon vertex.

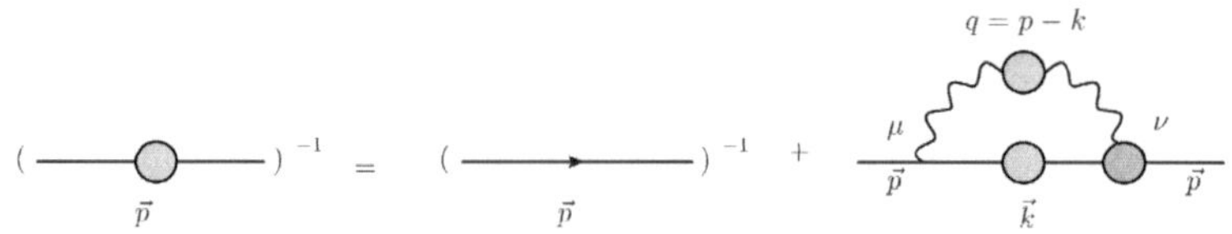

Figure 1. The quark SDE, given by Eq. (3). The grey blobs represent the full gluon and quark propagators and the complete quark-gluon vertex.

Following the momentum convention and the Lorentz indices indicated in Fig. 1, the renormalized SDE for the quark propagator is written as

$$S^{-1}(p) = Z_F(\not{p} - m_0) - Z_1 C_{\rm F} g^2 \int_k \gamma_\mu S(k)\Gamma_\nu(-p,k,q)\Delta^{\mu\nu}(q)\,, \tag{3}$$

where Γ_ν is the full quark-gluon vertex, $Z_1(\mu)$ and $Z_F(\mu)$ are the vertex and the quark wave-function renormalization constants, respectively. Here m_0 is the current bare mass and C_F is the Casimir eigenvalue for the fundamental representation. In addition, we have introduced the shorthand notation $\int_k \equiv (2\pi)^{-4} \int d^4k$.

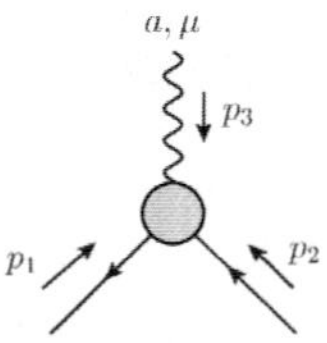

Figure 2. The full quark-gluon vertex.

An essential ingredient in the above equation is the full quark-gluon vertex, illustrated in Fig. 2, and written as

$$\Gamma^a_\mu(p_1, p_2, p_3) = g t^a \Gamma_\mu(p_1, p_2, p_3)\,,$$

where all momenta p_i are entering as showed in Fig. 2, and $t^a = \lambda^a/2$, where λ^a are the Gell-Mann matrices.

The vertex Γ_μ satisfies the fundamental Slavnov-Taylor identity (STI) [13]

$$q^\mu \Gamma_\mu(p_1, p_2, q) = F(q)[S^{-1}(-p_1)H(p_1, p_2, q) - \overline{H}(p_2, p_1, q)S^{-1}(p_2)]\,, \tag{4}$$

where $H(p_1, p_2, q)$ is the quark-ghost scattering amplitude represented in Fig. 3, and $F(q^2)$ is the dressing of the full ghost propagator, defined as

$$D(q^2) = \frac{F(q^2)}{q^2}. \tag{5}$$

The quark-ghost scattering amplitude $H(p_1, p_2, q)$ and its "conjugate" $\overline{H}(p_1, p_2, q)$ has the following Lorentz decomposition [6, 14]

$$\begin{aligned} H(p_1, p_2, q) &= X_0 \mathbb{I} + X_1 \not{p}_1 + X_2 \not{p}_2 + X_3 \tilde{\sigma}_{\mu\nu} p_1^\mu p_2^\nu\,, \\ \overline{H}(p_2, p_1, q) &= \overline{X}_0 \mathbb{I} - \overline{X}_2 \not{p}_1 - \overline{X}_1 \not{p}_2 + \overline{X}_3 \tilde{\sigma}_{\mu\nu} p_1^\mu p_2^\nu\,, \end{aligned} \tag{6}$$

where the form factors X_i are functions of the momenta, $X_i = X_i(p_1, p_2, q)$, $\overline{X}_i = \overline{X}_i(p_2, p_1, q)$, and $\tilde{\sigma}_{\mu\nu} \equiv \frac{1}{2}[\gamma_\mu, \gamma_\nu]$.

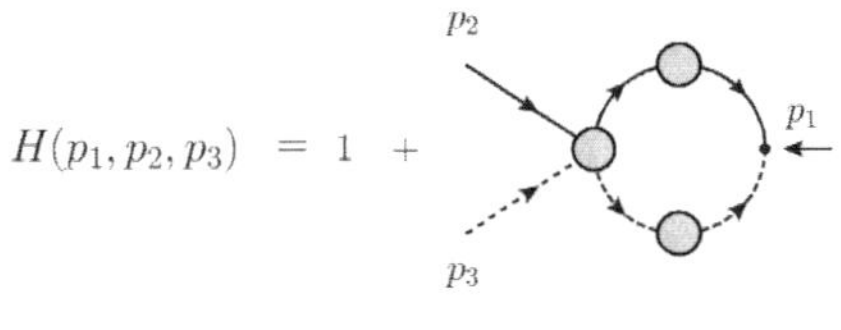

Figure 3. The quark-ghost scattering kernel $H(p_1, p_2, p_3)$.

Thus, we clearly see from Eqs (3) and (4) that most of the dynamics of the ghost sector will be introduced into the kernel of the gap equation by means of the full quark-gluon vertex.

It is interesting to notice that the STI, given by Eq. (4), enforces the all-older constraint [6]

$$Z_1 = Z_c^{-1} Z_F Z_H^{-1}, \tag{7}$$

where Z_c and Z_H are the renormalization constants of the ghost propagator and the quark-ghost scattering kernel, respectively. In the next section, these constants we will be fixed using the momentum subtraction scheme (MOM). Specifically, in this scheme the renormalized propagators and vertices should satisfy

$$\begin{aligned} F(q^2)|_{q^2=\mu^2} &= 1, \qquad S^{-1}(p)|_{p^2=\mu^2} = \not{\mu} - m(\mu), \\ \Delta^{-1}(q^2)|_{q^2=\mu^2} &= \mu^2, \qquad \Gamma_\mu(p_1,p_2,q)|_{p_1^2=p_2^2=q^2=\mu^2} = \gamma_\mu, \\ H(p_1,p_2,q)|_{p_1^2=p_2^2=q^2=\mu^2} &= 1, \end{aligned} \tag{8}$$

where μ is renormalization (subtraction) point.

3. The quark-gluon vertex and the gap equation

Let us now turn our attention to the quark-gluon vertex. This vertex can be separated into longitudinal and transverse parts in the following way [8, 9]

$$\Gamma_\mu(p_1,p_2,q) = \Gamma_\mu^L(p_1,p_2,q) + \Gamma_\mu^T(p_1,p_2,p_3), \tag{9}$$

where the transversal part $\Gamma_\mu^T(p_1,p_2,q)$ satisfies automatically the STI of Eq. (4), since $q^\mu \Gamma_\mu^T(p_1,p_2,q) = 0$.

The most general Lorentz decomposition for the longitudinal part of the vertex $\Gamma_\mu^L(p_1,p_2,q)$ can be written as [14]

$$\Gamma_\mu^L = L_1\gamma_\mu + L_2(\not{p}_1 - \not{p}_2)(p_1-p_2)_\mu + L_3(p_1-p_2)_\mu + L_4\tilde{\sigma}_{\mu\nu}(p_1-p_2)^\nu, \tag{10}$$

where $L_i = L_i(p_1,p_2,q)$ are the form factors, whose dependence on the momenta has been suppressed to keep a compact notation. Notice that the tree level expression is recovered by setting $L_1 = 1$ and $L_2 = L_3 = L_4 = 0$; then, $\Gamma_\mu^{[0]} = \gamma_\mu$.

Due to the fact that the behavior of the vertex Γ_μ^L is constrained by the STI of Eq. (4), the form factors L_i's appearing in the Eq. (10) are given in terms of the form factors X_is of Eq. (6). The full expressions for L_i's in terms of the form factors X_is can be found in Ref. [6]. Here, for the sake of simplicity, we will consider the case where only the scalar component of the quark-ghost scattering kernel is non-vanishing *i.e.* $X_0 \neq 0$ while $X_i = \overline{X}_i = 0$ for $i \geq 1$. In this limit, we obtain that

$$\begin{aligned} L_1 &= F(q)X_0(q)\left[\frac{A(p_1)+A(p_2)}{2}\right] \\ L_2 &= F(q)X_0(q)\left[\frac{A(p_1)-A(p_2)}{2(p_1^2-p_2^2)}\right] \\ L_3 &= F(q)X_0(q)\left[\frac{B(p_1)-B(p_2)}{(p_1^2-p_2^2)}\right] \\ L_4 &= 0 \end{aligned} \tag{11}$$

According the above expression, the form factors L_i display an explicit dependence on the product $F(q)X_0(q)$, thus carrying valuable information about the ghost infrared sector of the

Journal of Physics: Conference Series **706** (2016) 052018 doi:10.1088/1742-6596/706/5/052018

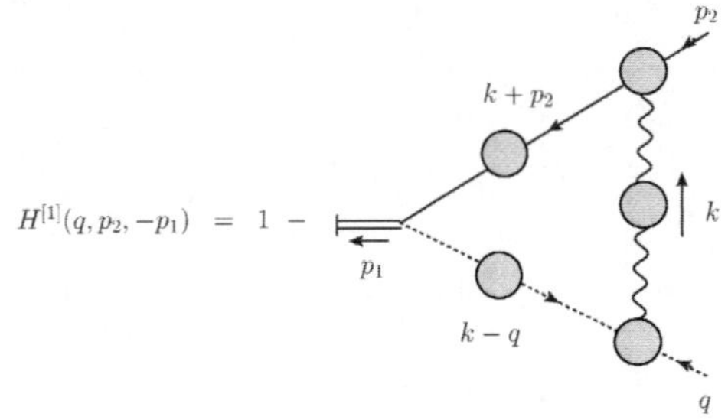

Figure 4. The quark-ghost scattering kernel at "one-loop dressed" approximation.

theory. It is interesting to notice that in the limit of $F(q)X_0(q) = 1$, the form factors of Eq. (11) reduce to the so-called Ball-Chiu (BC) vertex, Γ_μ^{BC} [15]. Therefore, our final Ansatz for the longitudinal part of the vertex is given by

$$\Gamma_\mu^L(p_1,p_2,q) = F(q)X_0(q)\Gamma_\mu^{BC}(p_1,p_2,q)\,, \tag{12}$$

Substituting the vertex Γ_μ^L defined in the Eqs. (10) and (11) into the gap equation (3), we arrive at the following coupled system for $A(p)$ and $B(p)$

$$\begin{aligned} A(p) &= Z_F + Z_c^{-1}g^2C_{\rm F}\int_k \frac{\Delta(q)F(q)X_0(q)}{A^2(k)k^2+B^2(k)}\mathcal{K}_A(p,k)\,,\\ B(p) &= Z_4 m + Z_c^{-1}g^2C_{\rm F}\int_k \frac{\Delta(q)F(q)X_0(q)}{A^2(k)k^2+B^2(k)}\mathcal{K}_B(p,k)\,, \end{aligned} \tag{13}$$

where $q = p - k$, $\mathcal{K}_A(p,k)$ and $\mathcal{K}_B(p,k)$ are kernels defined in Ref. [6].

In order to proceed further, now we derive the dynamical equation governing the behavior of the scalar form factor, $X_0(q)$. Our starting point is the Fig. 4, where we have the diagrammatic representation of the quark-ghost scattering kernel, $H^{[1]}$, at the "one-loop dressed" level. In this approximation $H^{[1]}$ is given by

$$H^{[1]} = Z_H - i\frac{g^2C_A}{2}\int_k D(q-k)G_\nu\Delta^{\mu\nu}(k)S(k+p_2)\Gamma_\mu(p_2,-p_2-k,k)\,. \tag{14}$$

To evaluate Eq. (14) further, we will use the tree-level expression for the gluon-ghost vertex $G_\nu = (q-k)_\nu$, whereas for the vertex Γ_μ we employ the following simpler Ansatz

$$\Gamma_\mu(p_2,-k-p_2,k) = F(k)X_0(k)\Gamma'_\mu(p_2,-k-p_2,k)\,, \tag{15}$$

with

$$\begin{aligned} \Gamma'_\mu(p_2,-k-p_2,k) &= \frac{1}{2}\left[A(p_2+k)+A(p_2)\right]\gamma_\mu \\ &+\frac{1}{2}\frac{k^\mu}{k^2}\Big[\left[A(p_2+k)-A(p_2)\right](2\,\not{p}_2+\not{k}) - 2\left[B(p_2+k)-B(p_2)\right]\Big]\,. \end{aligned} \tag{16}$$

Notice that the quark-gluon vertex given in Eq. (15) satisfies the STI of Eq. (4).

In this work we restrict ourself to the study of X_0 in the so-called *quark symmetric limit*, where the quark momenta have the same magnitude and opposite signs, *i.e.* $p_1 = -p/2$, $p_2 = p/2$

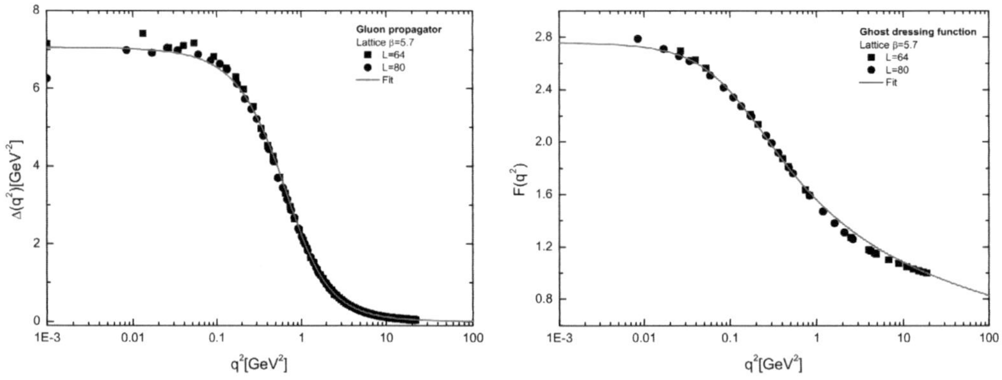

Figure 5. Lattice data (black dots) for the gluon propagator, $\Delta(q)$, (left panel) and the ghost dressing function, $F(q)$ (right panel) renormalized at $\mu = 4.3$ GeV [12]. The red continuous curve are the fits for the lattice data.

and $q = -p$ [6]. Taking the appropriate traces in Eq. (6), it is straightforward to derive from Eq. (14) the following linear integral equation for the form factor $X_0^{[1]}$ (in the Euclidean space)

$$X_0^{[1]}(p) = Z_H + \frac{g^2 C_A}{8} \int_k \frac{D(p+k)F(k)\Delta(k)A_2[A_2 + A_1]}{A_2^2(p/2+k)^2 + B_2^2} \left[p^2 - \frac{(p\cdot k)^2}{k^2}\right] X_0^{[1]}(k)\,. \quad (17)$$

where we have introduced the shorthand notation $A_1 = A(p/2)$, $A_2 = A(p/2+k)$, $B_1 = B(p/2)$, and $B_2 = B(p/2+k)$.

4. Numerical analysis

We clearly see that the coupled system for $A(p)$ and $B(p)$ given by Eq. (13) depends on the nonperturbative form of the three basic Greens functions, namely $\Delta(q)$, $F(q)$ and $X_0^{[1]}(p)$. For $\Delta(q)$ and $F(q)$ we use the lattice data obtained in Ref. [12], and shown in the Fig. 5. We clearly see that both lattice results for $\Delta(q)$, $F(q)$ are infrared finite. Such a feature can be associated to a purely non-perturbative effect that gives rise to a dynamical gluon mass [16, 17], which saturates the gluon propagator in the infrared.

Now, we are in position to solve the system formed by Eqs. (13) and (17). Substituting into Eqs. (13) the lattice data for $\Delta(q)$ and $F(q)$, with the modification $Z_c^{-1}\mathcal{K}_{A,B} \to \mathcal{K}_{A,B}F(q)$, to enforce the correct renormalization group behavior of the dynamical mass (see discussion in [6].), we determine numerically the unknown functions $A(p)$, $B(p)$ and $X_0^{[1]}(p)$ when $\alpha(\mu) = 0.29$ for $\mu = 4.3$ GeV.

The result for $X_0^{[1]}(p)$ is shown in Fig. 6. Notice that we have fixed the value of Z_H, appearing in Eq. (17), by enforcing that the renormalization condition, $X_0^{[1]}(\mu) = 1$.

On the left panel of Fig. 6, we present the case where $m_0 = 5$ MeV. As happened in the analysis presented in Ref. [6], $X_0^{[1]}(p)$ displays a maximum located 1 GeV, while in the ultraviolet and infrared regions $X_0^{[1]}(p) \to 1$ (black curve). Notice that the resulting peak is slightly higher than the one obtained by simply substituting $X_0^{[1]}(k) = 1$ on the rhs of (17) (green dashed curve) [6]. On the right panel, we show the behavior of $X_0^{[1]}(p)$ for the following current masses:

Journal of Physics: Conference Series **706** (2016) 052018 doi:10.1088/1742-6596/706/5/052018

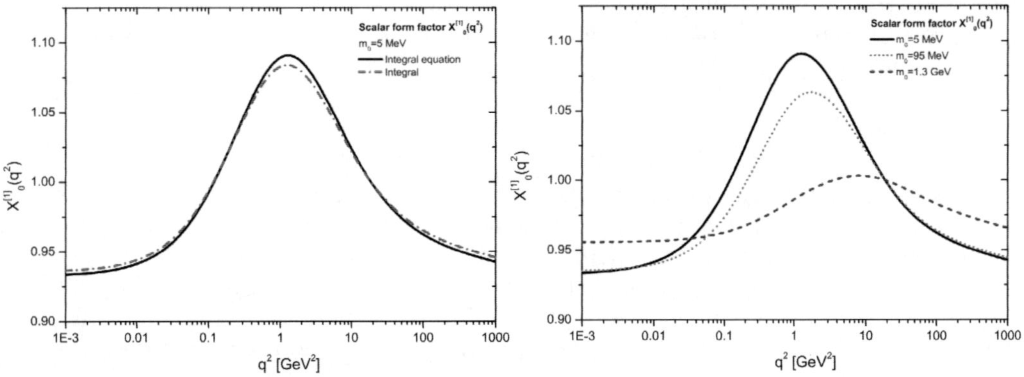

Figure 6. The numerical result for the form factor $X_0^{[1]}(q)$ given by Eq. (17).

$m_0 = 5$ MeV (black continuous), $m_0 = 95$ MeV (red dotted) and $m_0 = 1.3$ GeV (blue dashed). We clearly see a suppression as the current mass increases. Although this peak is not very pronounced, it turns out to be essential for providing to the kernel of the gap equation the enhancement required for the generation of phenomenologically compatible constituent quark masses.

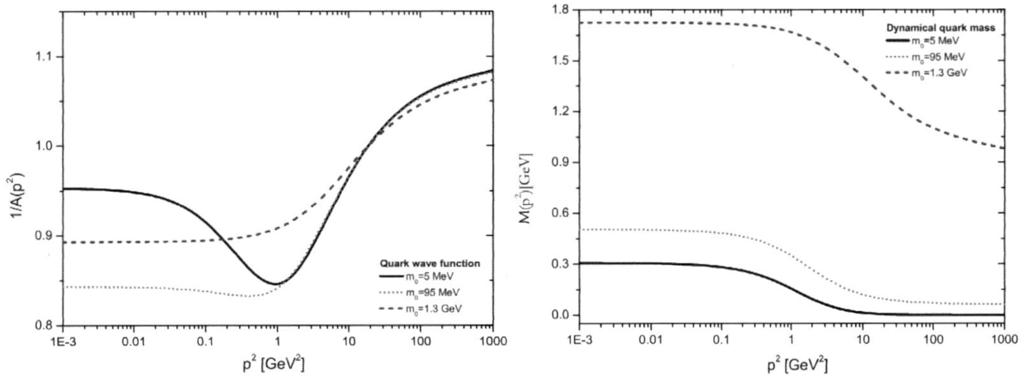

Figure 7. The numerical solution for the quark wave function $A^{-1}(p)$ (left panel). Dynamical quark mass $\mathcal{M}(p)$ (right panel).

Finally, in Fig. 7 we show the corresponding results for the quark wave function $A^{-1}(p)$ (left panel) and for the dynamical quark mass $\mathcal{M}(p)$ (right panel). From this plot one observes the effects due to the inclusion of a current mass term, which explicitly breaks the chiral symmetry. We present the results for three different current quark masses $m_0 = 5$ MeV (black continuous), $m_0 = 95$ MeV (red dotted) and $m_0 = 1.3$ GeV (blue dashed), corresponding to the up/down, strange and charm quarks. It is clear that the amount of dynamical mass generated in the case of light quarks is much more considerable than for the heavier ones. For example, for $m_0 = 5$ MeV, the dynamical mass produced is $\mathcal{M}(0) = 310$ MeV, for $m_0 = 95$ MeV we obtain $\mathcal{M}(0) = 503$ MeV, whereas for $m_0 = 1.3$ MeV we obtain $\mathcal{M}(0) = 1.72$ GeV. In addition, using the Pagels-Stokar formula [18], we have computed the pion decay constant for the case of $m_0 = 5$ MeV. The result obtained is 98.4 MeV, which is in good agreement with the experiment value of 93 MeV [19].

5. Conclusions

It has been shown that the quark gap equation gives rise to phenomenologically compatible results for the dynamical quark mass using an infrared finite gluon propagator. The crucial ingredient in this analysis is the structure of the non-abelian quark-gluon vertex and the quark-ghost scattering kernel. Both quantities are responsible for furnishing the additional strength in the intermediate region around 1 GeV, necessary for the breaking of the CS. In addition, light quarks are significantly more affected by the dynamical mass generation than the heavier ones.

Acknowledgments

The authors thank the organizers of the XIII International Workshop on Hadron Physics for their hospitality. This work was developed under the support of the National Council for Scientific and Technological Development-CNPq (JCC and ACA) and São Paulo Research Foundation - FAPESP (ACA).

References

[1] C. D. Roberts and A. G. Williams, Prog. Part. Nucl. Phys. **33**, 477 (1994).
[2] J. Papavassiliou, J. Phys. Conf. Ser. **631**, no. 1, 012006 (2015).
[3] I. C. Cloet and C. D. Roberts, Prog. Part. Nucl. Phys. **77**, 1 (2014).
[4] C. D. Roberts, arXiv:1509.02925 [nucl-th].
[5] J. Papavassiliou and J. M. Cornwall, Phys. Rev. D **44**, 1285 (1991).
[6] A. C. Aguilar and J. Papavassiliou, Phys. Rev. D **83**, 014013 (2011).
[7] C. S. Fischer and R. Alkofer, Phys. Rev. D **67**, 094020 (2003).
[8] A. C. Aguilar, D. Binosi, D. Ibañez and J. Papavassiliou, Phys. Rev. D **90**, no. 6, 065027 (2014).
[9] A. C. Aguilar, D. Binosi, J. C. Cardona and J. Papavassiliou, PoS ConfinementX , 103 (2012).
[10] A. Cucchieri and T. Mendes, PoS **LAT2007**, 297 (2007).
[11] A. Cucchieri and T. Mendes, Phys. Rev. D **81**, 016005 (2010).
[12] I. L. Bogolubsky, E. M. Ilgenfritz, M. Muller-Preussker and A. Sternbeck, PoS LAT **2007**, 290 (2007).
[13] W. J. Marciano and H. Pagels, Phys. Rept. **36**, 137 (1978).
[14] A. I. Davydychev, P. Osland and L. Saks, Phys. Rev. D **63**, 014022 (2001).
[15] J. S. Ball and T. W. Chiu, Phys. Rev. D **22**, 2542 (1980).
[16] A. C. Aguilar, D. Binosi and J. Papavassiliou, Phys. Rev. D **78**, 025010 (2008).
[17] A. C. Aguilar, D. Binosi and J. Papavassiliou, JHEP **1201**, 050 (2012).
[18] H. Pagels and S. Stokar, Phys. Rev. D **20**, 2947 (1979).
[19] K. A. Olive *et al.* [Particle Data Group Collaboration], Chin. Phys. C **38**, 090001 (2014).

XIII International Workshop on Hadron Physics IOP Publishing
Journal of Physics: Conference Series **706** (2016) 052019 doi:10.1088/1742-6596/706/5/052019

Scattering process in the Scalar Duffin-Kemmer-Petiau gauge theory

J Beltran[1], B M Pimentel[1], D E Soto[2]

[1] Instituto de Física Teorica, Universidade Estadual Paulista, Rua Dr. Bento Teobaldo Ferraz 271, S.P, 01140-070, Brazil.

[2] Facultad de Ciencias, Universidad Nacional de Ingeniera UNI, Avenida Tupac Amaru S/N apartado 31139 Lima, Per.

E-mail: jhosep@ift.unesp.br, pimentel@ift.unesp.br, dsotob@uni.edu.pe

Abstract. In this work we calculate the cross section of the scattering process of the Duffin-Kemmer-Petiau theory coupling with the Maxwell's electromagnetic field. Specifically, we find the propagator of the free theory, the scattering amplitudes and cross sections at Born level for the Moeller and Compton scattering process of this model. For this purpose we use the analytic representation for free propagators and take account the framework of the Causal Perturbation Theory of Epstein and Glaser.

1. Introduction

One of the main restriction to construct field equations is the Lorentz covariance and the causality principle. Thus, elementary particles are described by Lorentz-covariant field equations and they do not interact at a distance. Furthermore, in order to analyse the relativistic effects on the physical systems, relativistic wave equations, such the Dirac, Klein-Gordon-Fock (KGF) and Duffin-Kemmer-Petiau (DKP) equations, are constructed. The solutions of the relativistic wave equations with arbitrary spin have a long history. Besides the KGF and Dirac equations, the Dirac-like first order Lorentz invariant Kemmer equation supplemented with the β-matrices algebra, for both spin-0 and spin-1 particles was first derived by Kemmer in 1939 [1] and they are called the DKP[1] equations. Exact solutions of the DKP equation in the presence of an external field have been investigated by some authors, such as the quantum oscillator in the framework of the DKP theory [2, 3] during the last few decades.

It is known that in the free field case the DKP and KGF theories are equivalent, both in classical and quantum pictures. For instance, it was shown that both theories are equivalent in the classical level for the cases of minimal interaction with electromagnetic [4] and gravitational fields [5]. Strict proofs of equivalence between both theories were also given for cases of interaction of quantized scalar field with classical and quantized electromagnetic, Yang-Mills and external gravitational fields [6]. However, there are still no general equivalent proofs of equivalence between these theories when interactions and decays of unstable particles are taken into account.

Perhaps one of the most evident advantages in working with this theory is the fact that derivative couplings do not appear between DKP and the gauge field. This property has been

[1] There were two independent previous works given by Duffin and Petiau.

used by Gribov, who employed the spin-1 DKP theory to study the quark confinement problem [7]. Such property will result in manifestly covariant expressions for the interaction Hamiltonian and the vacuum expectation values of time ordered products of fields. Other advantages are the formal similarity with SQED and QED, what allows an unified treatment of the scalar and vector fields.

For spin zero particles it is known as the Duffin-Kemmer-Petiau theory for scalar fields (SDKP). One of the difficulties in working with SQED is the presence of a term of second order in the coupling constant in the interaction Hamiltonian, which causes trouble in proving gauge invariance. In SDKP theory it was achieved, by an effective approach, that this second order term does not contribute to the S-matrix, and thus it can be neglected when we construct the Feynman rules for the theory.

The Epstein-Glaser-Scharf causal perturbative method [8, 9] applied to SQED [10, 11] is a finite ultraviolet (UV) theory which differs from the conventional formalism because the perturbative construction starts from the first order interaction, and the quadratic term appears in the process of distribution splitting as a consequence of gauge invariance. Following the causal approach, J.T. Lunardi, B.M. Pimentel and J.S. Valverde [12] studied the SDKP coupled with the electromagnetic field and found the UV free radiative corrections.

Moreover, when the causal approach is supplemented with the analytical representation, this framework gives us a powerful method to approach formally Light-front Dynamics. In particular, it has been obtained a Light-Front Quantum Electrodynamics without prescriptions [13]. The same idea was applied for Generalized Quantum Electrodynamics [14], in the usual dynamics, to calculate the Podolsky's corrections to the Bhabha scattering at the Born level.

In this work we will approach the SDKP coupled with the electromagnetic field in the framework of the causal perturbative method supplemented with the analytical representation. Our goals are to calculate the free electromagnetic and scalar propagators and, to calculate the amplitude transitions and the cross section, at Born level, for the Moeller and Compton scattering.

2. The S-Matrix inductive Causal Program

The axiomatic construction of perturbative Quantum Field Theory is reviewed from a causal point of view introduced by Epstein and Glaser in 1973 [8]. In this axiomatic approach the scattering operator S can be written in the following formal perturbative series:

$$S[g] = 1 + \sum_{n=1}^{\infty} \frac{1}{n!} \int dx_1 dx_2 ... dx_n T_n(x_1, x_2, ..., x_n)\, g(x_1)\, g(x_2) ... g(x_n), \tag{1}$$

where we can identify the quantity T_n as an operator-valued distribution and $g^{\otimes n}$ its test function[16]. The operator S can be constructed using *Causality* and *Poincaré invariance* as the principal physical properties. From *Causality* we can prove the following property

$$T_n(x_1, ..., x_m, x_{m+1}, ..., x_n) = T_m(x_1, ..., x_m)\, T_{n-m}(x_{m+1}, ..., x_n), \tag{2}$$

if $\{x_1^0, ..., x_m^0\} > \{x_{m+1}^0, ..., x_n^0\}$.

The first step in the construction of S is to define $T_1(x)$. In the KGF Formalism we define $T_1^{KG}(x) \equiv i : \mathcal{L}_{\text{int}}^{1^o} :$, where $\mathcal{L}_{\text{int}}^{1^o}$ is the interacting lagrangian in first order on the coupling constant avoiding the e^2 term obtained by minimal coupling prescription for getting gauge invariance[11]. For DKP formalism the e^2 term does not exist, and we can follow the definition $T_1(x) \equiv i : \mathcal{L}_{\text{int}} :$ as for Quantum Electrodynamics (QED)[9].

The second step is to construct the *intermediary distributions* $A_2'(x_1, x_2) \equiv -T_1(x_1)T_1(x_2)$ and $R_2'(x_1, x_2) \equiv -T_1(x_2)T_1(x_1)$ with the goal to obtain advanced A_2 and retarded R_2

distributions as

$$A_2(x_1,x_2) \equiv -T_1(x_1)T_1(x_2) + T_2(x_1,x_2) = A'_2(x_1,x_2) + T_2(x_1,x_2), \tag{3}$$

$$R_2(x_1,x_2) \equiv -T_1(x_2)T_1(x_1) + T_2(x_1,x_2) = R'_2(x_1,x_2) + T_2(x_1,x_2). \tag{4}$$

We can see from (2) that the support of A_2 is the past of light cone centered in x_2, and for R_2 the support is the future part. Unlike the usual way to get the advanced and retarded part of propagators using the step function, the Epstein-Glasser's method does not multiply distributions in the same variable because such multiplication is not always well defined.

Now, from (3) and (4) we can compute explicitly the *causal distribution* $D_2(x_1,x_2) \equiv R'_2(x_1,x_2) - A'_2(x_1,x_2) = R_2(x_1,x_2) - A_2(x_1,x_2)$. In the computation of D_2 we can use Wick's theorem to obtain the following form for the causal distribution

$$D_2(x_1,x_2) = \sum_k : \left[\prod_j \phi^\dagger(x_j)\right] d_2^k(x_1,x_2) \left[\prod_l \phi^\dagger(x_l)\right] :: \left[\prod_m A(x_m)\right] :, \tag{5}$$

where $d_2^k(x_1,x_2)$ are numerical distributions with the light cone centered in x_2 as support. Using the distribution theory, we can compute R_2 as a retarded part of D_2. This process is called *causal splitting* and it is not a trivially process. In *causal splitting* we need to find the *order of singularity* ω of $d_2^k(x_1,x_2)$. In function of ω, we have two solutions for the retarded part $\hat{r}_2(p)$ of $d_2^k(x_1,x_2)$ in momentum space[17]

$$\hat{r}(p) = \frac{i}{2\pi} sgn(p^0) \int dt \frac{\hat{d}(tp)}{1-t+sgn(p^0)i0^+}, \quad \text{when } \omega < 0 \tag{6}$$

$$\hat{r}(p) = \left[\frac{i}{2\pi} sgn(p^0) \int dt \frac{\hat{d}(tp)}{t^{\omega+1}(1-t+sgn(p^0)i0^+)}\right] + \sum_{a=0}^{\omega} C_a p^a, \quad \text{when } \omega \geqslant 0 \tag{7}$$

The solution (6) is the trivial splitting using a step function $\Theta(x^0)$. The second solution (7) is carefully computed with the distribution theory and in that computation all steps are well defined. In the second solution, the constant C_a appears because of the uniqueness of the solution. Those constants will be computed using physical properties as gauge invariance, charge invariance, etc. Finally, we compute the second term $T_2(x_1,x_2)$ making the necessary substitutions. With $T_2(x_1,x_2)$ we compute all we need from the scattering operator until second order.

3. Free DKP causal propagators

The Epstein-Glasser causal approach follows the Heisenberg program[15]. For that reason we work with the causal propagators obtained from the free field theory. In this work we are going to use Wightman's formalism to compute those propagators[14][17]. Wightman's formalism tell us that if we know the green function $\hat{G}(k)$ associated to some free field equation, then the analytical representation of the positive and negative propagators $\hat{D}^{(\pm)}$ are

$$\left\langle \hat{D}^{(\pm)}, \varphi \right\rangle = (2\pi)^{-2} \oint_{c_\pm} \hat{G}(k)\varphi(k_0)dk_0. \tag{8}$$

where $c_{+(-)}$ is a counterclockwise closed path which contains only the positive (negative) poles of the Green function $\hat{G}(k)$, and $\varphi(k_0)$ are test functions of $\hat{D}^{(\pm)}$. Using (8), we may define the *causal* propagator:

$$\hat{D}(k) = \hat{D}^{(+)}(k) + \hat{D}^{(-)}(k), \tag{9}$$

furthermore, for local field theories, this propagator must have causal support in the configuration space

$$suppD(x_1, x_2) \subseteq \Gamma_2^+(x_2) \cup \Gamma_2^-(x_2). \tag{10}$$

This property must be proven explicitly for each case.

Now we are going to apply Wightman's formalism for DKP's fields. The free DKP theory is considered as a Lagrangian [4]

$$\mathcal{L}_{DKP} = \frac{i}{2}\bar{\psi}\beta^{\mu}\overleftrightarrow{\partial}_{\mu}\psi - m\bar{\psi}\psi, \tag{11}$$

where ψ is a multi-component wave function, $\bar{\psi} = \psi^{\dagger}\eta^0$, and $\eta^0 = 2\left(\beta^0\right)^2 - 1$. We have that $\{\beta^{\mu}\}$ are a set of matrices ($\mu = 0, 1, 2, 3$) satisfying the algebraic relations

$$\beta^{\mu}\beta^{\nu}\beta^{\rho} + \beta^{\rho}\beta^{\nu}\beta^{\mu} = \beta^{\mu}g^{\nu\rho} + \beta^{\rho}g^{\mu\nu}. \tag{12}$$

The equations of motion are then

$$\mathcal{D}(\partial)\psi = (i\beta.\partial - m)\psi = 0\ ,\ \bar{\psi}\overleftarrow{\mathcal{D}}(\partial) = \bar{\psi}\left(i\beta.\overleftarrow{\partial} + m\right) = 0. \tag{13}$$

It is known that the algebra of the β matrices has only three irreducible representations, whose degrees are 1, 5 and 10 [1]. The first one is trivial, having no physical content. The second and third ones correspond, respectively, to the scalar and vectorial representations. In this work we shall restrict us to the scalar case.

To obtain the causal propagator, we can find that the Green function of the DKP field equation (13) is given by:

$$\hat{G}(k) = \frac{1}{m}\frac{\beta.p(\beta.p + m)}{p^2 - m^2} - \frac{I}{m}, \tag{14}$$

then, using the analytical representation (8), the positive (PF) and negative (PF) frequency propagators are

$$\hat{S}^{(\pm)}(p) = \frac{1}{m}\beta.p(\beta.p + m)\hat{D}_m^{(\pm)}(p). \tag{15}$$

where $\hat{D}_m^{(\pm)}$ are the PF and NF scalar propagators[17]. These PF and NF propagators are related to the contractions of DKP fields as follows

$$\overbrace{\psi_a(x)\bar{\psi}_b(y)} \equiv \left[\psi_a^{(-)}(x), \bar{\psi}_b^{(+)}(y)\right] = \frac{1}{i}S_{ab}^{(+)}(x - y), \tag{16}$$

$$\overbrace{\bar{\psi}_a(x)\psi_b(y)} \equiv \left[\bar{\psi}_a^{(-)}(x), \psi_b^{(+)}(y)\right] = \frac{1}{i}S_{ba}^{(-)}(x - y), \tag{17}$$

Moreover, the *causal DKP propagator* is obtained as $\hat{S} = \hat{S}^{(+)} + \hat{S}^{(-)}$, with the following result

$$\hat{S}(p) = \frac{1}{m}\beta.p(\beta.p + m)\hat{D}_m(p), \tag{18}$$

where $\hat{D}_m$ is the known scalar massive Pauli-Jordan propagator given by

$$\hat{D}_m(p) = \hat{D}_m^{(+)} + \hat{D}_m^{(-)} = \frac{i}{2\pi}sgn(p_0)\delta\left(p^2 - m^2\right). \tag{19}$$

It is straightforward confirm that $\hat{S}(p)$ has a causal support as required.

For electromagnetic field quantization by Wightman's formalism we recommend the reference [17]. The most important result for our calculation is the contraction between two electromagnetic fields

$$\overbrace{A_\mu(x)\,A_\nu(y)} \equiv \left[A_\mu^{(-)}(x)\,,A_\nu^{(+)}(y)\right] = iD_{\mu\nu}^{(+)}(x-y)\,, \tag{20}$$

where $D_{\mu\nu}^{(+)}(x_1-x_2)$ is given by the following expression

$$D_{\mu\nu}^{(+)}(x) = g_{\mu\nu}D_0^{(+)}(x)\,, \tag{21}$$

and D_0 is given by (19) with $m = 0$.

4. Moeller and Compton scattering

As we said in section 2. for DKP gauge theory we define $T_1(x) \equiv ie : \bar{\psi}(x)\beta^\mu\psi(x) : A_\mu(x)$. Then, in the Epstein-Glasser approach the causal distribution D_2 has the following structure

$$D_2(x,y) = D_2^{(1)} + D_2^{(2)} + D_2^{(3)} + D_2^{(4)} + D_2^{(5)} \tag{22}$$

where each term in (22) could be associated with different processes. In this work we are interested in Moeller and Compton scattering $D^{(1)}$ and $D^{(2)}$ respectively. Those terms explicitly are

$$D_2^{(1)} = -ie^2 : \overline{\psi}(x_2)\beta^\mu\psi(x_2)\overline{\psi}(x_1)\beta^\nu\psi(x_1) : D_{\mu\nu}(x_1-x_2) \tag{23}$$

$$\begin{aligned} D_2^{(2)} =& ie^2 : \overline{\psi}(x_1)\beta^\nu S(x_1-x_2)\beta^\mu\psi(x_2) :: A_\mu(x_2)A_\nu(x_1) : \\ &- ie^2 : \overline{\psi}(x_2)\beta^\mu S(x_2-x_1)\beta^\nu\psi(x_1) :: A_\mu(x_2)A_\nu(x_1) : \end{aligned} \tag{24}$$

4.1. Moeller Scattering

The differential cross section for Moeller scattering is obtained from (23). For *splitting* (23) the computation of the order of singularity gives us $\omega = -2$, what means that the retarded part is given by the solution (6). Making the necessary substitutions we get for $T_2^{(1)}$

$$T_2^{(1)}(x_1,x_2) = -e^2 i g_{\mu\nu}\beta_{ab}^\mu\beta_{cd}^\nu : \overline{\psi}_a(x_2)\psi_b(x_2)\overline{\psi}_c(x_1)\psi_d(x) : D_0^F(x_1-x_2), \tag{25}$$

where $D_0^F(x_1-x_2)$ is the massless Feynman propagator which in the momentum space is

$$D_0^F(k) = -(2\pi)^{-2}\frac{1}{k^2+i0^+} \tag{26}$$

Because the distributional nature of S-matrix, we make our computation of cross section smearing S-matrix with the wave packets

$$|\psi_i\rangle = \int d^3p_1 d^3q_1 \psi_i(\mathbf{p}_1,\mathbf{q}_1)\,a^\dagger(\mathbf{p}_1)\,a^\dagger(\mathbf{q}_1)\,|\Omega\rangle\,, \tag{27}$$

$$|\psi_f\rangle = \int d^3p_2 d^3q_2 \psi_f(\mathbf{p}_2,\mathbf{q}_2)\,a^\dagger(\mathbf{p}_2)\,a^\dagger(\mathbf{q}_2)\,|\Omega\rangle\,, \tag{28}$$

where $a^\dagger(\mathbf{p}_l)$ is a creation operator of a DKP particle with 3-momentum $\mathbf{p}_l$ and if $\mathbf{p}_i$ and $\mathbf{p}_f$ are the ingoing and outgoing 3-momentum of the DKP's particles, then $|\psi_i\rangle$ and $|\psi_f\rangle$ are the wave-packet functions sharply peaked in $\mathbf{p}_i$ and $\mathbf{p}_f$ respectively [9]. Then, by definition we have for the transition amplitude $\mathcal{P}_{fi}$

$$\mathcal{P}_{fi} \equiv \left(|\psi_f\rangle, |\psi_i\rangle\right)^2 = |S_{fi}|^2 = \left|\left\langle \psi_f \left| S_2^{(1)} + \ldots \right| \psi_i \right\rangle\right|^2 . \tag{29}$$

In the laboratory frame, the incoming particles are a beam of particles with a cylinder form of radius R and velocity $\mathbf{v}$, then summing over all final states we obtain [9]

$$\sum_f \mathcal{P}_{fi}(R) = \frac{1}{\pi R^2}\left[(2\pi)^2 \frac{1}{|\mathbf{v}|}\int d^3p_2 d^3q_2 \left|\mathcal{M}(p_i, q_i, p_2, q_2)\right|^2 \delta(p_2 + q_2 - p_i - q_i)\right], \tag{30}$$

where $\mathcal{M}$ is a distributional quantity related to the *S-Matrix* as it follows

$$S_{fi} = ie^2\delta(p_f + q_f - p_i - q_i)\mathcal{M}. \tag{31}$$

The scattering cross section in the laboratory frame is then given by

$$\sigma_{\text{lab}} \equiv \lim_{R\to\infty} \pi R^2 \overline{\sum_f P_{if}(R)}. \tag{32}$$

where $\overline{\sum_f P_{if}(R)}$ is the average of $\sum_f P_{if}(R)$ over the cylinder of radius R defined by the incoming particles.

Finally, after some no trivial computation, we end with the following differential cross section

$$\frac{d\sigma_{c.m}}{d\Omega} = \frac{\alpha^2}{4s}\left|\frac{s-t}{u} + \frac{s-u}{t}\right|^2, \tag{33}$$

where α is the fine-structure constant and $\{s, t, u\}$ the Mandelstam's variables. The solution (33) is in accordance with the one obtained using the KGF equation by the *causal approach* [11] and the one obtained using the usual approach[19][18].

4.2. Compton Scattering

For computing the differential cross section for Compton scattering we use (24). The order of singularity of (24) is $\omega = 0$. This value tell us that to obtain the retarded part in the causal splitting process we must use the solution (7) where a non fixed constants will appear. Then, making the necessary substitutions we get the 2-points distribution $T_2^{(2)}$

$$\begin{aligned} T_2^{(2)} &= e^2 i : \overline{\psi}_c(x)\beta^{\nu}_{cd}(-S^F_{da}(x-y) + C_{da}\delta(x-y))\beta^{\mu}_{ab}\psi_b(y) :: A_\mu(y)A_\nu(x) : \\ &+ e^2 i : \overline{\psi}_a(y)\beta^{\mu}_{ab}(-S^F_{bc}(y-x) - C'_{bc}\delta(x-y)\beta^{\nu}_{cd}\psi_d(x) :: A_\mu(y)A_\nu(x) : \end{aligned} \tag{34}$$

where C_{da} and C'_{bc} can not be fixed by causality or causal splitting process. To obtain them we need to use other physical properties or symmetries of S-Matrix. Using ***charge invariance*** we get $C = C'$ and using ***Gauge invariance*** we get $C = \frac{\mathrm{I}}{m}$. We want to emphasize here that if we make the integration in Delta Dirac distribution in the S-matrix we are going to obtain the multiplication of two DKP fields and two electromagnetic fields in the same variable. This result is consistent with the term e^2 placed in the usual approach to get gauge invariance by minimal coupling prescription.

In order to compute the differential cross section we could follow the same steps as for the Moeller scattering, but here we need to smear the S-Matrix with the following wave packets

$$|\psi_i\rangle = \int d^3p_1 d^3k_1 \psi_i(\mathbf{p}_1, \mathbf{q}_1)\, a^\dagger(\mathbf{p}_1)\, \varepsilon_{i\nu} \mathrm{b}^\dagger_\nu(\mathbf{k}_1)\, |\Omega\rangle, \tag{35}$$

$$|\psi_f\rangle = \int d^3p_2 d^3k_2 \psi_f(\mathbf{p}_2, \mathbf{q}_2)\, a^\dagger(\mathbf{p}_2)\, \varepsilon_{f\nu} \mathrm{b}^\dagger_\nu(\mathbf{k}_2)\, |\Omega\rangle, \tag{36}$$

where $\varepsilon_{i\nu}$ and $\varepsilon_{f\nu}$ are the 4-polarization vectors for the incoming and outgoing photons, and the $\mathrm{b}_\nu^\dagger(\mathbf{k}_l)$ are the creation operator valued distributions for photons. The other steps in the computation are the same which were used for Moeler scattering.

With the help of the polarization conditions

$$\varepsilon_i k_i = 0, \quad \varepsilon_f k_f = 0 \tag{37}$$

and using the reference system where

$$p_i \varepsilon_i = 0, \quad p_i \varepsilon_f = 0 \tag{38}$$

we get

$$\begin{aligned} \frac{d\sigma_{\text{lab}}}{d\Omega} &= \frac{e^4}{16\pi^2} \frac{\omega_f^2}{m^2\omega_i^2} \left[\varepsilon_f^\mu(\mathbf{k}_f)\varepsilon_{i\mu}(\mathbf{k}_i)\right]^2 \\ &= \frac{\alpha^2\omega_f^2}{m^2\omega_i^2} \left[\varepsilon_f^\mu(\mathbf{k}_f)\varepsilon_{i\mu}(\mathbf{k}_i)\right]^2 . \end{aligned} \tag{39}$$

where α is the fine-structure. The solution (39) is in accordance with the one obtained using the KGF equation by the Causal Approach [11] and the one obtained using the usual approach[19].

5. Conclusions

We could note that the combination of the analytical representation technique with the Epstein-Glasser causal approach gives us a well define theory for the study of the SDKP.

We have calculated the DKP free propagator using analytical representation, and the cross section for the Moeller and Compton scattering process for the SDKP.

For the Moeller and Compton scattering process we find, at Born level, whose the results of the differential cross sections are identical to that obtained in the usual treatment of Scalar Quantum Electrodynamics. These results provide us an evidence that the SDKP and the KGF theory are equivalent, at Born level, even when they are coupled with the electromagnetic field for these processes.

For future works, we will calculate the radiative corrections and determine some important general results such as the renormalizability of the theory.

Acknowledgments

J Beltrán thanks CAPES for full support. B M Pimentel thanks CAPES and CNPq for partial support. D E Soto thanks CNPq for partial support.

References

[1] Petiau G 1936 *Acad. Roy. de Belg. A. Sci. Mem. Collect.* **16** ; Duffin R Y 1938 *Phys. Rev.* **54**, 1114; Kemmer N 1939 *Proc. R. Soc. London* **137A**, 91.

[2] Cardoso T R, Castro L B, Castro A S 2010 *J. Phys. A: MAth. Theor.* **43** 055306.

[3] Nedjadi Y and Barrett R C 1994 *J. Phys. A: Math. Gen.* **27** 4301. Nedjadi Y and Barrett R C 1998 *J. Phys. A: Math. Gen.* **31** 6717.

[4] Lunardi J T, Pimentel B M, Texeira R G and Valverde J S 2000 *Phys. Lett.* **A288**, 165; Nowakowski M 1998 Phys. Lett. **A244**, 329.

[5] Lunardi J T, Pimentel B M and Texeira R G 2001 in *Procs. Workshop on Geometrical Aspects of Quantum Fields*, eds; Bytsenko A A, Gonçalves A E and Pimentel B M (World Scientific, Singapore), p111.

[6] Fainberg V Ya and Pimentel B M 2000 *Theor. Math. Phys.* **124**, 445; Phys. Lett. **A271**, 16.

[7] Gribov V 1999 *Eur. Phys. J.* **C10**, 71.

[8] Epstein H and Glaser V 1973 *Ann. Inst. H. Poincaré A* **19** 211.

[9] Scharf G 1995 Finite Quantum Electrodynamics The Causal Approach, 2nd ed., (Berlin Heidelberg New York: Springer-Verlag).
[10] Dütsch M, Krahe F, Scharf G 1993 Scalar QED Revisited *Nuovo Cimento* **106** 277.
[11] Beltran J 2014 Eletrodinâmica Quântica Escalar via Teoria de Perturbação Causal: um estudo [Master's thesis] *São Paulo: Instituto de Física Teorica Universidade Estadual Paulista.*
[12] Lunardi J T, Pimentel B M and Valverde J S 2002 *Int. J. Mod. Phys.* A17, 205.
[13] Bufalo R , Pimentel B M, Soto D E 2014 *Annals of Physics* **351** 10341061; *Annals of Physics* **351** 10621084.
[14] Bufalo R, Pimentel B M, Soto D E 2014 The Causal approach for the electron-positron scattering in the Generalized Quantum Electrodynamics, *Phys. Rev. D* **90**, 085012.
[15] Heisenberg W 1943 *Z. Physik* 120, 513.
[16] Bogoliubov N N and Shirkov D V 1980 Introduction to the Theory of Quantized Fields (New York: John Wiley & Sons) 3rd ed.
[17] Soto D E 2014 Eletrodinâmica Quântica Generalizada a la Teoria de Perturbação Causal PhD[Thesis] *São Paulo: Instituto de Física Teorica Universidade Estadual Paulista.*
[18] Akhiezer A I and Berestetski V B 1965 Quantum Electrodynamics (New York: Inter.) 2nd ed.
[19] Itzykson C and Zuber J-C 1980 Quantum Field Theory (McGraw-Hill).

Study of the $Z_b(10610)$ and $Z_b(10650)$ states through $B\bar{B}^*$ and $B^*\bar{B}^*$ interactions using local hidden gauge approach

J. M. Dias[1,2], **F. Aceti**[2] **and E. Oset**[2]

[1] - Instituto de Física, Universidade de São Paulo, Caixa Postal 66318, 05389 – 970, São Paulo, São Paulo Brazil
[2] - Departamento de Física Teórica, Universidad de Valencia and IFIC, Centro Mixto Universidad de Valencia - CSIC, Institutos de Investigación de Paterna, Apartado 22085, 46071 Valencia, Spain

E-mail: jdias@if.usp.br

Abstract. We have studied the $B\bar{B}^*$ and $B^*\bar{B}^*$ interactions for isospin $I = 1$ using the Local Hidden gauge approach. Since both interactions via one light meson exchange are not allowed by Okubo-Zweig-Iizuka (OZI) rule, we investigated the contributions for those interactions coming from two pions, interacting and noninteracting among themselves, and also due to the heavy vector meson exchange, in which the OZI rule no longer holds. From the amplitudes calculated by these mechanism, we determine an effective potential which is used as a kernel of the Bethe-Salpeter equation. Our goal is look for poles in the T-matrix in attemp to relate them with the charged $Z_b(10610)$ and $Z_b(10650)$ states observed by Belle Collaboration. For the $B\bar{B}^*$ interaction, we find a loosely bound state with the mass in the range $10587 - 10601$ MeV, which is close to the experimental value observed by Belle Collaboration. For the $B^*\bar{B}^*$ case, we find a cusp at 10650 MeV.

1. Introduction

Since the discovery of the $X(3872)$ by Belle Collaboration [1], many others Collaborations like BaBar, CDF, D0 and BESIII are measuring a series of new structure in the Charmonium mass region. They are called XYZ Charmoniumlike states. If they were a simple Charmonium, they would decay into a pair of charm mesons, that is, mesons which are made of one quark (anti-quark) charm and a light quark. Instead, they decay into J/ψ plus pions which is an unusual feature for a simple Charmonium state. Furthermore, properties such as mass and decay modes predicted by the potential models do not fit with that observed by the experiment. All these striking behaviour led to many theoretical and experimental efforts to understand the quark content of the XYZ. Most of the theoretical work done so far attempts to accommodate them in a exotic picture. By exotic we mean a more complex quark structure beyond the known quark anti-quark (mesons) and three quarks (baryons) configuration like hybrid, meson molecule, glueballs and hadrocharmonium. It is a challenge to understand these new charmoniumlike states as exotic since using the models mentioned above it is relatively simple to reproduce the masses of those states. The same challenges also concern the bottomoniumlike states. Among them, the $Z_b(10650)$ and $Z_b(10610)$ are very interesting. They were observed by the

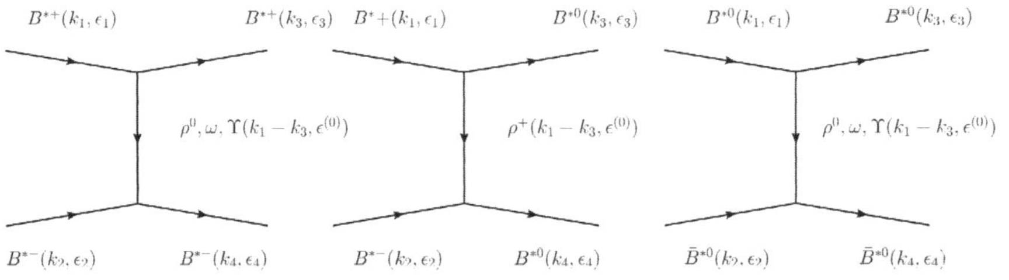

Figure 1. Vector exchange diagrams contributing to the process $B^*\bar{B}^* \to B^*\bar{B}^*$.

Belle collaboration in $\pi^\pm h_b(nP)$ and $\pi^\pm \Upsilon(mS)$, with the $n = 1,2$ and $m = 1,2,3$, invariant mass distribution of the $\Upsilon(5S)$ decay channel [2]. As a result of the measurements, Belle reported: $M_{Z_b(10610)} = (10608.4 \pm 2.0)$ MeV, $\Gamma_{Z_b(10610)} = (15.6 \pm 2.5)$ MeV and for $Z_b(10650)$, $M_{Z_b(10650)} = (10653.2 \pm 1.5)$ MeV and $\Gamma_{Z_b(10650)} = (14.4 \pm 3.2)$ MeV. The quantum numbers are reported as $J^P = 1^+$ and positive G parity. Therefore, in this work we use the Lagrangians from the Local Hidden Gauge approach in order to study the interactions $B\bar{B}^*$ and $B^*\bar{B}^*$. Our aim is look into the possibility of creating states from those interactions for $I = 1$ exchanging mesons. Since the exchange of a light mesons is not allowed by OZI rule, we should circumvent the OZI restriction and investigate the interactions by means of a heavy vector meson exchange and also due to the two pions, interacting and noninteracting among themselves, exchange. In view of this, we follow the procedure done by the authors of Refs. [6, 8] and perform an extension to the bottom sector in order to look on $B\bar{B}^*$ and $B^*\bar{B}^*$ interactions.

2. Formalism

2.1. The heavy vector meson exchange

In order to study the $B\bar{B}^*$ and $B^*\bar{B}^*$ interactions by means of heavy vector exchange, we need Lagrangians which can describe the VPP and VVV vertices. They are given by

$$\mathcal{L}_{VPP} = -ig\langle V^\mu[P, \partial_\mu P]\rangle \ , \tag{1}$$

$$\mathcal{L}_{VVV} = ig\langle (V^\mu \partial_\nu V_\mu - \partial_\nu V_\mu V^\mu)V^\nu\rangle \ . \tag{2}$$

The coupling g is given by $g = M_V/2f_\pi$, being $f_\pi = 93$ MeV the pion decay constant, while M_V is the vector meson mass. In Eqs. (1) and (2), the symbol $\langle\ \rangle$ stands for the trace of SU(4). The vector field V_μ is represented by the SU(4) matrix, which is parametrised by 16 vector mesons including the 15-plet and singlet of SU(4) [9]. while P is a matrix containing the 15-plet of the pseudoscalar mesons written in the physical basis in which η, η' mixing is taken into account [3],

The channels we are interested in are those with $B = 0$, $S = 0$ and isospin $I = 1$. In the $B^*\bar{B}^*$ case, they are $B^*\bar{B}^*$ and $\rho\Upsilon$. In the case of $B\bar{B}^*$ we are only interested in the positive G-parity combination, namely $(B\bar{B}^* + cc)/\sqrt{2}$ and also $\eta_b\,\rho$ and $\pi\,\Upsilon$.

2.1.1. The $B\bar{B}^*$ case For this case, the most important diagrams are depicted in Fig. 1. In order to calculate the vertices of those diagrams we can use the Lagrangians defined earlier. Then, taking into account all particles involved in the exchange, we obtain for the $I = 1$ combination the following expression

$$t^{I=1,S=0,1,2}_{B^*\bar{B}^*\to B^*\bar{B}^*} = -g^2 + g^2\left[\frac{2M_\rho^2 M_\omega^2 + M_\Upsilon^2(-M_\omega^2 + M_\rho^2)}{4M_\Upsilon^2 M_\omega^2 M_\rho^2}\right](4M_{B^*}^2 - 3s)\,, \tag{3}$$

where s stands for the center of mass energy of the $B^*\bar{B}^*$ system.

Consider now the other channel, $B^*\bar{B}^* \to \rho\Upsilon$. The relevant diagrams are depicted in Fig 2 [9]. The procedure to get the amplitude for this channel is analogous to what we have done earlier. Thus, the amplitude in isospin $I = 1$ basis for the spin $S = 0, 2$ states in s-wave, corresponding to all diagrams of Fig. 2 plus the contact term is given by

$$t^{I=1,S=0,2}_{B^*\bar{B}^*\to\rho\Upsilon} = -2g^2 + g^2\left[\frac{2M^2_{B^*} + M^2_{\Upsilon} + M^2_{\rho} - 3s}{M^2_{B^*}}\right]. \tag{4}$$

The interaction in $S = 1$ vanishes as a consequence of a cancellation of terms where the ρ^0 and Υ are interchanged in the diagrams. The diagonal $\rho\Upsilon \to \rho\Upsilon$ transition is again OZI forbidden and null in this approach.

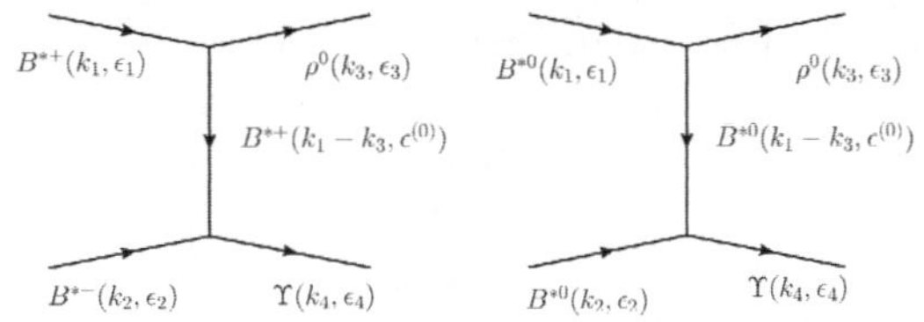

Figure 2. Vector exchange diagrams contributing for the $B^*\bar{B}^* \to \rho\,\Upsilon$ channel.

Eqs. (3) and (4) will be used as a kernel of the Bethe-Salpeter equation as we shall discuss it later.

2.1.2. The $B^*\bar{B}^*$ case In this case, the Lagrangians defined in Eqs. (1) and (2) can also be used to provide the vertices of the $PV \to PV$ interaction through exchange of a heavy vector. In Refs. [5, 4] these amplitudes were already calculated in s-wave, where the authors were concerned with axial-vector resonances dynamically generated. Besides, in Ref. [6] the same equation for the amplitude is used for the $D\bar{D}^*$ interaction. In our case, we extend these amplitudes for the $B\bar{B}^*$ interaction in the isospin $I = 1$ channel, with the result

$$V_{ij}(s) = -\frac{\vec{\epsilon}\,\vec{\epsilon}\,'}{8f_\pi^2}\,\mathcal{C}_{ij}\left[3s - (M^2 + m^2 + M'^2 + m'^2) - \frac{1}{s}(M^2 - m^2)(M'^2 - m'^2)\right], \tag{5}$$

where the masses M (M') and m (m') in Eq. (5) correspond to the initial (final) vector meson and pseudoscalar meson, respectively. The indices i and j represent the initial and final VP channels $(B\bar{B}^* + cc)/\sqrt{2}$, $\eta_b\,\rho$ and $\pi\,\Upsilon$.

The $\mathcal{C}_{ij}$ are constants that can be displayed in a 3×3 matrix, which for the positive G-parity of the $B\bar{B}^*$ combination, is

$$\mathcal{C}_{ij} = \begin{pmatrix} -\psi & \sqrt{2}\gamma & \sqrt{2}\gamma \\ \sqrt{2}\gamma & 0 & 0 \\ \sqrt{2}\gamma & 0 & 0 \end{pmatrix}, \tag{6}$$

where $\gamma = \left(\frac{m_L}{m_H}\right)^2$ and $\psi = \left(\frac{m_L}{m_{H'}}\right)^2$. Those factors are defined in this way in order to take into account the suppression due to the exchange of a heavy vector meson. Concerning the parameters m_L, m_H and $m_{H'}$, we choose their values in order to have the same order of magnitude of the light and heavy vector meson masses: $m_L = 800$, $m_H = 5000$ MeV and $m_{H'} = 9000$ MeV. These masses stand for the ρ or ω, B^* and Υ respectively.

3. T-matrix

The amplitudes obtained above provide the kernel of the Bethe-Salpeter equation in coupled channels,

$$T = (1 - VG)^{-1} V \ , \tag{7}$$

where V is the potential. In the $B\bar{B}^*$ case, it is a 3×3 matrix given by the Eq. (5) with the C_{ij} defined by Eq. (6), which is associated with the channels: $B\bar{B}^*$, $\eta_b \rho$ and $\pi\Upsilon$. In the $B^*\bar{B}^*$ case, V is a 2×2 matrix whose elements are the amplitudes given by Eqs. (3) and (4) corresponding to the channels $B^*\bar{B}^*$ and $\rho\Upsilon$, respectively. In Eq. (7), G is a diagonal matrix whose elements are associated with the two meson loops, G_l, for each channel l. Explicitly, we have

$$G_l = i \int \frac{d^4 q}{(2\pi)^4} \frac{1}{q^2 - m^2 + i\epsilon} \frac{1}{(q-P)^2 - M^2 + i\epsilon} \ , \tag{8}$$

where m is the mass of the pseudoscalar (in the $B\bar{B}^*$ case) or vector (in the $B^*\bar{B}^*$ case), while M is the vector meson mass involved in the loop in the channel l. In Eq. (8) P means the total four-momentum of the mesons. The integral of Eq. (8) is logarithmically divergent and it can be regularized with a cut off in the momentum space or dimensional regularization [6, 8, 9].

3.1. $B\bar{B}^*$ and $B^*\bar{B}^*$ interactions by means of two interacting pion exchange

Following the idea of the authors of Refs. [6, 8], where the authors investigated the $D\bar{D}^*$ and $D^*\bar{D}^*$ interactions by means of two interacting pions exchange, we extent these mechanism to the bottom case, studying the $B\bar{B}^*$ and $B^*\bar{B}^*$ interactions.

3.1.1. The $B\bar{B}^*$ case For this case, the relevant diagrams contributing to this interaction are depicted in Fig. 3.

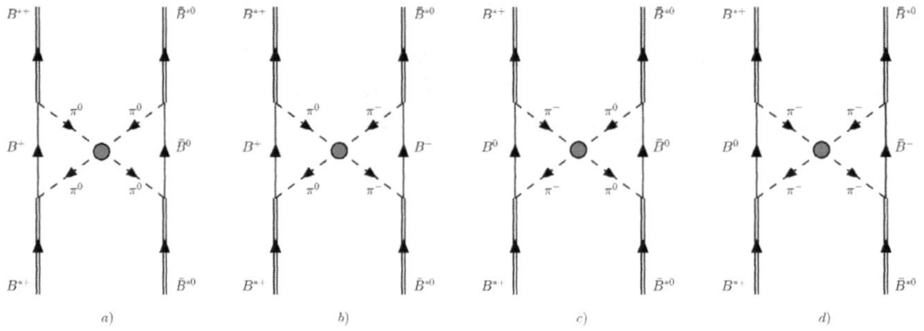

Figure 3. Diagrams contributing to the two pions interaction in lowest order in $I = 1$ for the $B^*\bar{B}^* \to B^*\bar{B}^*$ process [9].

Since we are following the approach of the Ref. [8], we can start from the amplitude obtained by its authors and replace the masses of the D and D^* mesons by the B and B^* mesons, respectively. Therefore, we get

$$t^{\sigma}_{B^*\bar{B}^*}(\vec{q}) = a^2 \frac{3}{2} \left[\frac{1}{f^2} \frac{\vec{q}^2 + \frac{m_\pi^2}{2}}{1 - G(-\vec{q}^2)\frac{1}{f^2}(\vec{q}^2 + \frac{m_\pi^2}{2})} \right] . \tag{9}$$

Next, we shall consider the same mechanism for the $B^*\bar{B}^*$ interaction.

Journal of Physics: Conference Series **706** (2016) 052020 doi:10.1088/1742-6596/706/5/052020

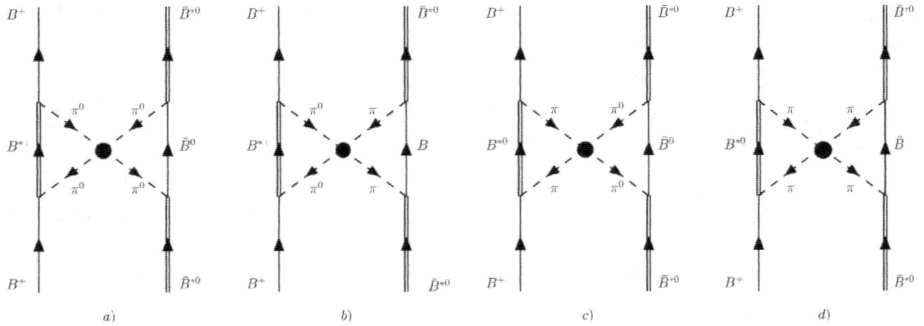

Figure 4. Diagrams contributing to the two pion exchange interaction in lowest order for the $B\bar{B}^* \to B\bar{B}^*$ process in $I = 1$ [9].

3.1.2. The $B^*\bar{B}^*$ case The diagrams contributing to this process are illustrated in Fig. 4.

For this case, the only difference to the former case is the appearing of another triangular loop (see the details in Ref. [9]).

3.2. Noninteracting two pions exchange
In this case, the contributions to the $B\bar{B}^*$ and $B^*\bar{B}^*$ interactions only come from the diagrams (a) and (d) of Figs. 3 and 4, respectively. All the details related to the derivation of the amplitudes $t^{\pi\pi}_{B\bar{B}^*}$ and $t^{\pi\pi}_{B^*\bar{B}^*}$ can be found in Ref. [8]. Explicitly, the amplitude $t^{\pi\pi}_{B^*\bar{B}^*}$ is given by

$$\begin{aligned} t^{\pi\pi}_{B^*\bar{B}^*} &= \tfrac{5}{4}\, g_B^4 \tfrac{A}{15} \int \tfrac{d^3p}{(2\pi)^3} (4\vec{p}^2 - \tfrac{\vec{q}^2}{4})^2\, F^2\, \tfrac{1}{\omega_1+\omega_2}\, \tfrac{1}{2\omega_1\omega_2}\, \tfrac{1}{4E_B^2}\, \tfrac{1}{p_1^0-\omega_1-E_B+i\epsilon} \\ &\times \tfrac{1}{p_1^0-\omega_2-E_B+i\epsilon} \left(1 + \tfrac{E_B+\omega_1+\omega_2-p_1^0}{p_1^0-\omega_1-E_B+i\epsilon} + \tfrac{E_B+\omega_1+\omega_2-p_1^0}{p_1^0-\omega_2-E_B+i\epsilon}\right) , \end{aligned} \tag{10}$$

where $A = 5$ is associated with spin $J = 0$, while $A = 2$ is related to the $J = 2$ case, $\omega_1 = \sqrt{(\vec{p}+\vec{q}/2)^2 + m_\pi^2}$, $\omega_2 = \sqrt{(\vec{p}-\vec{q}/2)^2 + m_\pi^2}$ are the energies of the pions and $E_B(\vec{p}) = \sqrt{\vec{p}^{\,2} + m_B^2}$ is the energy of the B meson. $F(\vec{q})$ is a form factor of the type

$$F = F_1(\vec{p}+\frac{\vec{q}}{2})\, F_2(\vec{p}-\frac{\vec{q}}{2}) = \frac{\Lambda^2}{\Lambda^2 + (\vec{p}+\frac{\vec{q}}{2})^2}\, \frac{\Lambda^2}{\Lambda^2 + (\vec{p}-\frac{\vec{q}}{2})^2} , \tag{11}$$

with $\Lambda = 700$ GeV, which is used to improve the convergence. On the other hand, for the $t^{\pi\pi}_{B\bar{B}^*}$ amplitude we have

$$\begin{aligned} t^{\pi\pi}_{B\bar{B}^*} &= -\tfrac{5}{4} g_B^4\, \tfrac{1}{2}\, \vec{\epsilon}\,'\, \vec{\epsilon}\,'' \int \tfrac{d^3p}{(2\pi)^3} (\vec{p}^{\,2} - \vec{q}^{\,2}) \left[(4\vec{p}^{\,2} - \tfrac{\vec{q}^{\,2}}{4}) - \tfrac{1}{\vec{q}^{\,2}}\left[(2\vec{p}\vec{q})^2 - \tfrac{\vec{q}^{\,4}}{4}\right]\right] \tfrac{F^2}{\omega_1+\omega_2}\, \tfrac{1}{2\omega_1\omega_2} \\ &\times \tfrac{1}{2E_B}\, \tfrac{1}{2E_V} [\omega_1^2 + \omega_2^2 + \omega_1\omega_2 - (\omega_1+\omega_2)(2p_1^0 - E_{B^*} - E_B) + (p_1^0 - E_{B^*})(p_1^0 - E_B)] \\ &\times \tfrac{1}{p_1^0-\omega_1-E_{B^*}+i\epsilon}\, \tfrac{1}{p_1^0-\omega_1-E_B+i\epsilon}\, \tfrac{1}{p_1^0-\omega_2-E_{B^*}+i\epsilon}\, \tfrac{1}{p_1^0-\omega_2-E_B+i\epsilon} , \end{aligned} \tag{12}$$

where $E_{B^*}(\vec{p}) = \sqrt{\vec{p}^{\,2} + m_{B^*}^2}$ is the energy of the B^* meson.

4. Results

Once we have obtained all the amplitudes which contribute to the $B\bar{B}^*$ and $B^*\bar{B}^*$ interactions, we determine an effective potential to be used as a kernel of the Bethe-Salpeter equation, as follows: we calculate the integral $\int d^3q\, V(|\vec{q}|)$ related to the strength of all the potentials associated with the pions and convert them in a potential of the form

$$V_{eff}\theta(q_{max} - |\vec{q}\,|)\theta(q_{max} - |\vec{q}\,'|)\,, \tag{13}$$

where q_{max} is the maximum momentum used in the loops in Eq. (8), such that $\int_{q<q_{max}} d^3q\, V_{eff}$ is equal to the sum of $\int d^3q\, V_i\,(q)$. Then, we take as effective potential the sum of the Eq. (13) plus the one coming from the heavy vector exchange. Both have the same form of the Eq. (13) and can be used in the Bethe-Salpeter equation with the same G loop function regularized with the cut-off $|\vec{q}_{max}|$.

On the other hand, the value of the strength depends on the value of the upper limit of the integral $\int d^3q\, V(q)$. For this reason we calculated the effective potential V_{eff} using values of this limit for the light meson exchange potential varying from 700 to 1100 MeV for both $B^*\bar{B}^*$ and $B\bar{B}^*$ interactions. Changing the upper limit in $\int d^3q\, V_i\,(q)$ introduces large uncertainties in the approach concerning the final potential. The strength of the final potential, summing V_{eff} and the vector exchange, can be a factor $2.4 - 14.5$ times the one of the vector exchange alone for the case of $B^*\bar{B}^*$ with $J = 0$, while for $J = 2$ we find a factor $1.2 - 5.2$. For the case of $B\bar{B}^*$ the factor varies between 30 and 64.

In the following we study the shape of the $|T_{11}|^2$ which stands for diagonal transitions, $B\bar{B}^*$ or $B^*\bar{B}^*$ to themselves, depending on the case.

4.1. The $B\bar{B}^*$ case

For this case, we have calculated the T-matrix considering the channels: $B\bar{B}^*$, $\eta_b\,\rho$ and $\pi\,\Upsilon$. The T-matrix was evaluated between those channels for the $\sqrt{s}$ values around 10600 MeV. For the G loop function, we have use the dimensional regularization formula with the following values for the subtraction constants: $\alpha_{B\bar{B}^*} = -2.79$, $\alpha_{\eta_b\rho} = -3.56$ and $\alpha_{\pi\Upsilon} = -3.78$. This is equivalent to use a cut-off equal to $|\vec{q}_{max}| = 700$ MeV. In Fig. 5 the shape of $|T_{11}|^2$, the component of

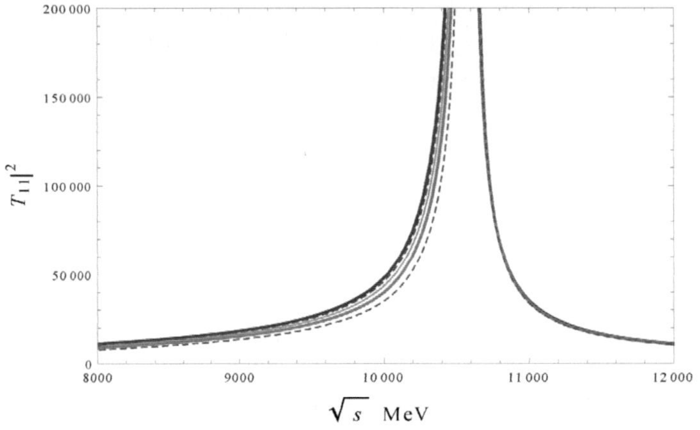

Figure 5. $|T_{11}|^2$ as a function of the $\sqrt{s}$ center of mass energy for the case of $B\bar{B}^*$. Each curve is associated with a value of the integration limit: 700 MeV, 800 MeV, 900 MeV, 1000 MeV, 1100 MeV. The peak moves from right to left as the integration limit increases [9].

the T matrix that describes the transition $B\bar{B}^* \to B\bar{B}^*$, for different values of the integration limit, is depicted. As can be seen, even choosing values of the limit between 700 and 1100 MeV, the effect on the binding and the width is small. As a result, we find that the position of the peak moves slightly to higher energies for decreasing values of the upper limit and it is seen in the range of 10587 – 10601 MeV. These values are very close to what was found by the Belle collaboration, $M_{Z_b(10610)} = (10608.4 \pm 2.0)$ MeV.

4.2. The $B^*\bar{B}^*$ case

For this case, we have two channels: $B^*\bar{B}^*$ and $\rho\,\Upsilon$. Again, we use the dimensional regularization form of the loop function G with the subtraction constants $\alpha_{B^*\bar{B}^*} = -2.79$ and $\alpha_{\rho\Upsilon} = -3.56$, corresponding to a cut off value equal to $|\vec{q}_{max}| = 700$ MeV.

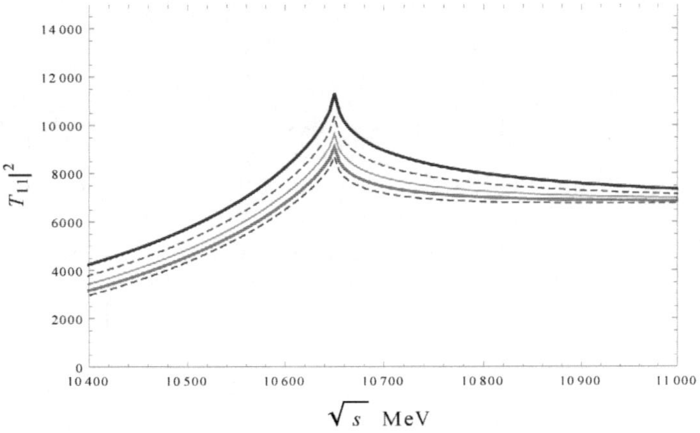

Figure 6. $|T_{11}|^2$ as a function of the $\sqrt{s}$ center of mass energy for the case of $B^*\bar{B}^*$ for $J = 0$. Each curve is associated with a value of the integration limit: 700 MeV, 800 MeV, 900 MeV, 1000 MeV, 1100 MeV. The peak moves from bottom to top as the integration limit increases [9].

Fig. 6 shows the shape of $|T_{11}|^2$, which means the component of the T matrix that describes the transition from $B^*\bar{B}^*$ to itself, for different values of the integration limit plotted as a function of the center of mass energy, $\sqrt{s}$, of the system. This peak corresponds to spin $J = 0$. In Fig 7, we show the shape of $|T_{11}|^2$ for the $J = 2$ case, again for different values of the integration limit. It is important to emphasize that, according to Eq. (4), there is no contribution in the transition matrix T from $B^*\bar{B}^*$ to $\rho\,\Upsilon$ channel for spin $J = 1$. In this case, $B^*\bar{B}^*$ stands as a single channel.

From these figures we noted that it is always found a structure for the peak of $|T_{11}|^2$, which corresponds clearly to a cusp. Whether to call this a resonant state or not it is a question of criterion. We should however note that the $a_0(980)$ appears in the experiments (or in the theories) [10, 11] as a cusp and is universally accepted as a resonance. Our findings, obtained a cusp for the $|T_{11}|^2$ amplitude in this case, would come to support the claims of the former works [12, 13].

5. Conclusions

We have studied the $B\bar{B}^*$ and $B^*\bar{B}^*$ interactions for $I = 1$ using the Lagrangians from the Hidden Local Symmetry approach. We stressed that from the point of view of the light meson exchange, those interactions are not allowed by the OZI rule. Therefore, we investigated the same interactions through the exchange of heavy meson and also due to the two pions, interacting and noninteracting among themselves, exchange. Our goal was to look for poles in the T-matrix, solving the Bethe-Salpeter equation using an effective potential obtained from the last meson

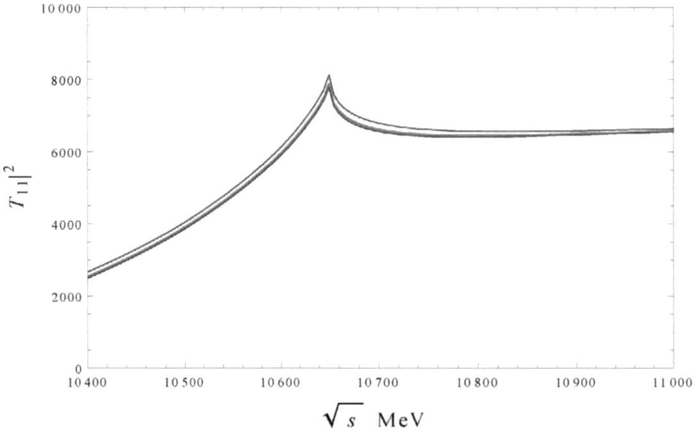

Figure 7. $|T_{11}|^2$ as a function of the $\sqrt{s}$ center of mass energy for the case of $B^*\bar{B}^*$ for $J = 2$. Each curve is associated with a value of the integration limit: 700 MeV, 800 MeV, 900 MeV, 1000 MeV, 1100 MeV. The peak moves sligthtly from bottom to top as the integration limit increases [9].

exchange mechanism, in order to relate them with the charged states $Z_b(10610)$ and $Z_b(10650)$ observed by Belle Collaboration. As a result, for the $B\bar{B}^*$ case, we found a bound state with mass is in the range $10587 - 10601$ MeV which is very close to the experimental mass value of the $Z_b(10610)$ state. For the $B^*\bar{B}^*$ interaction, we have found a cusp at 10650 MeV for the $J = 0$ and also $J = 2$ spin cases.

5.1. Acknowledgments

This work was supported by Brazilian funding agency FAPESP and by the Spanish Ministerio de Economia y Competitividad and European FEDER funds under the Contract No. FIS2011-28853-C02-01, and the Generalitat Valenciana in the program Promoteo II, 2014/068. We also acknowledge the support of the European Community-Research Infrastructure Integrating Activity Study of Strongly Interacting Matter (acronym HadronPhysics3, Grant Agreement No. 283286) under the Seventh Framework Programme of the EU.

References

[1] S. K. Choi *et al.* [Belle Collaboration], Phys. Rev. Lett. **91**, 262001 (2003) [hep-ex/0309032].
[2] A. Bondar *et al.* [Belle Collaboration], Phys. Rev. Lett. **108**, 122001 (2012) [arXiv:1110.2251 [hep-ex]].
[3] D. Gamermann, E. Oset and B. S. Zou, Eur. Phys. J. A **41**, 85 (2009)
[4] D. Gamermann and E. Oset, Eur. Phys. J. A **33**, 119 (2007).
[5] L. Roca, E. Oset and J. Singh, Phys. Rev. D **72**, 014002 (2005).
[6] F. Aceti, M. Bayar, E. Oset, A. M. Torres, K. P. Khemchandani, F. S. Navarra and M. Nielsen, Phys. Rev. D **90**, 016003 (2014).
[7] E. Oset, H. Toki, M. Mizobe and T. T. Takahashi, Prog. Theor. Phys. **103**, 351 (2000).
[8] F. Aceti, M. Bayar, J. M. Dias and E. Oset, Eur. Phys. J. A **50**, 103 (2014).
[9] J. M. Dias, F. Aceti and E. Oset, Phys. Rev. D **91**, 076001 (2015).
[10] P. Rubin *et al.* [CLEO Collaboration], Phys. Rev. Lett. **93**, 111801 (2004) [hep-ex/0405011].
[11] J. A. Oller, E. Oset and J. R. Pelaez, Phys. Rev. D **59**, 074001 (1999) [Erratum-ibid. D **60**, 099906 (1999)] [Erratum-ibid. D **75**, 099903 (2007)] [hep-ph/9804209].
[12] D. V. Bugg, Europhys. Lett. **96**, 11002 (2011) [arXiv:1105.5492 [hep-ph]].
[13] E. S. Swanson, arXiv:1409.3291 [hep-ph].

Journal of Physics: Conference Series **706** (2016) 052021 doi:10.1088/1742-6596/706/5/052021

Phase transitions in neutron star equation of state induced by the delta resonances matter

J C T Oliveira[1], H Rodrigues[2] and S B Duarte[3]

[1] Departamento de Física, Universidade Federal de Roraima, Campus do Paricarana s/n, 69310-270, Roraima, Brazil
[2] Centro Federal de Educação Tecnológica do Rio de Janeiro, Av. Maracanã 229, 20271-110, Rio de Janeiro, RJ, Brazil
[3] Centro Brasileiro de Pesquisas Físicas, Rua Dr. Xavier Sigaud 150, 22290-180, Rio de Janeiro, RJ, Brazil

E-mail: jose.oliveira@ufrr.br

Abstract. In the present work we determine the equation of state and the population of baryons and leptons, and also we discuss the implication of changes in the baryon-meson coupling constants to the formation of delta matter in the stellar medium. And also in this work the phase transition is explored with respect to the domain of the delta-mesons coupling constants.

1. Introduction

With the high production of delta-resonances (~30% of baryon population) in the dense phase of relativistic heavy ion collisions [1-3] arises a great interest in the study of the delta matter formation in neutron star structure. In this work the delta-rich hadronic condensate is studied in the context of a relativistic mean field approximation for hadrons in beta-equilibrium with the leptons present in the medium. The baryonic sector includes the complete 1/2-spin octet, and delta resonances with 3/2-spin, which interact by exchange of scalar, vector and isovector mesons σ, ω and ρ, respectively.

2. The model

In this work we adopted the nonlinear Walecka (NLW) model in the mean field approximation, including the 1/2-spin baryon octet (n, p, Λ^0, Σ^-, Σ^0, Σ^+, Ξ^-, Ξ^0), the 3/2-spin baryonic resonances represented by the delta-matter (Δ^-, Δ^0, Δ^+, Δ^{++}) and by Ω^- in the baryonic sector [4,5]. In the leptonic sector are considered electrons and muons. The Lagrangian density of this hadronic and leptonic dense medium [6,7] is given by $\mathcal{L} = \mathcal{L}_F + \mathcal{L}_l$, where

$$\begin{aligned}\mathcal{L}_F = {} & \sum_B \Psi_B (i\gamma_\mu \partial_\mu - m_B)\Psi_B + \sum_\zeta R_\zeta \nu (i\gamma_\mu \partial_\mu - m_\zeta)\, R\nu_\zeta \\ & + 1/2\,(\partial_\mu \sigma\, \partial^\mu \sigma - m_\sigma^2 \sigma^2) - b/3\; m_N (g_{\sigma N}\sigma)^3 - c/4\,(g_{\sigma N}\sigma)^4 - 1/4\; \omega_{\mu\nu}\,\omega^{\mu\nu} \\ & + 1/2\; m_\omega^2\, \omega_\mu\, \omega^\mu - 1/4\; \rho_{\mu\nu} \,.\, \rho^{\mu\nu} + 1/2\; m_\rho^2 \rho_\mu . \rho^\mu + \sum_\lambda \Psi_\lambda (i\gamma_\mu \partial_\mu - m_\lambda)\Psi_\lambda \,.\end{aligned}$$

is the Lagrangian density for free baryons, electrons and muons ($\lambda = e^-, \mu^-$), and σ, ω and ρ are meson fields. In addition, the interaction Lagrangian density is given by the following expression

$$\mathcal{L}_I = \sum_B \Psi_B \,(g_{\sigma B}\,\sigma - g_{\omega B}\,\gamma_\mu\,\omega_\mu - 1/2\; g_{\rho B}\,\gamma_\mu\,\tau.\rho^\mu)\Psi_B$$
$$+ \sum_\zeta R_\zeta \nu \,(g_{\sigma\zeta}\,\sigma - g_{\omega\zeta}\,\gamma_\mu\,\omega_\mu - 1/2\; g_{\rho\,\zeta}\,\gamma_\mu\,\tau.\rho^\mu) R\nu_\zeta\,.$$

In the equations above the operator Ψ_B represents the Dirac spinor describing the 1/2-spin baryon-octet, with $B = n, p, \Lambda^0, \Sigma^-, \Sigma^0, \Sigma^+, \Xi^-, \Xi^0$. The operator $R_\zeta\nu$ is the Rarita-Schwinger spinor for the 3/2-spin resonances [8], with $\zeta = \Delta^-, \Delta^0, \Delta^+, \Delta^{++}, \Omega^-$.

3. Results and conclusions

In the present work we have used the results of the finite density QCD Sum Rules (QCDSR) to determine the possible values of delta-meson coupling constants [9], where we have taken different values for the quantities $\alpha = g_{\omega\Delta}/g_{\omega N}$, $\beta = g_{\sigma\Delta}/g_{\sigma N}$ and $\gamma = g_{\rho\Delta}/g_{\rho N}$. For the NLW model we have used the set of coupling constants given in Ref. [10], namely $(g_{\sigma N}/m_\sigma)^2 = 9.927\, fm^2$, $(g_{\omega N}/m_\omega)^2 = 4.820\, fm^2$, $(g_{\rho N}/m_\rho)^2 = 4.791\, fm^2$, $b = 0.008659$ and $c = -0.002421$.

In figure 1 we show the results for the population of baryons and leptons versus density for the set of universal coupling values ($\alpha = \beta = \gamma = 1.0$), and for the case ($\alpha = 0.6$, $\beta = \gamma = 1.0$), respectively. In figure 2 we show the behavior of the equation of state for different values of delta-meson coupling constants. Notice that in the figure 2 has $dP/d\rho < 0$ for some ranges of densities. Indeed, what is happening in this intervals is a phase transition, which must be treated by making a Maxwell construction. In figure 3 we show the behavior of the pressure versus density after the Maxwell construction. We note that the delta resonance interaction induces a liquid-gas type phase transition accompanying delta matter formation at densities values of neutron star interior.

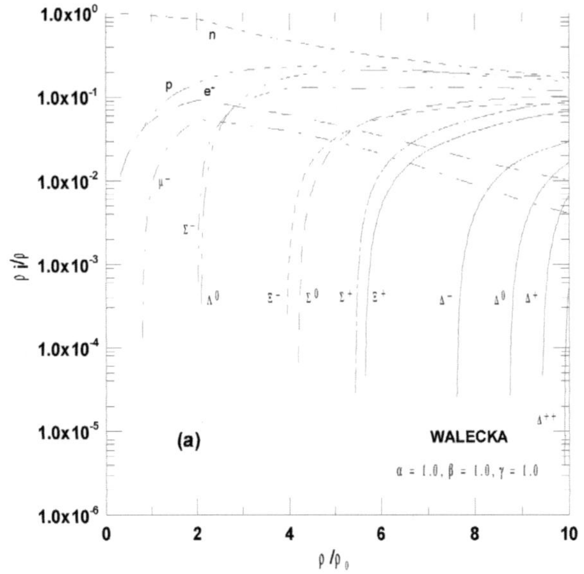

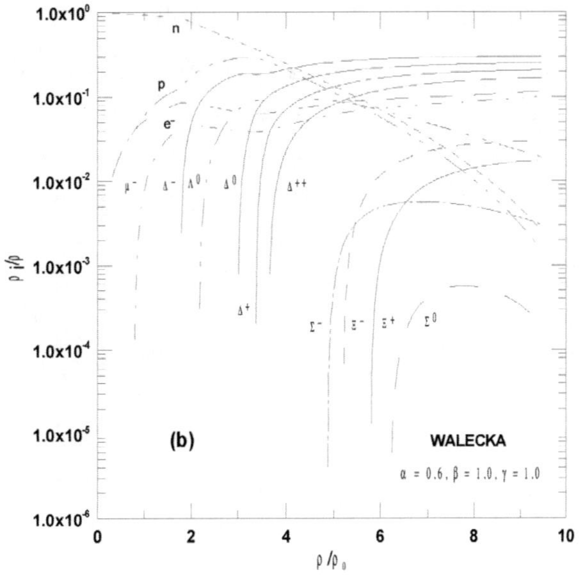

Figure 1. Population of baryons and leptons versus density for the set of universal coupling values ($\alpha = \beta = \gamma = 1.0$) in the part (a), and for the case ($\alpha = 0.6$, $\beta = \gamma = 1.0$) in the part (b).

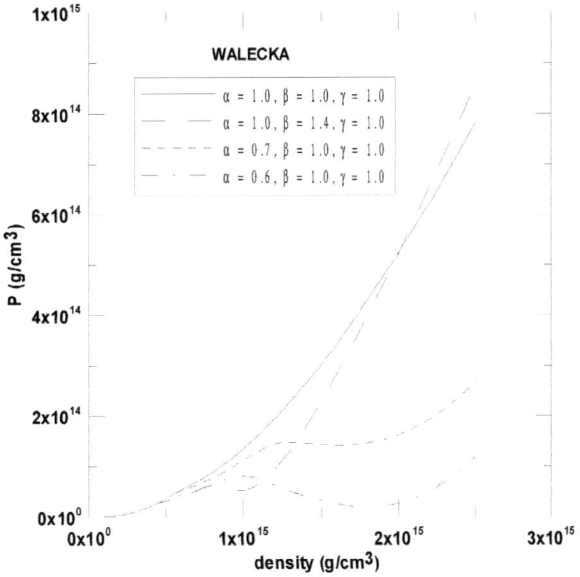

Figure 2. Pressure versus density for different values of delta-meson coupling constants.

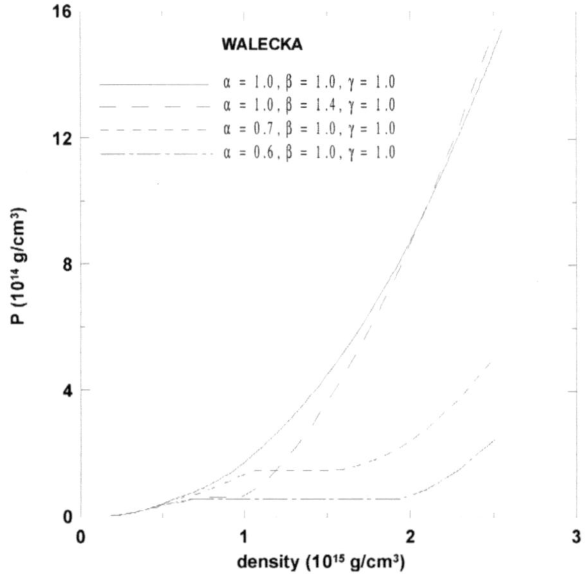

Figure 3. Pressure versus density after the Maxwell construction.

References

[1] Hofmann M et al 1995 Phys. Rev. C **51** 2095
[2] Hjort E L et al 1997 Phys. Rev. Lett. **79** 4345
[3] Hong B et al 1997 Phys. Lett. B **407** 115
[4] Oliveira J C T, Kyotoku M, Chiapparini M, Rodrigues H and Duarte S B 2000 Mod. Phys. Lett. A **15** 1529
[5] Oliveira J C T, Duarte S B, Rodrigues H, Chiapparini M and Kyotoku M 2007 Int. J. Mod. Phys. D **16** 175
[6] Walecka J D 1974 Ann. Phys. **83** 497
[7] Serot B D and Walecka J D 1986 Adv. Nucl. Phys. **16** 1
[8] Rarita W and Schwinger J 1941 Phys. Rev. **60** 61
[9] Kosov D S, Fuchs C, Martemyanov B V and Faessler A 1998 Phys. Lett. B **421** 37
[10] Glendenning N K and Moszkowski S A 1991 Phys. Rev. Lett. **67** 2414

Radial oscillations of charged strange stars

J D V Arbañil and M Malheiro

Departamento de Física, Instituto Tecnológico de Aeronáutica, Centro Técnico Aeroespacial, 12228-900 São José dos Campos, SP, Brazil.

E-mail: arbanil@ita.br

Abstract. The radial oscillations of charged strange quark stars is investigated. It is considered that the fluid pressure follows the MIT bag model equation of state and the charge density to be proportional to the energy density, $\rho_e = \alpha\rho$ (where α is proportionality constant). The modified equations of radial oscillations to the introduction of the electric charge are integrated to determine the fundamental mode. It is found that the stability of the charged object decreases with the increment of the central energy density and with the growth of the charge fraction.

1. Introduction

Strange quark stars or strange stars (for short) are objects that contain approximately equal number of up, down and strange quarks together with smaller number of electrons. In strange stars the electrons are distributed o its surface, forming and electron surface width of several fermis. With the aim to known as the electric charge affects the equilibrium configurations of the strange stars, in [1, 2] Negreiros and collaborators studied the charged strange quark stars considering a star composed by strange matter where radial pressure follows the MIT bag model equation of state and a Gaussian distribution of the electric charge on the surface of the star. The authors found that the distribution of charge on the surface of the object produce significant changes in the structure of the object.

An important point that has not been analyzed in [1, 2] was the stability of charged strange stars. The stability of the charged stars are studied by solving the equation of radial oscillations for charged stars [3].

In this work, it is studied the stability of charged strange stars. We consider that the fluid pressure follows the MIT bag model equation of state and the charge density to be a function of the energy density of the form $\rho_e = \alpha\rho$ (where α is proportionality constant) [4]. We use the units $c = 1 = G$.

2. Radial oscillation equations for charged stars

To the study of radial oscillations of charged strange stars we use the equations presented in Brillante and Mishustin [3]

$$\frac{d\xi}{dr} = \frac{\xi}{2}\frac{d\nu}{dr} - \frac{1}{r}\left(3\xi + \frac{\Delta p}{p\Gamma}\right), \tag{1}$$

$$\frac{d\Delta p}{dr} = e^{\lambda-\nu}(p+\rho)\omega^2\xi r + \left(\frac{d\nu}{dr}\right)^2 \frac{(p+\rho)\xi r}{4} - 4\xi\left(\frac{dp}{dr}\right)$$
$$-8\pi(p+\rho)\xi r e^{\lambda}\left(p+\frac{q^2}{8\pi r^4}\right) - \left(\frac{1}{2}\frac{d\nu}{dr} + 4\pi r e^{\lambda}(p+\rho)\right)\Delta p, \quad (2)$$

being ω the eigenfrequency. The quantities $\xi \equiv \Delta r/r$ and Δp are assumed to have time dependence $e^{i\omega t}$. The parameter $\Gamma = (1+\rho/p)\,(dp/d\rho)$.

In order to solve Eqs. (1) and (2), in the center of the star $r = 0$ is required that for $r \to 0$ the coefficient of the $1/r$ in Eq. (1) must vanish. In other words, we require $(\Delta p)_{\mathrm{center}} = -3(\xi\Gamma p)_{\mathrm{center}}$. The surface of the star surface $r = R$ is determined by the condition $p \to 0$. This implies $(\Delta p)_{\mathrm{surface}} = 0$.

To solve the radial oscillation equations we use the shooting method. As a first step we integrate the hydrostatic equilibrium equation [5, 6, 7] for different values of α and ρ_c. After obtaining the coefficient of the radial oscillations for a given α and ρ_c, the pulsation equations are integrated from the center to the surface of the star. This process starts considering a trial value of ω^2. If after the integration the boundary condition $(\Delta p)_{\mathrm{surface}} = 0$ is not satisfied, the trial value ω^2 is corrected in order to match this condition at the surface of the object.

3. Results and final remarks

In Figs. 1 is plotted the total mass normalized in solar masses as a function of the central energy density for four values of α. In this figure we can note that the mass of the spherical object grows with the charge fraction. The increase of the mass is due to the charge grows with the increment of the charge fraction. From this it can be understood that the charge acts as an affective pressure and helps the radial pressure to counteract more mass against the collapse.

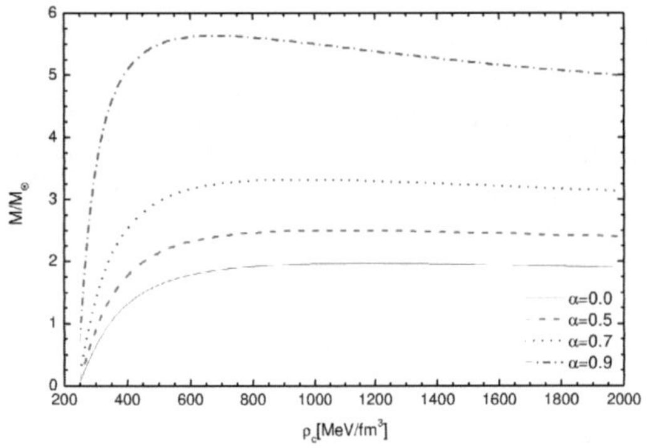

Figure 1. Total mass of the star, in solar masses, against the central energy density for some values of α.

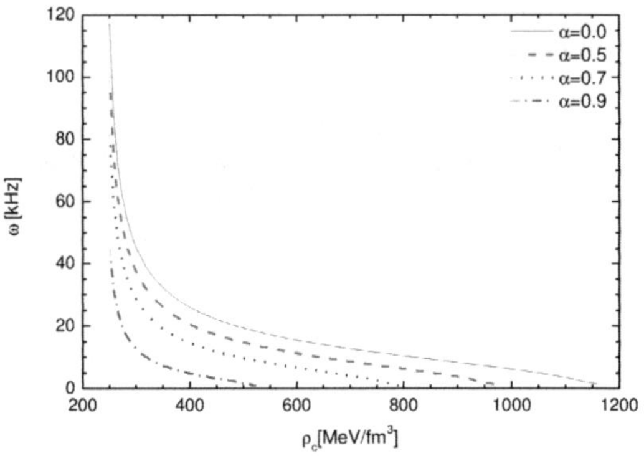

Figure 2. The eigenfrequency of oscillation of the fundamental mode as a function of the central energy density for some values of α.

The behavior of the eigenfrequency of the fundamental mode with the central energy density, for some values of α, it is shown in Fig. 2. Note that the eigenfrequency of oscillation decreases with the increment of the central energy density indicating that the stability of the object decreases with the increment of ρ_c. We can observe also that the increment of the charge fraction affects the stability of the charged star. The stability of the charged object decreases with the increment of α.

Observing Figs. 1 and 2 we can note that the central energy density used to find the maximum mass points are not the same central energy densities used to find the zero eigenfrequencies.

This means that, when is fixing the charge fraction, the condition $dM/d\rho_c > 0$ ($dM/d\rho_c < 0$) is a necessary but not a sufficient condition to determine regions made of stable (unstable) equilibrium configurations.

A detailed analysis about radial oscillations of charged strange quark stars will be reported in a future work [14].

References

[1] Negreiros R P, Weber F, Malheiro M, and Usov V 2009 *Phys. Rev. D* **80** 083006
[2] Malheiro M, Negreiros R P, Weber F, and Usov V 2011 *J. Phys. Conf. Ser.* **312** 042018
[3] Brillante A and Mishustin I N, 2014 *Europhys. Lett.* **105** 39001
[4] Ray S, Espíndola A L, Malheiro M, Lemos J P S, and Zanchin V T 2003 *Phys. Rev. D* **68** 084004
[5] Bekenstein J 1971 *Phys. Rev. D* **4** 2185
[6] Arbañil J D V, Lemos J P S, and Zanchin V T 2013 *Phys. Rev. D* **88** 084023
[7] Arbañil J D V, Lemos J P S, and Zanchin V T 2014 *Phys. Rev. D* **89** 104054
[8] Arbañil J D V and Malheiro M 2015 Equilibrium and stability of charged strange quark stars (in preparation)

A very simple statistical model to the quarks asymmetries

Luis Augusto Trevisan[1] **and Carlos Mirez**[2]

[1]Departament of Mathematics and Statistics, Universidade Estadual de Ponta Grossa, Av. Carlos Cavalcante 4748
Ponta Grossa, Paraná 84.030-900, Brazil
[2]Instituto de Ciência, Engenharia e Tecnologia, Universidade Federal dos Vales do Jequitinhonha e Mucuri, Campus do Mucuri, Rua do Cruzeiro 01, Jardim São Paulo, 39803-371, Teófilo Otoni, Minas Gerais, Brazil.

E-mail: luisaugustotrevisan@yahoo.com.br

Abstract. A simple statistical model is developed with the Fock states being the meson-hadron fluctuations. As expected, a insight about the violation of the Gottfried sum rule is obtained, and also a small difference between the strangeness amount in proton and neutron is explained.

1. Introduction

During the last decades, many collaborations, such as NMC,BCDMS,SLAC, have been performed to describe the nucleons structure function. These experiments have been executed with a large range of energy (Q^2) and from that works, the particle data functions (PDF) have been obtained, (see for instance the NNPDF collaboration at https://nnpdf.hepforge.org/ and MRS-MRST-MSTW collaborations at http://durpdg.dur.ac.uk/hepdata/mrs.html.)

Among the interesting results of the collaborations, is the violation of the Gottfried sum rule [1], that was not predict considering only the QCD phenomenology. One popular explanation is the meson cloud around the nucleon - there are more π^+ meson around the proton than π^- ,because the fluctuation $udd(\bar{d}u) >$ is more probable in the Fock space.

The observed violation of the GSR is depending on Q^2 [2][3], but there are few phenomenological works to explain such behavior. Abbate [2] shows a decrease of the difference $F_2^p - F_2^n$ till $Q^2 = 10GeV^2$ and a small increase at $Q^2 = 30GeV^2$. Sohaily et al. [3] made a correlation among the scattering energy and the variables (Temperature, Volume, Chemical potential) of the statistical model by Zhang [4].

The polarized structure functions were also studied in the experimental way by E142,E143,E154,SMC,HERMES and COMPASS and it was found that there is a difference between the contribution of the strange polarized quarks to the total polarized structure function in proton and neutron.Lets consider

$$n_{s\uparrow} - n_{s\downarrow} + n_{\bar{s}\uparrow} - n_{\bar{s}\downarrow} = \Delta s$$

where $n_{s\uparrow}$ means the amount of strange quark with positive spin, and so on, the result for the proton is:

$$\Delta s = -0.10 \pm 0.03$$

and for the neutron

$$\Delta s = -0.06 \pm 0.04$$

. Here we adopted the same results of ref [5]. We studied the relation among all these facts and present a naive statistical model to provide a physical insight about it. This work is structured as follows: we review some concepts and facts about the structure function in the next section, after the statistical model is presented, the results are in section 4 and a discussion is in the last section.

2. Review on the structure function and quark sea asymmetries

On the present work, we use the following notation: $q_p(x)$ is the a quark (u, d, s) distribution function inside the proton, and naturally, $q_n(x)$ refers to the neutron case. We also suppose the symmetries below:

$$u_p(x) = d_n(x) = u_v(x) + u_s(x)$$
$$d_p(x) = u_n(x) = d_v(x) + d_s(x)$$

where the new index v and s refers to valence quark and sea quark, respectively.

The structure function F_2^p for the proton is given by:

$$F_2^p(x) = \frac{4}{9}\left(u_p(x) + \bar{u}_p(x)\right) + \frac{1}{9}\left(d_p(x) + \bar{d}_p(x)\right) + \frac{1}{9}\left(s_p(x) + \bar{s}_p(x)\right) + \mathrm{S} \tag{1}$$

and, F_2^n

$$F_2^n(x) = \frac{4}{9}\left(u_n(x) + \bar{u}_n(x)\right) + \frac{1}{9}\left(d_n(x) + \bar{d}_n(x)\right) + \frac{1}{9}\left(s_n(x) + \bar{s}_n(x)\right) + \mathrm{S} \tag{2}$$

where S is the contribution of another particles of the sea, without the strange quark. The strange quarks must remain with the index , because this is the crucial point of the present work. The antiquarks have the bar and dont need the index. Writing again F_2^p and F_2^n we have:

$$F_2^p(x) = \frac{4}{9}\left(u_v(x) + u_s(x) + \bar{u}(x)\right) + \frac{1}{9}\left(d_v(x) + d_s + \bar{d}(x)\right) + \frac{1}{9}\left(s_p(x) + \bar{s}_p(x)\right) + \mathrm{S} \tag{3}$$

and

$$F_2^n(x) = \frac{1}{9}\left(u_v(x) + u_s u_n(x)\right) + \frac{4}{9}\left(d_v(x) + d_s(x) + \bar{d}(x)\right) + \frac{1}{9}\left(s_n(x) + s_n(x)\right) + \mathrm{S} \tag{4}$$

Therefore, making the difference $(F_2^p - F_2^n)$, and integrating, the result is:

$$\int_0^1 F_2^p(x) - F_2^n(x) = \frac{1}{3} + \frac{2}{3}\int_0^1 u_s(x) - d_s(x)dx + \frac{2}{9}\int_0^1 s_p(x) - s_n(x)dx \tag{5}$$

The dependence with Q^2 was studied by Abate [2], that obtained the table 1:

In the Fock representation of hadron, besides the three quarks, the fluctuation with quark-antiquark may be represented as a state, and the combination of a valence quark with a antiquark may originates a meson-hadron state. Considering the creation of the pairs $d\bar{d}$, $u\bar{u}$ and $s\bar{s}$, one may have, for the proton, the states:

$$|uud> = |uud> + |udd(u\bar{d})> + |uds(u\bar{s})> + |uuu(d\bar{u}> ... + \text{othercombinations}$$

Table 1. The amounts of the Gottfried sum rule versus different values of Q^2

$Q^2(GeV^2)$	S_G
1	0.2849
2	0.2548
3	0.2479
4	0.2415
5	0.2329
10	0.2279
30	0.2450

Table 2. The main fluctuations from proton that causes the violation of the Gottfried sum rule and have strangeness

state	energy (MeV)	total
$\|p>$	939	939
$\|n\pi^+>$	939+139.6	1078.6
$\|\Lambda K^+>$	1115.683 +493.7	1609.383
$\|\Sigma^0 K^+>$	1192.6 + 493.7	1686.3
$\|\Sigma^+ K^0>$	1189.3+497.7	1687
$\|p\pi^0>$	939+135	1074

with some possible configurations,

$$|p>=|p>+|n\pi^+>+|\Lambda K^+>+|\Sigma^0 K^+>+|\Delta^+ + \pi^->+...$$

and for the neutron

$$|udd>=|udd>+|uud(d\bar{u})>+|uds(d\bar{s})>+|ddd(u\bar{d}>...+\text{oc}$$

that may be

$$|n>=|n>+|p\pi^->+|\Lambda K^0>+|\Sigma^- K^+>+|\Delta^-\pi^+>+...\text{oc}$$

The difference in the probability of the state with $d\bar{d}$ and $u\bar{u}$ originates the violation of the Gottfried sum rule. The model presented in the next section describe the behavior of such asymmetry and gives an insight about the difference of the amount of strangeness in the proton and neutron.

3. A simple statistical model

The phenomena describe above are analysed with a simple statistical model. Each Fock state has a energy level, that is the sum of masses of the hadron with the meson, therefore

The Fermi-Dirac distribution gives the probability p_i for each state i, at some temperature T

$$p_i = \frac{1}{1+\exp[(E_i - \mu_{p,n})/T]} \tag{6}$$

where $\mu_{p,n}$ is the chemical potential for the proton or neutron. As usual

$$_{i=1}^{n}\sum p_i = 1$$

Table 3. The main fluctuations from neutron that causes the violation of the Gottfried sum rule and have strangeness

state	energy (MeV)	total
$\|n>$	939	939
$\|p\pi^->$	939+139.6	1078.6
$\|\Lambda K^0>$	1115.683+497.7	1613.383
$\|\Sigma^0 K^0>$	1192.6+497.7	1690.3
$\|\Sigma^- K^+>$	1197.4+493.7	1691.1
$\|n\pi^0>$	939+135	1074

Table 4. The dependence of states with the temperature.

T	$\|n\pi^+>$	$\sum$ states with s_p -$\sum$ states with s_n
10	8.60928776E-05	2.52272171E-28
25	3.68270986E-02	3.71425330E-12
50	0.13331053	4.15262150E-07

and the thermal mass is defined as (if n states are considered):

$$M_T = \sum_{i=1}^{n} p_i E_i$$

.

The thermal mass is a parameter for the energy system, that is, the mean energy. The comparison with the states energies help us to understand the physical meaning of the model.

4. Results

On the table 4 we show the dependence of the statistical weight and the temperature (T). The second column shows the probability of oscillation $n\pi^+>$, that is the more important contribution for the $\bar{u}$ and $\bar{d}$ asymmetry. Also, in the same table, the difference among the four states for neutron and the four states for proton that have some strange contribution. The strange contribution to proton and neutron structure function may have, therefore, different amounts, and the difference between them increases with the energy.

On despite that few states were considered, the main conclusion that comes from the model will not change with further improvements.

5. Discussion and conclusion

In the present work we review some features of the structure functions for proton and neutron, as the dependence with the energy in the scattering. The asymmetry $\bar{u}$ -$\bar{d}$ increases, after has a small decrease. We also observe that must have a small difference in the amount of the strangeness in the proton and neutron and that this difference is more evident if the energy rises.

To explain these facts in a qualitative way, we implement a simple statistical model, with a Fermi-Dirac distribution for the Fock states. The physical explanation for the observed data, that comes from the model, is that with low scattering energy -that is, low temperature- , the states that generates the $\bar{d}$ and $\bar{u}$ asymmetry are more frequent , while, if the energy increases, the states with strange particles (or states that do not contribute to the $\bar{u} - \bar{d}$ asymmetry

)gain statistical weight. Also, due the small difference of mass between the states with strange quarks in proton and neutron, there is a small difference on the strange contribution. Therefore, the model may explain two different phenomena. As a final remark, we notice that the small difference in the strangeness may be multiplied in a nuclear medium, where the number of protons and neutrons may be different, and this may affect some physical properties in the system.

References

[1] Gottfried K.,(1967)*Phys.Rev.Lett* **18**, 1174
[2] Abbate R.,Forte S., (2005) *Phys. Rev.* D**72** http://link.aps.org/doi/10.1103/PhysRevD.72.117503
[3] S. Sohaily , M. M. Yazdanpanah and A. Mirjalili ,(2012) *International Journal of Modern Physics* A, **V27**, 16, doi = 10.1142/S0217751X12500832
[4] Zhang Y, Shao L., Ma B.Q.,(2009) *Phys. Lett.* B **671** 30
[5] Bhalerao R.S. and Kelkar N.G. and Ram B. (2000) *Physics Letters* B , **476** 285 doi = "http://dx.doi.org/10.1016/S0370-2693(00)00158-1",

Pions near the chiral critical point

M Hippert[1], E S Fraga[1] and E M Santos[2]

[1] Instituto de Física, Universidade Federal do Rio de Janeiro, Caixa Postal 68528, 21941-972, Rio de Janeiro, RJ, Brazil

[2] Instituto de Física, Universidade de São Paulo, Caixa Postal 66318, 05314-970, São Paulo, SP, Brazil.

E-mail: hippert@if.ufrj.br

Abstract. It is an exciting possibility that the QCD critical point can be found in ultrarelativistic heavy-ion collision experiments (HICs). While quantities such as some event-by-event moments of specific observables should display strong non-monotonic behavior near the critical point and could, hence, be used as signatures of criticality, it is not clear that this behavior could effectively be observed in the highly non-ideal scenario of HICs. We here employ Monte Carlo simulations to test second-order moments of pion observables as possible signatures of the critical point while taking into account some realistic ingredients, similar to the ones found in HICs. We make use of simplified models to introduce spurious contributions and dynamical effects.

1. Introduction

The QCD chiral phase diagram might present a second-order critical endpoint (CEP) accessible to current heavy-ion collision experiments (HICs) [1, 2, 3, 4, 5]. The experimental discovery of such a feature of the QCD phase diagram would be of huge importance and a large amount of literature has been dedicated to the study of the effects of the CEP in experimentally accessible observables. Among possible signatures for searching the CEP, it has been proposed that critical fluctuations in its neighborhood could have a large impact upon moments of the event-by-event distribution of observables [6, 7, 8, 9, 10, 11, 12, 13, 14, 15]. This would result in a non-monotonic increase of these moments as experimental conditions probe this region of the QCD phase diagram.

However, it is not at all obvious that the non-monotonic increase in these quantities could be visible in the environment of the small amount of plasma produced in HICs [16, 17, 18], especially if effects from dynamics and spurious contributions are considered. In the following, we study the behavior of second-order moments of pions near criticality within a simplified model [19]. In order to introduce some of the expected experimental limitations of a real experiments, we apply Monte Carlo techniques.

2. Effective theory

Our choice of approach for studying the critical behavior of the observables of interest is taken from Refs. [6, 8]. In this context, the chiral field σ is used as an order parameter for the chiral phase transition. In the second-order CEP, its mass, m_σ, is expected to vanish, thus giving rise to strong long-wavelength fluctuations and the divergence of the corresponding correlation length $\xi \approx 1/m_\sigma$. In order to describe this behavior, a classical treatment is applied along with a homogeneous approximation $\sigma(x) \approx \sigma_0$. The probability distribution for σ_0 is given by

$$P[\sigma_0] = e^{-\Omega[\sigma_0]/T}\,, \tag{1}$$

where T is the temperature and the effective potential Ω is expanded in powers of σ_0,

$$\Omega[\sigma_0] \approx V\,\frac{1}{2}\,m_\sigma^2\,\sigma_0^2 + \mathcal{O}(\sigma_0^3)\,. \tag{2}$$

Interaction with pions is only introduced as a shift in the pion mass squared, m_π^2,

$$\mathcal{L}_{int} = -G\,\sigma_0\,\vec{\pi}\cdot\vec{\pi} + \mathcal{O}(\phi^4)\,, \tag{3}$$

or equivalently,

$$m_\pi^2 = m_\pi^{(0)\,2} + 2\,G\,\sigma_0\,, \tag{4}$$

where a value of $G \approx 300$ MeV near the CEP is taken from [6].

Eqs. (1), (2) and (4) yield a probability distribution for m_π^2. This distribution turns out to be Gaussian within the present approximations, with a width $2\,G\,\xi\,\sqrt{T/V}$, where V is the system volume. This picture enables us to have a grasp on critical correlations among pions while essentially regarding them as free particles. As the pion mass fluctuates, it collectively affects the pion multiplicity and transverse momentum distribution, generating event-by-event correlations among observables. In particular, it can be shown that

$$\langle \Delta n_p\,\Delta n_k\rangle \approx \frac{(G\xi)^2}{TV}\,\frac{1}{\omega_p}\frac{1}{\omega_k} f_p(1+f_p)\,f_k(1+f_k) + \mathcal{O}((\delta m_\pi^2)^4)\,, \tag{5}$$

where k and p denote distinct (free) one-particle pionic states, of energy ω_p and ω_k and occupation numbers n_p and n_k, respectively, and $f_p = (e^{\omega_p/T}-1)^{-1}$. The correlation in Eq. (5) can be used to calculate critical contributions to second-order moments of observables of pions. Its quadratic dependence in ξ indicates an enhancement of these contributions as the CEP is approached and ξ increases.

The picture outlined above enables the construction of a very simple Monte Carlo algorithm for generating distributions of pions displaying the desired critical correlations [19]:

(i) draw m_π^2 from a Gaussian distribution of width $2\,G\,\xi\sqrt{T/V}$;

(ii) for each mode p, draw an occupation number n_p according to the Boltzmann factor $e^{-\omega_p\,n_p/T}$.

In order to have a finite number of modes, it is necessary to impose boundary conditions. We impose Dirichlet boundary conditions on a sphere, partially introducing finite-size effects into our treatment.

3. Heavy-ion collisions

Since ξ increases near the CEP, Eq. (5) shows that second-order moments of pions should exhibit non-monotonic behavior as the freeze-out conditions of the plasma formed in HICs approach and depart from its neighborhood. Our goal is to test whether this behavior is sufficiently strong to be experimentally detected in realistic conditions, providing signatures of the critical point.

Our simulations aim at reproducing freeze-out conditions from the RHIC Beam Energy Scan program. We choose parameters inspired on results from STAR regarding $Au + Au$ collisions at center-of-mass energy $\sqrt{s_{NN}} = 7.7$ GeV in the $0-5\%$ centrality class and pretend the CEP is near the corresponding freeze-out parameters. For this centrality and this center-of-mass energy, the freeze-out temperature and plasma radius are in the ranges $T_{f.o.} \approx 110 - 150$ MeV and $R_{f.o.} \approx 6 - 8$ fm, taking different values for chemical and kinetic freeze-out [20, 21]. We take $T = 130$ MeV and $R = 6.8$ fm and restrict our analysis to pions of transverse momentum $p_T \leq 1$ GeV and pseudo-rapidity in the window $|y| \leq 0.5$. As to introduce some realism, we include limitations in the growth of ξ due to critical dynamics and fluctuations of the freeze-out parameters [19].

3.1. Fluctuations of freeze-out parameters

The experimental study of event-by-event correlations implies a sample of events. However, in HICs, it is not at all possible to prepare or post-select identical events. While centrality binning (and possibly other kinds of post-selection) can be used to reduce this effect, experimental events in a given sample will always vary in initial conditions and evolution, so that freeze-out parameters will inevitably fluctuate among them. Since these fluctuations affect the resulting distribution of particles as a whole, they can generate collective fluctuations that provide a background to critical fluctuations in this kind of analysis.

The final volume and geometry of the plasma, for instance, depend on initial conditions and should fluctuate accordingly. As a simplified model, we consider the colliding nuclei as disks, due to Lorentz contraction, and suppose the plasma to be a sphere of volume proportional to the intersection area between the nuclei at the instant of collision. A probability distribution for the freeze-out radius $R_{f.o.}$ is then obtained by considering a linear probability distribution for the impact parameter $\mathcal{P}(b) \propto b$. The intersection area can be easily found as a function of the impact parameter b and the nuclear radius R_N,

$$A(b, R_N) = 2R_N^2 \cos^{-1}\left(\frac{b}{2R_N}\right) - b\sqrt{R_N^2 - \frac{b^2}{4}}, \tag{6}$$

while the proportionality constant can be found by fixing the plasma radius for a given centrality class (we choose $R_{f.o.} = 6.8$ fm for $0 - 5\%$ centrality) [19]. Our estimate of R_N is the measured parameter $r_0 = 6.38$ fm of the Woods-Saxon nuclear density profile for the gold nucleus [22, 23].

While lacking a simplified model for temperature fluctuations and its relation to volume fluctuations, we just assume the freeze-out temperature to fluctuate according to a Gaussian distribution of 5% width, which we believe to be a moderate choice. Since pions are not directly sensitive to the baryonic chemical potential μ_B, fluctuations of its freeze-out value are not taken into account.

3.2. Critical slowing down

As explicit in Eq. (5), second order correlations exhibit an approximately quadratic dependence on ξ. While in equilibrium and in the thermodynamic limit $\xi \to \infty$ as the CEP is approached, this cannot be the case for a real system with finite size and lifetime. Since ξ should not grow arbitrarily fast, as its equilibrium value increases towards infinity, so do the equilibration times, in what is known as critical slowing down [24]. This effect can significantly affect the proposed signatures of criticality [25].

In order to take this limitation into account, we use the treatment of Ref. [25] to find the time-dependence of ξ near the CEP, including several simplifying approximations and assumptions. With the purpose of using the Ising universality class of QCD, the system is considered to be in equilibrium until a sufficiently large value ξ_0 of the correlation length is reached, so that universality arguments apply. The system is also taken to cool down at a constant rate dT/dt, at constant baryonic chemical potential and exactly over the critical point, while the reduced temperature is taken to be proportional to the Ising reduced magnetic field when mapping the problem to the Ising Model [19].

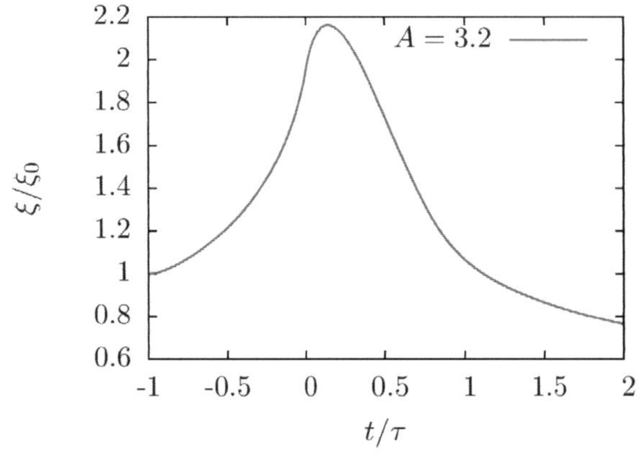

Figure 1. Evolution of the correlation length ξ in the most favorable scenario, with $A = A_{max} = 3.2$. Its value never exceeds 2.2 ξ_0, being most likely under 1.8 ξ_0, depending on at which instant t freeze-out occurs.

Besides universal parameters and the dimensionless constant A, there are two necessary scales that determine ξ as a function of time t: the cooling time $\tau := \frac{T_0 - T_E}{|dT/dt|}$, where T_E is the critical temperature and T_0 is the temperature above which the treatment can be applied, and the initial correlation length ξ_0 [19]. Since $\xi_0 > 1/T_E$, we take $\xi_0 = 1/120$ MeV $= 1.6$ fm and, based on two-pion interferometry results for the lifetime of the system, we estimate the time scale between reaching ξ_0 and crossing the critical point as $\tau = 5.5$ fm [21]. The longer the correlation length has to respond to criticality and the smaller its initial value, the higher the factor $\xi(t)/\xi_0$ can get. We therefore consider these values to be very optimistic. While the dimensionless constant A is not estimated in Ref [25], it can be constrained by asking that the correlation length never increases above the speed of light. The greater its value, the faster ξ can grow. It can thus be shown that $A \leq A_{max} = 3.2$ [19].

Fig. 1 shows $\xi(t)$ for $A = A_{max}$ within this approach, with the system crossing the CEP at $t = 0$. It is clear that ξ should not grow by a factor larger than 2.2, reaching up to ~ 3.5 fm. Since this value is smaller than the typical system radius, we believe dynamical effects to be more restraining than the finite size of the system.

4. Results

Finally, we make simulations of 10^6 events for various values of ξ, ranging from $\xi_0 = 1.6$ fm to 2.2×1.6 fm $= 3.5$ fm, and calculate, for each of them, the second-order moments $\langle(\Delta N)^2\rangle$ and $\langle(\Delta\bar{p}_T)^2\rangle$ for charged pions, where N is the particle multiplicity and $\bar{p}_T$ is the mean transverse momentum among particles of a given event [19]. These moments are normalized by the appropriate power of $\langle N\rangle$ to cancel out system-size dependence.

Our results are shown in Fig. 2, as a function of the combination $\xi_\chi^2 := (G\,\xi)^2$. We choose to display them as relative variations with respect to the values for $\xi = 1.6$ fm, revealing that the tested signatures can increase up to $\sim 10\%$ in the case of $\langle(\Delta N)^2\rangle$. A larger signal could be found for the mixed moment $\langle\Delta N\,\Delta\bar{p}_T\rangle$. However, since chemical and kinetic freeze-out occur separately, there should be no correlation between N and $\bar{p}_T$, so we believe this result to be artificial.

It is also interesting to find how these signatures depend on the instant of freeze-out. For that purpose, a linear fit can be made in Fig. 2 to extract the dependence of $\langle(\Delta N)^2\rangle/\langle(\Delta N)^2\rangle_0$ in ξ_χ^2, which can be used along with the time dependence $\xi(t)$ obtained in Section 3.2. The result is shown in Fig. 3, with the time variable replaced by the difference between the critical and the freeze-out temperatures using $T_0 - T_E = 44$ MeV. It should not be interpreted as the non-monotonic behavior expected as the beam energy of an experiment is varied, but rather as

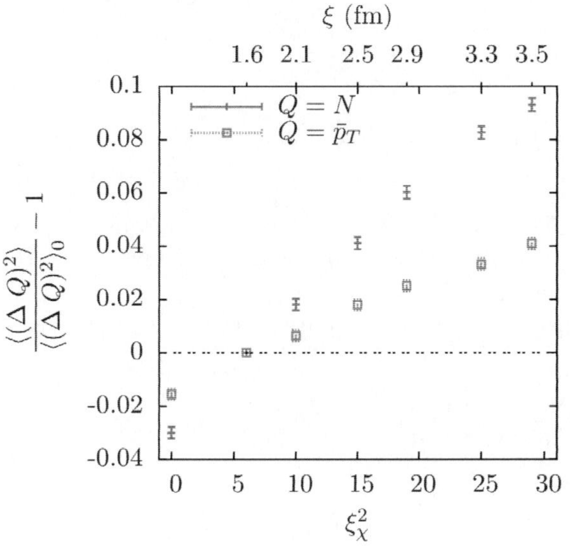

Figure 2. Signal as a function of ξ_χ^2 and ξ, in proportion to the reference value, taken at $\xi = 1.6$ fm. The variances of the charged pion multiplicity, N, and the average transverse momentum of charged pions for a single event, $\bar{p}_T$, are shown.

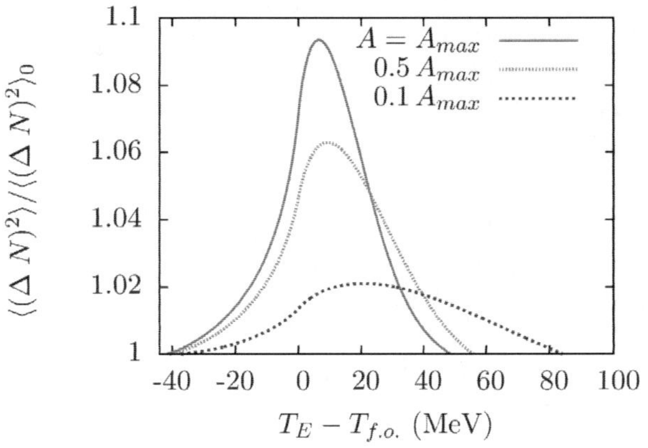

Figure 3. Behavior of the signal to baseline ratio of the variance of N as the freeze-out temperature is varied over a cooling trajectory which crosses the CEP at $T = T_E$, for different choices of the parameter A.

displaying the influence of the amount by which the system cools before freeze-out given that cooling occurs over the critical point. Different values of the parameter A are used and one should bear in mind that changing it has the same effect of changing the cooling time τ by the same factor.

5. Final remarks

In this work, we have analyzed second-order event-by-event moments of pions as signatures of the chiral critical point in HICs. It should be regarded as a first step towards a more realistic understanding of the impact of experimental limitations upon signals of the CEP in HICs. As such, it can and should be enhanced in many ways such as by including contributions from resonance decays, re-scatering and hydrodynamic flow. Since we have assumed perfect equilibrium (apart from Section 3.2), used very simple models and chosen optimistic values whenever estimates were necessary, our results should be regarded as limiting how optimistic one could be about the performance of tested signatures [19]. Results for higher-order moments and moments of protons, which should both yield stronger signatures [8, 9], are left for future developments.

Acknowledgments
This work was partially supported by CAPES, CNPq and FAPERJ.

References

[1] Stephanov M 2006 *PoS* **LAT2006** 024 (*Preprint* hep-lat/0701002)

[2] Rajagopal K 1995 (*Preprint* hep-ph/9504310)

[3] Stephanov M A 2004 *Prog.Theor.Phys.Suppl.* **153** 139–156 (*Preprint* hep-ph/0402115)

[4] Hatta Y and Ikeda T 2003 *Phys. Rev.* D **67**(1) 014028 URL http://link.aps.org/doi/10.1103/PhysRevD.67.014028

[5] Fraga E S, Palhares L F and Sorensen P 2011 *Phys.Rev.* C **84** 011903 (*Preprint* 1104.3755)

[6] Stephanov M A, Rajagopal K and Shuryak E V 1999 *Phys.Rev.* D **60** 114028 (*Preprint* hep-ph/9903292)

[7] Stephanov M A 2002 *Phys.Rev.* D **65** 096008 (*Preprint* hep-ph/0110077)

[8] Stephanov M 2009 *Phys.Rev.Lett.* **102** 032301 (*Preprint* 0809.3450)

[9] Athanasiou C, Rajagopal K and Stephanov M 2010 *Phys.Rev.* D **82** 074008 (*Preprint* 1006.4636)

[10] Mukherjee S, Venugopalan R and Yin Y 2015 (*Preprint* 1506.00645)

[11] Antoniou N G, Diakonos F K and Saridakis E N 2007 *Nucl. Phys.* A **784** 536–550 (*Preprint* hep-ph/0610382)

[12] Antoniou N G, Diakonos F K and Saridakis E N 2008 *Phys. Rev.* C **78** 024908 (*Preprint* 0709.0339)

[13] Luo X (STAR) 2013 *Nucl. Phys.* A **904-905** 911c–914c [Central Eur. J. Phys.10,1372(2012)] (*Preprint* 1210.5573)

[14] Adamczyk L *et al.* (STAR Collaboration) 2013 (*Preprint* 1309.5681)

[15] Luo X (STAR) 2013 *PoS* **CPOD2013** 019 (*Preprint* 1306.3106)

[16] Palhares L, Fraga E and Kodama T 2010 *J.Phys.* G **37** 094031

[17] Kiriyama O, Kodama T and Koide T 2006 (*Preprint* hep-ph/0602086)

[18] Braun J, Klein B, Pirner H J and Rezaeian A 2006 *Phys.Rev.* D **73** 074010 (*Preprint* hep-ph/0512274)

[19] Hippert M, Fraga E S and Santos E M 2015 (*Preprint* 1507.04764)

[20] Chatterjee S, Das S, Kumar L, Mishra D, Mohanty B, Sahoo R and Sharma N 2014 *Adv.High Energy Phys.* **2014** 349013 URL http://www.hindawi.com/journals/ahep/aa/349013/ref/

[21] Adamczyk L *et al.* (STAR Collaboration) 2014 (*Preprint* 1403.4972)

[22] Vries H D, Jager C D and Vries C D 1987 *Atomic Data and Nuclear Data Tables* **36** 495 – 536 ISSN 0092-640X URL http://www.sciencedirect.com/science/article/pii/0092640X87900131

[23] Jager C D, Vries H D and Vries C D 1974 *Atomic Data and Nuclear Data Tables* **14** 479 – 508 ISSN 0092-640X nuclear Charge and Moment Distributions URL http://www.sciencedirect.com/science/article/pii/S0092640X74800021

[24] Hohenberg P and Halperin B 1977 *Rev.Mod.Phys.* **49** 435–479

[25] Berdnikov B and Rajagopal K 2000 *Phys.Rev.* D **61** 105017 (*Preprint* hep-ph/9912274)

Journal of Physics: Conference Series **706** (2016) 052025 doi:10.1088/1742-6596/706/5/052025

Phase structure of cold magnetized color superconducting quark matter

PG Allen[1], AG Grunfeld[1,2] and NN Scoccola[1,2,3]

[1] Department of Theoretical Physics, Comisión Nacional de Energía Atómica, Av.Libertador 8250, 1429 Buenos Aires, Argentina

[2] CONICET, Rivadavia 1917, 1033 Buenos Aires, Argentina

[3] Universidad Favaloro, Solís 453, 1078 Buenos Aires, Argentina

E-mail: `allen@tandar.cnea.gov.ar`

Abstract. The influence of intense magnetic fields on the behavior of color superconducting cold quark matter is investigated using an $SU(2)_f$ NJL-type model for which a novel regulation scheme is introduced. In such a scheme the contributions which are explicitly dependent on the magnetic field turn out to be finite and, thus, do not require to be regularized. As a result of this, non-physical oscillations that arise from regularizing magnetic field dependent terms are naturally removed, and oscillations that are actually physical can be better appreciated. The phase diagrams in the $\tilde{e}B - \mu$ plane are presented for different values of the diquark coupling.

1. Introduction

At asymptotically large chemical potentials, the fact that cold quark matter behaves as a color superconductor can be shown by using perturbative methods in the context of quantum chromodynamics (QCD)[1]. However, such methods cannot be applied in the range of moderate densities relevant for, amongst others, the astrophysics of strongly magnetized compact stellar objects[2]. Moreover, since the well-known sign problem prevents lattice QCD calculations from being performed at sufficiently low temperatures and finite chemical potential, one has to rely on effective models to analyze the behavior of magnetized quark matter in this region. One particular model that has been extensively used for this purpose is the Nambu-Jona-Lasinio (NJL) model[3]. This is an effective model originally devised to study the dynamics of chiral symmetry breaking, in which gluon degrees of freedom are integrated out and interactions are described by local four-quark interactions. The incorporation of additional diquark interactions into the model allows for the description of color superconducting matter[4].

In this context, the effect of a constant magnetic field has been analyzed by several authors [5, 6, 7, 8, 9, 10]. However, the Landau level (LL) structure acquired by the quark dispersion relations introduces complications in the regularization procedure necessary to treat the divergent vacuum integrals characteristic of the local interactions in NJL-type models. In the superconducting case, the only prescriptions used until now applied regulating functions on each Landau level integral separately [6, 7, 8, 9, 10]. This procedure, however, might introduce non-physical oscillations [11, 12, 13], making the interpretation of the results unclear. On the other hand, another way of regularizing the NJL model under magnetic fields was reported in Ref.[14] in the absence of color superconductivity. In this procedure, a sum over all Landau levels in the vacuum term is performed, and the divergence is isolated into a term that has no explicit dependence on the magnetic field, and can be regularized in the standard fashion. The main purpose of this work is to introduce a regularization scheme which can be considered an extension of this procedure to the superconducting case. The results obtained within this "Magnetic field independent regularization" (MFIR) will be compared to the ones obtained with a few regularizations which have been used in the literature, showing that the aforementioned non-physical oscillations are removed. Then, the behavior of the model under magnetic field and varying coupling strength will be studied.

2. Magnetized cold quark matter within the SU(2)$_f$ NJL model in the presence of color pairing interactions

2.1. The thermodynamical potential in the mean field approximation

We consider a NJL-type SU(2)$_f$ Lagrangian density which includes scalar-pseudoscalar and color pairing interactions. In the presence of an external magnetic field and chemical potential it reads:

$$\begin{aligned}\mathcal{L} &= \bar\psi\left[i\,\tilde{\not{D}} - m_c + \mu\,\gamma^0\right]\psi \\ &\quad + G\left[(\bar\psi\psi)^2 + (\bar\psi i\gamma_5\vec\tau\psi)^2\right] + H\left[(i\bar\psi^C\,\epsilon_f\epsilon_c^3\gamma_5\psi)(i\bar\psi\,\epsilon_f\epsilon_c^3\gamma_5\psi^C)\right]. \end{aligned} \tag{1}$$

Here, G and H are coupling constants, $\psi = (u,d)^T$ represents a quark field with two flavors, $\psi^C = C\bar\psi^T$ and $\bar\psi^C = \psi^T C$, with $C = i\gamma^2\gamma^0$, are charge-conjugate spinors and $\vec\tau = (\tau_1,\tau_2,\tau_3)$ are Pauli matrices. Moreover, $(\epsilon_c^3)^{ab} = (\epsilon_c)^{3ab}$ and $(\epsilon_f)^{ij}$ are antisymmetric matrices in color and flavor space respectively. Furthermore, m_c is the (current) quark mass that we take to be the same for both flavors and μ is the quark chemical potential. The coupling of the quarks to the electromagnetic field $\tilde{\mathcal{A}}_\mu$ is implemented through the covariant derivative $\tilde D_\mu = \partial_\mu - i\tilde e\tilde Q\tilde{\mathcal{A}}_\mu$. Note that here we are dealing with "rotated" fields. In fact, as is well known, in the presence of a non-vanishing superconducting gap Δ, the photon acquires a finite mass. However, as shown in Ref.[15], there is a linear combination of the photon and the eighth component of the gluon field that leads to a massless rotated $U(1)$ field. The associated rotated charge matrix $\tilde Q$ is given by

$$\tilde Q = Q_f\otimes 1_c - 1_f\otimes\left(\frac{\lambda_8}{2\sqrt3}\right) \tag{2}$$

where $Q_f = \mathrm{diag}(2/3,-1/3)$ and λ^8 is the color quark matrix $\lambda_8 = \mathrm{diag}(1,1,-2)/\sqrt3$. Then, in a six dimensional flavor-color representation $(u_r,u_g,u_b,d_r,d_g,d_b)$, the rotated $\tilde q$ for different quarks are: $u_r = 1/2, u_g = 1/2, u_b = 1, d_r = -1/2, d_g = -1/2, d_b = 0$. The rotated unit charge $\tilde e$ is given by $\tilde e = e\cos\theta$, where θ is the mixing angle which is estimated to be $\simeq 1/20$[17]. In the present work we consider a static and constant magnetic field in the 3-direction, $\tilde{\mathcal{A}}_\mu = \delta_{\mu 2}x_1 B$, which in fact is also a mixture of the electromagnetic field and color fields.

In what follows we work in the mean field approximation (MFA), assuming that the only non-vanishing expectation values are $<\bar\psi\psi>= -(M-m_c)/2G$ and $< i\bar\psi^C\,\epsilon_f\epsilon_c^3\gamma_5\psi>= -\Delta/2H$, which can be chosen to be real. Here, M and Δ are the so-called dressed quark mass and superconducting gap, respectively. The resulting MFA thermodynamic potential at vanishing temperature reads

$$\Omega_{\mathrm{MFA}} = \frac{(M-m_c)^2}{4G} + \frac{\Delta^2}{4H} - \sum_{|\tilde q|=0,\frac12,1} P_{|\tilde q|} \tag{3}$$

where

$$P_{|\tilde q|=0} = \int\frac{d^3p}{(2\pi)^3}\left[E_0^+ + |E_0^-|\right], \tag{4}$$

$$P_{|\tilde{q}|=1} \;=\; \frac{\tilde{e}B}{8\pi^2}\sum_{k=0}^{\infty}\alpha_k\int_{-\infty}^{\infty}dp_z\left[E_1^+ + \left|E_1^-\right|\right], \tag{5}$$

$$P_{|\tilde{q}|=1/2} \;=\; \frac{\tilde{e}B}{4\pi^2}\sum_{k=0}^{\infty}\alpha_k\int_{-\infty}^{\infty}dp_z\left[E_{1/2}^+ + E_{1/2}^-\right]. \tag{6}$$

Here, we have introduced $\alpha_k = 2 - \delta_{k0}$ and

$$\begin{aligned} E_0^{\pm} &= \sqrt{p^2 + M^2} \pm \mu \\ E_1^{\pm} &= \sqrt{p_z^2 + 2k\tilde{e}B + M^2} \pm \mu \\ E_{1/2}^{\pm} &= \sqrt{\left[\sqrt{p_z^2 + k\tilde{e}B + M^2}\;\pm\;\mu\right]^2 + \Delta^2}. \end{aligned} \tag{7}$$

Clearly, Eqs.(4-6) are divergent and, thus, require to be regularized. Some alternative schemes to achieve this will be discussed in the following subsection. Given the corresponding regularized form $\Omega^{reg}_{\rm MFA}$, the associated gap equations for M and Δ then read

$$\frac{\partial\Omega^{reg}_{\rm MFA}}{\partial(M,\Delta)} = 0\ . \tag{8}$$

For each value of μ and $\tilde{e}B$, several solutions of these equations will generally exist, corresponding to different possible phases, and the most stable solution is that associated to the absolute minimum of the thermodynamic potential.

To regularize the integrals, several prescriptions were introduced in which some cutoff function $h_\Lambda(q)$ is introduced, with $q = p$ in the case of Eq.(4) and $q = \omega_{p_z,k} \equiv \sqrt{p_z^2 + 2k|\tilde{q}|B}$ for Eqs.(5,6). The obvious and simplest choice would be to take $h_\Lambda(q) = \Theta(\Lambda - q)$, scheme which will be referred to as "sharp function regularization" (ShFR). In this case, the integral in Eq.(4), which is magnetic field independent, is cut off when $p = \Lambda$. The contributions coming from Eqs.(5-6) include a sum over Landau levels, and the integral in each of these is cut off for the p_z that satisfies $\Lambda = \sqrt{p_z^2 + 2k|\tilde{q}|B}$, that is, such that the momentum and magnetic field contribution to the quark dispersion relation does not exceed the value of the cut-off. This would seem like a natural way of extending the 3D sharp cutoff zero magnetic field regularization to the finite $\tilde{e}B$ case. However, the magnetic field dependence of the prescription brings in strong unphysical oscillations. In an attempt to minimize the effects of this magnetic field dependence, the Heaviside function can be replaced with a smooth regulator, prescription that will be referred to as "smooth function regularization" (SmFR). For definiteness in this work, we consider the function $h_\Lambda(q) = 1/(1 + \exp[(q/\Lambda - 1)/a])$. We have verified that other possible choices lead to similar results. To choose the value of the constant a that determines the regulator smoothness one is limited by the fact that a too steep function does not improve over the ShFR results and that a too smooth function leads to values of the quark condensate in absence of the magnetic field which are quite above the phenomenological range. Here, we follow Refs.[8, 9] and consider $a = 0.05$. We will see that this procedure softens the oscillations but does not remove them.

To fully get rid of the above mentioned regularization artifacts, we introduce in what follows an alternative scheme in which the contributions that are explicitly dependent on the magnetic field turn out to be finite and thus do not need to be regularized. We will refer to this regularization scheme as the "magnetic field independent regularization" (MFIR). We start by considering the $P_{|\tilde{q}|=0}$ contribution. Since it is independent of the magnetic field, it can be treated in the usual way [3]. Introducing a sharp 3D cutoff we get

$$P_{|\tilde{q}|=0} = \frac{1}{\pi^2}\int_0^\Lambda dp\, p^2 \sqrt{p^2+M^2} + \frac{\Theta(\mu-M)}{\pi^2}\left[\frac{\mu(\mu^2-M^2)^{3/2}}{3} - \frac{(\mu^2-M^2)^2}{8} h\left(\frac{M}{\sqrt{\mu^2-M^2}}\right)\right], \tag{9}$$

where $h(z) = (2+z^2)\sqrt{1+z^2} + z^4 \ln[z/\sqrt{1+z^2}]$. In the case of $P_{|\tilde{q}|=1}$ we note that, except for the value of the quark charge, the expression coincides with that analyzed in Ref.[14] where no color pairing interactions were considered. Following the steps in that reference, we get

$$P_{|\tilde{q}|=1} = \frac{1}{\pi^2}\int_0^\Lambda dp\, p^2 \sqrt{p^2+M^2} + \frac{\tilde{e}B}{4\pi^2}\sum_{k=0}^{k_{max}} \alpha_k \left[\mu\sqrt{\mu^2-s_k^2} - s_k^2 \ln\left(\frac{\mu+\sqrt{\mu^2-s_k^2}}{s_k}\right)\right] + \frac{(\tilde{e}B)^2}{2\pi^2}\left[\xi'(-1,x) + \frac{x-x^2}{2}\ln x + \frac{x^2}{4}\right] \tag{10}$$

where $x = M^2/(2\tilde{e}B)$, $k_{max} = \text{Floor}[(\mu^2-M^2)/(2\tilde{e}B)]$ and $s_k = \sqrt{M^2+2k\tilde{e}B}$. In Eqs.(9-10), the first term is a vacuum contribution which does not explicitly depend on the magnetic field and the second term is the matter contribution. The last term in Eq.(10) is the explicit magnetic field contribution to the vacuum, which has been isolated into a finite term. The case of $|\tilde{q}| = 1/2$ is more involved. However, it can be cast into the form (see Appendix of [18] for details)

$$P_{|\tilde{q}|=1/2} = \frac{2}{\pi^2}\int_0^\Lambda dp\, p^2 \left(E_\Delta^+ + E_\Delta^-\right) + \frac{(\tilde{e}B)^2}{2\pi^2}\left[\xi'(-1,y) + \frac{y-y^2}{2}\ln y + \frac{y^2}{4}\right] + \frac{(\tilde{e}B)^2}{2\pi^2}\int_0^\infty dp \left[\sum_{k=0}^{\infty}\alpha_k\, f(p^2+k) - 2\int_0^\infty dx\, f(p^2+x)\right] \tag{11}$$

where

$$E_\Delta^\pm = \sqrt{(\sqrt{p^2+M^2}\pm\mu)^2+\Delta^2} \tag{12}$$

$$y = (M^2+\Delta^2)/(\tilde{e}B) \tag{13}$$

and

$$f(z) = \sum_{s=\pm1}\left[\sqrt{(\sqrt{z+2x}+s\,\mu/\sqrt{\tilde{e}B})^2+y-2x} - \sqrt{z+y}\right]. \tag{14}$$

In this expression a 3D sharp cutoff has been introduced to regularize the first term, i.e. the one that contains contributions from the vacuum and matter which do not explicitly depend on the magnetic field. Note that, as in the case of vanishing $\tilde{e}B$ discussed in e.g. Ref.[19], these cannot be disentangled into two terms unless $\Delta = 0$. The second term is the vacuum magnetic contribution analogous to the $|\tilde{q}| = 1$ case. Finally, the third term is an additional explicitly magnetic field dependent matter contribution which turns out to be finite.

In this work, the parameters will be set to the following values: $m_c = 5.595$ MeV, $\Lambda = 620.9$ MeV^2 and a coupling constant G such that $G\Lambda^2 = 2.212$. This parameter choice leads $M_0 = 340$ MeV within the MFIR regularization, where M_0 represents the vacuum quark effective mass in the absence of external magnetic fields.

3. Numerical results

3.1. Comparison between regularization schemes

The behavior of the dressed mass as a function of magnetic field is displayed in Fig.1, for the three regularizations and for $\mu = 0$. The results for ShFR and SmFR exhibit non-physical oscillations, whose origin lies in the magnetic field dependence of the regularization in the vacuum term, that causes the contribution of a given LL to be larger for lower magnetic fields. In the ShFR, which is the most extreme case, the only LLs participating in the sum are those for which $\Lambda^2 \geq 2k|\tilde{q}|B$. Therefore, depending on the magnetic field, more or less terms appear and each time the relation is satisfied for a given k, there will be a discontinuity in the derivative of the thermodynamic potential. This singularity, hence, does not correspond to a phase transition. The soft regulator, which could be regarded as a way to handle this problem and remove sharp oscillations, still contains this pathology, because in this case the contribution of a given LL also depends on the magnetic field through the Fermi-type regulator function. So, even though the smooth integrals partly conceal this problem, the oscillations are still present and we can actually see that for ShFR and SmFR they are in phase. On the other hand, in the MFIR scheme the mass increases steadily with the magnetic field displaying the usual "magnetic catalysis effect" as in e.g. Refs.[14, 20, 21, 22].

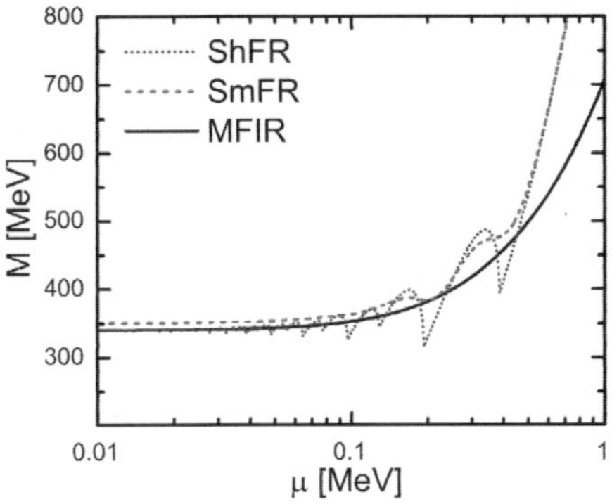

Figure 1. M vs $\tilde{e}B$ in the B phase, for the three regularization schemes considered and $H/G = 0.75$. Note that $\Delta = 0$ in this phase.

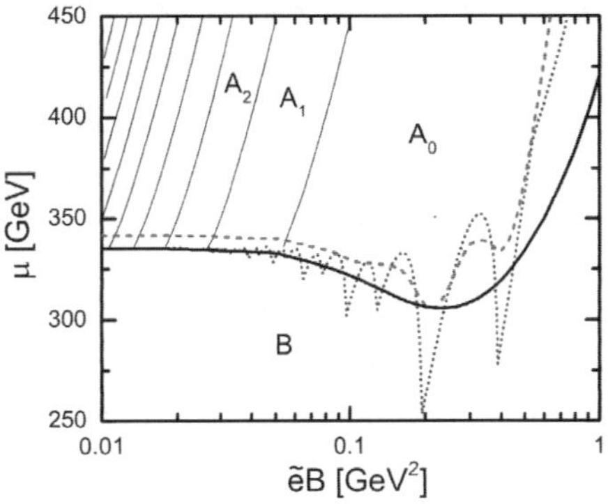

Figure 2. Phase diagram in the $\tilde{e}B - \mu$ plane, for the three regularization schemes considered and $H/G = 0.75$. All transitions seen in this diagram are first order.

The phase diagrams in the $\tilde{e}B - \mu$ plane are present in Fig.2. Despite the differences owing to the regularization, a few general properties are common to all prescriptions. For low enough chemical potential, there is always a vacuum phase (B) in which chiral symmetry is broken and the superconducting gap is zero. When chemical potential is raised, there is a first order transition to a region in which the quark population is non-zero, the superconducting gap is finite and chiral symmetry is restored (strictly speaking, restoration is only approximate due to the small but non-zero current quark mass). The near vertical lines, which are known as vAdH transitions, are first order transitions that separate the superconducting region into several phases which are denoted as A_i. In a given A_i phase, Landau levels from zero to i are populated, and through any given vAdH transition, the value of i changes in one unit. The vAdH transitions coincide almost exactly for all three regularizations. We should also emphasize that these transitions are effectively physical because quark density is discontinuous at these, and that they in turn induce physical oscillations in the order parameters as a function of the magnetic field. Despite these general similarities in the phase diagrams for the three regularizations, we can once again see that the critical chemical potential for the transition from the B phase to the A-type phases exhibits strong oscillations in the ShFR case, and that these continue to exist in SmFR, even though they are slightly suppressed. These are small for low magnetic fields, but become larger in the intermediate $\tilde{e}B$ range and once again make the phase diagram hard to interpret. In the MFIR scheme, we note that the critical chemical potential is approximately independent of $\tilde{e}B$ for values below 0.07 GeV^2, then it decreases until it reaches a minimum near $\tilde{e}B = 0.2$ GeV^2 and after this value, it increases back again, giving rise to the usual well-shaped curve related to the "inverse magnetic catalysis effect"[23]. Due to the regulation artifacts, this feature is much less evident in the ShFR and SmFR.

3.2. MFIR results for different coupling ratios

Given that the MFIR is seen to make an improvement over the other displayed regularizations, we will now concentrate on analyzing within this scheme the phase diagrams and their dependence on the coupling constant. In the previous section, only $H/G = 0.75$ was considered. However, given that the value of this ratio is subject to certain degree of uncertainty, it is worthwhile to explore the consequences of varying it within a reasonable range. Thus, in what follows, the representative values $H/G = 0.5, 0.75$ and 1 will be considered. Note that values $H/G > 1$ are quite unlikely to be realized in QCD.

The phase diagrams in the $\tilde{e}B - \mu$ plane for the three coupling ratios considered are displayed in Fig.3. As was seen in the previous section for $H/G = 0.75$, the system is in the B phase for low μ, where chiral symmetry is broken, and in one of the possible A-type phases for a high enough μ value, where chiral symmetry is approximately restored. However, if the coupling ratio is changed, other phases may appear. The first transition encountered if the phase diagram is traversed in the direction of increasing μ will be referred to as the "main transition". It has approximately the same shape in all displayed phase diagrams and it connects the B phase to populated phases in general. As H/G increases, the main transition is displaced downwards in its entirety, and the depth of the "inverse magnetic catalysis well" diminishes. Also, the superconducting gap in the A-type phases becomes larger. For $H/G = 0.5$ (left panel), an intermediate phase that will be referred to as a C-type phase exists. It is enclosed by the main transition from below and by another first order transitions from above. Both transitions are horizontal in a large magnetic field range so that the phase extends in an approximately horizontal band up to $\tilde{e}B \simeq 0.11$ GeV^2, where it is bounded by a crossover type transition, signalled by the peak of the chiral susceptibility. In the C-type phase, chiral symmetry is partially restored: the mass is lower than in the vacuum phase but higher than in the chirally restored phases, and density is finite for the $|\tilde{q}| = 0$ and $|\tilde{q}| = 1$ quarks (only the lowest LL being occupied for the latter species). It is important to point out that the existence of the C-type phase is not a consequence of the diquark pairing channels. Actually, superconducting effects are still small for $H/G = 0.5$ and the phase diagram remains basically unchanged with respect to the $H/G = 0$ case. In particular, the superconducting gap is always below 1 MeV in the C-phase.

As the coupling constant ratio is increased from the value $H/G = 0.5$, the upper phase transition is displaced downwards. This causes the C-type phase to shrink until it eventually disappears around $H/G \sim 0.65$, so that a relatively simple phase structure results, in which a single phase transition connects the B phase to the A-type phases. For an even larger value of H/G (~ 0.94), the phase transition splits into two once again, so that for $H/G = 1$ (right panel) there is an intermediate phase, which will be referred to as a D phase. This phase is qualitatively different from the one found in $H/G = 0.5$. To begin with, the transition from vacuum to this phase is second order. Also, both the dressed mass and the superconducting gap are appreciably large. This type of phase is sometimes referred to as a "mixed phase" [26], even though other meanings may be found in the literature for this term [27]. There is no quark

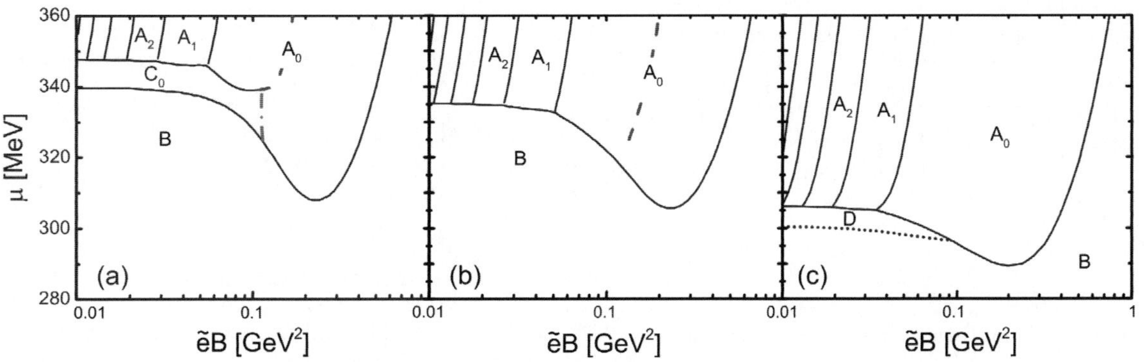

Figure 3. Phase diagrams within the MFIR scheme for different H/G ratios: (a) $H/G = 0.5$, (b) $H/G = 0.75$, (c) $H/G = 1$. Full black lines correspond to first order transitions and dotted black lines to second order transitions. The dash-dotted red lines represent chiral susceptibility crossovers. Blue dashed lines represent quark number susceptibility crossovers related to $|\tilde{q}| = 1/2$ quark which are too weak to be actually considered to separate distinct phases.

population for the $|\tilde{q}| = 0$ and $|\tilde{q}| = 1$ species, but the finiteness of Δ induces a non-zero density for $|\tilde{q}| = 1/2$ quarks. The transition leading to the A-type phases is first order as in the previous cases. The existence of this phase is a consequence of the diquark pairing alone and hence it exists for $\tilde{e}B = 0$. It is interesting to note, however, that it extends along the magnetic field axis of the phase diagram for an appreciable range.

4. Concluding Remarks

In the present work we explored the effects of magnetic field on cold color superconducting quark matter in the framework of the NJL-type model, using a regularization scheme in which the contributions which are explicitly dependent on the magnetic field turn out to be finite and, thus, do not require to be regularized. Such a "magnetic field independent regularization" (MFIR) scheme can be considered an extension of the method described in e.g. Ref.[14] to the case in which color pairing interactions are present. We compared the corresponding results with those obtained through the regularization methods used in previous works [6, 7, 8, 9, 10] and concluded that the MFIR scheme removes the unphysical oscillations originating in the regularization procedure. Taking this into account, we studied the phase structure in the $\tilde{e}B - \mu$ plane using three coupling ratio values which lead to qualitatively different results. We found that up to $H/G = 0.5$ superconducting effects are still small, and the phase diagram remains basically unchanged. In particular, there is an intermediate phase in which quark population exists but chiral symmetry is still strongly broken. The diquark gap in this phase is finite but extremely small. It is connected to other phases by first order transitions and a chiral crossover. As the coupling ratio is increased, the two transitions surrounding this phase move close to each other, causing it to disappear at $H/G \sim 0.65$. A relatively simple phase diagram hence exists for a narrow range around the standard value H/G =0.75. If the coupling ratio is increased beyond 0.94, a mixed phase is present in both parameter sets, where both condensates are large. Finally, we found that the inverse catalysis phenomenon is present, although it becomes less pronounced as H/G increases. Even though charge neutrality and β equilibrium conditions have not been taken into account in this work, they are certainly necessary for a correct description of the physics of compact stars. We expect to report on this in the near future.

Journal of Physics: Conference Series **706** (2016) 052025 doi:10.1088/1742-6596/706/5/052025

References

[1] Bailin D and Love A 1984 *Phys. Rept.* **107**, 325.
[2] Duncan R C and Thompson C 1992 *Astrophys. J.* **392**, L9;
[3] Vogl U and Weise W 1991 *Prog. Part. Nucl. Phys.* **27**, 195 ; Klevansky S 1992 *Rev. Mod. Phys.* **64**, 649; Hatsuda T and Kunihiro T 1994 *Phys. Rep.* **247**, 221.
[4] Buballa M 2005 *Phys. Rept.* **407**, 205 [hep-ph/0402234].
[5] Ferrer E J, de la Incera V and Manuel C 2005 *Phys. Rev. Lett.* **95**, 152002 [hep-ph/0503162];
[6] Fukushima K and Warringa H J 2008 *Phys. Rev. Lett.* **100**, 032007 [arXiv:0707.3785 [hep-ph]].
[7] Noronha J L and Shovkovy I A, 2007 *Phys. Rev.* D **76**, 105030 [2012 *Phys. Rev.* D **86**, 049901] [arXiv:0708.0307 [hep-ph]].
[8] Fayazbakhsh S and Sadooghi N 2010 *Phys. Rev.* D **82**, 045010 [arXiv:1005.5022 [hep-ph]].
[9] Fayazbakhsh S and Sadooghi N 2011 *Phys. Rev.* D **83**, 025026 [arXiv:1009.6125 [hep-ph]].
[10] Mandal T and Jaikumar P 2013 *Phys. Rev.* C **87**, 045208 [arXiv:1209.2432 [nucl-th]].
[11] Campanelli L and Ruggieri M 2009 *Phys. Rev.* D **80**, 034014 [arXiv:0905.0853 [hep-ph]].
[12] Frasca M and Ruggieri M 2011 *Phys. Rev.* D **83**, 094024 [arXiv:1103.1194 [hep-ph]].
[13] Gatto R and Ruggieri M 2013 *Lect. Notes Phys.* **871**, 87 [arXiv:1207.3190 [hep-ph]].
[14] Menezes D P, Benghi Pinto M, Avancini S S, Perez Martinez A and Providencia C 2009 *Phys. Rev.* C **79**, 035807 [arXiv:0811.3361 [nucl-th]].
[15] Alford M G, Berges J and Rajagopal K 2000 *Nucl. Phys.* B **571**, 269 [hep-ph/9910254].
[16] Gorbar E V 2000 *Phys. Rev.* D **62**, 014007 [hep-ph/0001211].
[17] Gorbar E V 2000 *Phys. Rev.* D **62**, 014007 [hep-ph/0001211].
[18] P. G. Allen, A. G. Grunfeld and N. N. Scoccola, arXiv:1508.04724 [hep-ph].
[19] Blaschke D, Volkov M K and Yudichev V L 2003 *Eur. Phys. J.* A **17**, 103 [hep-ph/0301065].
[20] Ebert D and Klimenko K G 2003 *Nucl. Phys.* A **728**, 203 [hep-ph/0305149].
[21] Allen P G, Pagura V P and Scoccola N N 2015 *Phys. Rev.* D **91**, no. 11, 114024 [arXiv:1502.00572 [hep-ph]].
[22] Boomsma J K and Boer D 2010 *Phys. Rev.* D **81**, 074005 [arXiv:0911.2164 [hep-ph]].
[23] Preis F, Rebhan A and Schmitt A 2011 *JHEP* **1103**, 033 [arXiv:1012.4785 [hep-th]].
[24] Allen P G and Scoccola N N 2013 *Phys. Rev.* D **88**, 094005 [arXiv:1309.2258 [hep-ph]].
[25] Grunfeld A G, Menezes D P, Pinto M B and Scoccola N N 2014 *Phys. Rev.* D **90**, no. 4, 044024 [arXiv:1402.4731 [hep-ph]].
[26] Huang M, Zhuang P f and Chao W q 2002 *Phys. Rev.* D **65**, 076012 [hep-ph/0112124].
[27] Neumann F, Buballa M and Oertel M 2003 *Nucl. Phys.* A **714**, 481 [hep-ph/0210078].

Cold quark matter phase diagram under strong magnetic fields within a generalized SU(2) NJL model

PG Allen[a], VP Pagura[b,c,d] and NN Scoccola[a,b,e]

[a] Department of Theoretical Physics, Comisión Nacional de Energía Atómica, Av.Libertador 8250, 1429 Buenos Aires, Argentina
[b]CONICET, Rivadavia 1917, 1033 Buenos Aires, Argentina
[c]IFLP, CONICET-Dpto. de Física, Univ.Nac. de La Plata, C.C.67, (1900) La Plata, Argentina
[d] Dpto. de Física Teórica-IFIC, Univ. de Valencia-CSIC, E-46100 Burjassot, Valencia, Spain
[e] Universidad Favaloro, Solís 453, 1078 Buenos Aires, Argentina

E-mail: allen@tandar.cnea.gov.ar

Abstract. We study the effect of intense magnetic fields on the phase diagram of cold, strongly interacting matter within an extended version of the Nambu-Jona-Lasinio model that includes flavor mixing effects and vector interactions. Different values of the relevant model parameters in acceptable ranges are considered. Charge neutrality and beta equilibrium effects, which are specially relevant to the study of compact stars, are also taken into account. In this case the behavior of leptons is discussed.

1. Introduction

The influence of intense magnetic fields on the properties of strongly interacting matter has become an issue of increasing interest in recent years [1]. This is mostly motivated by the realization that in some relevant physical situations, like high energy non-central heavy ion collisions [2] and compact stellar objects called magnetars [3], very strong magnetic fields may be produced. Since in these systems extreme temperatures and/or densities may be found, it is interesting to investigate which modifications are induced by the presence of strong magnetic fields on the whole QCD phase diagram. Unfortunately, even in the absence of those fields, the present knowledge of such phase diagram is only schematic due to the well-known difficulty given by the so-called sign problem which affects lattice calculations at finite chemical potential [4]. Of course, the presence of strong magnetic fields makes the situation even more complex. Thus, most of our present knowledge of their effect comes from investigations performed in the framework of effective models [5]. In this contribution we present some results of a study of the phase diagram of cold quark matter subject to intense magnetic fields in the framework of a generalized Nambu-Jona-Lasinio (NJL) model. The NJL-type models are effective relativistic quark models for non-perturbative QCD, where gluon degrees of freedom are integrated out and interactions are modelled through point-like interactions. In its simplest version [6] it only includes scalar and pseudo-scalar interactions that describe chiral symmetry breaking effectively. As is well known, however, a more detailed description of the low-energy quark dynamics requires that other channels like flavor-mixing and vector-meson interactions are taken into account [7]. In fact, some aspects of the effect of those interactions on the magnetized quark matter have already been investigated [8, 9]. The purpose of this work is to extend those analyzes by performing a detailed study of the resulting cold matter phase diagrams, including their dependence on the parameters that regulate the strength of these interactions. Moreover, the behavior of cold magnetized quark matter under $\beta-$equilibrium with leptons will be considered, condition relevant for the physics of compact stars.

This contribution is organized as follows. In Sec. 2 we provide some details of the model and its parametrizations as well as the way to deal with an external constant magnetic field. In Sec. 3 we discuss our results for symmetric quark matter. The situation for neutral matter is analyzed in Sec. 4. Our conclusions are given in Sec. 5.

2. Formalism

We consider the most general SU(2) Lagrangian density which includes a scalar-pseudoscalar interaction, vector-axial vector and the t´Hooft determinant interaction [7]. In the presence of an external magnetic field and chemical potential it reads:

$$\mathcal{L} = \bar{\psi}\left(i\not{D} - m_c + \hat{\mu}\,\gamma^0\right)\psi + \mathcal{L}_{int} \tag{1}$$

where

$$\begin{aligned}\mathcal{L}_{int} &= G_1 \sum_{a=0}^{3}\left[\left(\bar{\psi}\tau_a\psi\right)^2 + \left(\bar{\psi}i\gamma_5\tau_a\psi\right)^2\right] + G_2 \sum_{a=0}^{3}\left[\left(\bar{\psi}\gamma_\mu\tau_a\psi\right)^2 + \left(\bar{\psi}\gamma_\mu\gamma_5\tau_a\psi\right)^2\right] \\ &\quad + G_3\left[\left(\bar{\psi}\gamma_\mu\psi\right)^2 + \left(\bar{\psi}\gamma_\mu\gamma_5\psi\right)^2\right] + G_4\left[\left(\bar{\psi}\gamma_\mu\psi\right)^2 - \left(\bar{\psi}\gamma_\mu\gamma_5\psi\right)^2\right] + 2G_D\left(d_+ + d_-\right)\end{aligned} \tag{2}$$

Here, G_i with $i = 1, 4$ and G_D are coupling constants, $\psi = (u, d)^T$ represents a two-flavor quark field with two flavors, $d_\pm = det\left[\bar{\psi}\left(1 \pm \gamma_5\right)\psi\right]$, $\hat{\mu} = \mathrm{diag}\left(\mu_u, \mu_d\right)$ the quark chemical potentials, m_c is the (current) mass matrix that we take to be the same for both flavors , $\tau_0 = I_{2\times 2}$, and τ_a, $0 < a \leq 3$ denote the Pauli matrices. The coupling of the quarks to the electromagnetic field $\mathcal{A}_\mu$ is implemented through the covariant derivative $D_\mu = \partial_\mu - i\hat{q}\mathcal{A}_\mu$ where $\hat{q}$ represents the quark electric charge matrix $\hat{q} = \mathrm{diag}\left(q_u, q_d\right)$ where $q_u/2 = -q_d = e/3$. We consider a static and constant magnetic field in the 3-direction, $\mathcal{A}_\mu = \delta_{\mu 2} x_1 B$. In the mean-field approximation the grand-canonical thermodynamical potential reads

$$\Omega^{\mathrm{MFA}} = -\sum_{f=u,d}\theta_f + \Omega_{pot} \tag{3}$$

where θ_f gives the contribution from the gas of quasi-particles of each flavor $f = u, d$ and can be written as the sum of 3 contributions [10]

$$\begin{aligned}\theta_f^{\mathrm{vac}} &= \frac{N_c}{8\pi^2}\left\{\Lambda\left(2\Lambda^2 + M_f^2\right)\sqrt{\Lambda^2 + M_f^2} - M_f^4 \ln\left[\frac{(\Lambda + \sqrt{\Lambda^2 + M_f^2})}{M_f}\right]\right\}, \\ \theta_f^{\mathrm{mag}} &= \frac{N_c}{2\pi^2}\left(|q_f|B\right)^2\left[\zeta^{(1,0)}(-1, x_f) - \frac{1}{2}(x_f^2 - x_f)\ln x_f + \frac{x_f^2}{4}\right], \\ \theta_f^{\mathrm{med}} &= \frac{N_c}{4\pi^2}|q_f|B\sum_{\nu=0}^{\nu_f^{max}}\alpha_\nu\left[\tilde{\mu}_f\sqrt{\tilde{\mu}_f^2 - s_f(\nu,B)^2}\right. \\ &\qquad \left. - s_f(\nu,B)^2 \ln\left(\frac{\tilde{\mu}_f + \sqrt{\tilde{\mu}_f^2 - s_f(\nu,B)^2}}{s_f(\nu,B)}\right)\right],\end{aligned} \tag{4}$$

where $M_f = m_c + \sigma_f$ and $\tilde{\mu}_f = \mu_f - \omega_f$, with σ_f and ω_f being the mean field values of the scalar and vector meson fields, respectively. Λ represents a non covariant ultraviolet cutoff and $\zeta^{(1,0)}(-1, x_f) = d\zeta(z, x_f)/dz|_{z=-1}$ where $\zeta(z, x_f)$ is the Riemann-Hurwitz zeta function. In addition, $s_f(\nu, B) = \sqrt{M_f^2 + 2|q_f|B\nu}$ while $x_f = M_f^2/(2|q_f|B)$. In θ_f^{med}, the sum is over the Landau levels (LLs), represented by ν, while $\alpha_\nu = 2 - \delta_{\nu 0}$ is a degeneracy factor and ν_f^{max} is the largest integer that satisfies $\nu_f^{max} \leq (\tilde{\mu}_f^2 - M_f^2)/(2|q_f|B)$.

The Ω_{pot} contribution reads

$$\Omega_{pot} = \frac{(1-c_s)(\sigma_u^2 + \sigma_d^2) - 2c_s\,\sigma_u\,\sigma_d}{8g_s(1-2c_s)} - \frac{(1-2c_v)(\omega_u^2 + \omega_d^2) + c_v\,\omega_u\,\omega_d}{8g_v(1-2c_v)} \tag{5}$$

where we have introduced a convenient parametrization of the coupling constants in terms of the quantities g_s, c_s, g_v and c_v given by

$$g_s = G_1 + G_D, \quad g_v = G_2 + G_3 + G_4, \quad c_s = \frac{G_D}{G_1 + G_D}, \quad c_v = \frac{G_3 + G_4}{2(G_2 + G_3 + G_4)}. \tag{6}$$

The relevant gap equations are given by

$$\frac{\partial\Omega^{\mathrm{MFA}}(\sigma_u, \sigma_d, \omega_u, \omega_d)}{\partial(\sigma_u, \sigma_d, \omega_u, \omega_d)} = 0\,. \tag{7}$$

In our calculations we will consider first the simpler case of symmetric matter where both quarks carry the same chemical potential μ. Afterwards, we will analyze the case of neutral matter in which leptons are also present and β-equilibrium and charge neutrality are imposed. In this case the chemical potential for each quark, μ_f, is a function of μ and the lepton chemical potentials which have to be selfconsistently determined.

We will consider two SU(2) NJL model parametrizations: set 1, with $m_c = 5.595$ MeV, $g_s\Lambda^2 = 2.212$, $\Lambda = 620.9$ MeV, corresponds to that leading to $M_0 = 340$ MeV; and set 2, with $m_c = 5.833$ MeV, $g_s\Lambda^2 = 2.44$, $\Lambda = 587.9$ MeV, to that leading to $M_0 = 400$ MeV. Here M_0 is the vacuum quark effective mass in the absence of external magnetic field.

The presence of the t'Hooft determinant interaction reflects the $U_A(1)-$anomaly of QCD. Its strength, and consequently the amount of flavor mixing induced by this term, is controlled by the parameter c_s. An estimate for its value can be obtained from the $\eta - \eta'$ mass splitting within the 3$-$flavor NJL model [11]. This leads to $c_s \simeq 0.2$ [12]. In any case, to obtain a full understanding of the effects of flavor mixing we will vary the value of c_s in a range going from 0, which corresponds to a situation in which the two flavors are completely decoupled, to 0.5, being this the case of maximum flavor mixing described in Ref. [13]. Regarding the vector coupling constant g_v, we recall that one can obtain naturally the vector axial-vector interaction if one starts from a QCD-inspired color current-current interaction and then performs a Fierz transform into color-singlet channels, and that in this case the relation between coupling strengths is $g_v = g_s/2$ [7]. Yet, the value of g_v cannot be accurately determined from experiments nor from lattice QCD simulations and this is why it has been taken as a free parameter in most works. In the present work we take $0 < g_v/g_s < 0.5$.

3. Results for symmetric matter

In this section we present the results for the case of symmetric matter, obtained by solving the set of coupled "gap equations", Eqs. (7), for different values of magnetic field and chemical potential. We will consider in Sec. 3.1 the effect of flavor mixing on quark behavior in the absence of vector interactions and turn to the study of the role of those interactions in Sec. 3.2.

3.1. Effect of the flavor mixing interactions

Neglecting vector interactions implies taking $G_2 = G_3 = G_4 = 0$ in Eq. (2), while $G_1 \neq 0$ and $G_D \neq 0$. Therefore, Eqs. (7) will become a set of two coupled equations that must be solved for the independent variables M_u and M_d. The parameter c_s acts as a coupling between both flavors and we study how the phase transitions are modified as we vary this parameter.

The case $c_s = 0.5$ corresponds to ordinary NJL described in detail in Ref [13], where both flavors have identical behavior. If $c_s < 0.5$, then both masses will be independent variables, and transitions for each flavor might occur simultaneously or not in different regions of the phase diagrams. It is common to all these cases that chiral symmetry is completely broken for chemical potentials well below M_0 and that restoration occurs for high enough chemical potentials, usually accompanied by a large drop in the dressed mass. The inclusion of a constant magnetic field modifies the quark dispersion relation, introducing LL's into its spectrum. As a consequence, chiral symmetry restoration occurs in several steps, each of which is a transition where quark population appears on previously unoccupied LL's, giving rise to a potentially complex phase diagram, whose precise form depends on the parameter set and magnetic field.

We will start by analyzing the phase transitions for the case with no flavor mixing, i.e. $c_s = 0$, previously studied in Ref. [8]. The phase diagrams for both parameter sets can be seen in the upper two panels of Fig. 1. In a sense, each flavor will have its own independent phase diagram because the gap equations are decoupled. Here, each line corresponds to a transition where LL population of a single quark flavor occurs (red lines for down quarks, blue lines for up quarks). Both flavors will coincide for $eB = 0$ where $SU(2)$ symmetry is recovered and behave differently as eB increases, due to their different electric charges. Both diagrams are related through a scale change $eB \rightarrow 2eB$.

When c_s becomes finite (second row onwards in Fig. 1), flavors become coupled and a complex pattern is created, where transitions move closer together in some regions and even merge together in others, that is, the LL population changes simultaneously for both flavors (these are represented by black lines). We can see that already for $c_s = 0.03$, the transitions occur together for low magnetic fields, and then separate for $B = 0.02$ GeV^2 in set 1 and 0.1 GeV^2 in set 2.

For a better understanding of the meaning of the transition lines, we present in Fig. 2 the dressed masses for both flavors for set 2, $c_s = 0.03$ and $eB = 0.11$ GeV^2. When the first discontinuity is encountered at $\mu = 370$ MeV, M_u jumps to half its value and its lowest LL (LLL) becomes populated. On the other hand, the down flavor remains in vacuum. Actually, its mass presents a small discontinuity caused by the weak coupling to the up quark which is not to be confused with a down transition, that effectively occurs at $\mu = 374$ MeV. When both flavors are in the vacuum phase, $M_u > M_d$, while in the populated phase it happens the other way around, which is consistent with the phenomena of catalysis/anticatalysis discussed in Ref. [13].

For $c_s = 0.2$, the transitions for both flavors occur together, so the qualitative behavior is very similar to the full flavor mixing case $c_s = 0.5$. In fact, the model tends to full mixing quite quickly and only for $c_s \lesssim 0.1$ are relevant mixture effects (or more precisely, the absence of it) actually seen. On the other hand, cross over transitions do not have this property and tend much more slowly to the $c_s = 0.5$ behavior.

3.2. Effect of the vector interactions

In this section we analyze the effect of vector interaction terms. Since for $c_s \gtrsim 0.1$ the phase diagrams do not present qualitative variations we fix $c_s = 0.2$, which is also in the range of realistic values suggested in Ref. [12]. Although our results correspond to $c_v = 0$, we have shown that only small quantitative differences occur when varying this parameter from 0 to 1.

In Fig. 3 we present a series of phase diagrams obtained for different values of the ratio g_v/g_s. For set 1 we observe that as g_v/g_s increases the two lower and main transitions separate and several new transitions appear in the low eB region of the diagram. For set 2 at some small

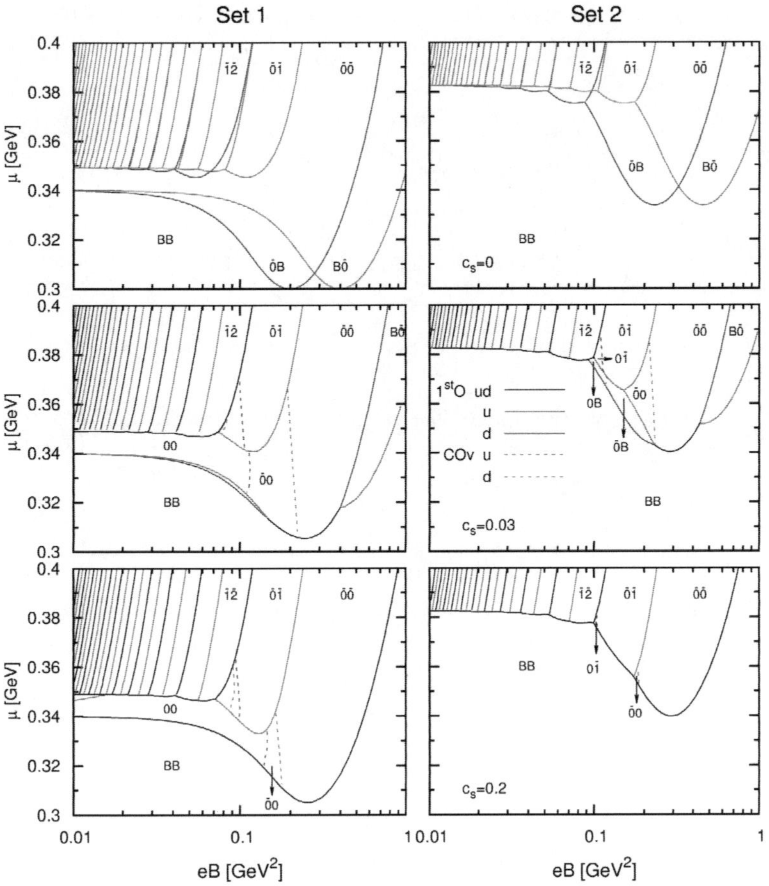

Figure 1. (Color online) Phase diagrams in the $eB - \mu$ plane for the case of symmetric matter with different values of flavor-mixing parameter c_s and no vector interactions. To simplify the figure we have introduced a compact notation to indicate the phases. The pair of integers mn corresponds to the C_{mn} phase and the pair $\bar{m}\bar{n}$ to the A_{mn} phase. The case in which one quark is in a C-type phase and the other in the A-type phase is indicated by putting a bar on top of the integer associated with the A-type phase.

$g_v/g_s \simeq 0.1$ the lower transition first splits into two, leaving a 00 phase in between which was not present before. For larger values of g_v/g_s the behavior is similar to that of set 1. In fact, increasing the vector interaction coupling has the same effect as going to a parameter set that reproduces a lower value of current mass M_0, described in Ref. [13]. Notice that the phase diagram for set 2 and $g_v/g_s = 0.3$ is in this sense very similar to set 1 and $g_v/g_s = 0$. The well-shaped behavior of the lowest transition is a manifestation of inverse magnetic catalysis (IMC), effect that is reduced with increasing g_v.

4. Results for neutral matter under $\beta-$equilibrium

We now turn our attention to stellar matter, fixing as in the previous section $c_s = 0.2$. If charge neutrality is imposed on our system, there will be a fixed relation between the densities of the up and down quarks, which will be necessarily different unless they are both zero. So, even though

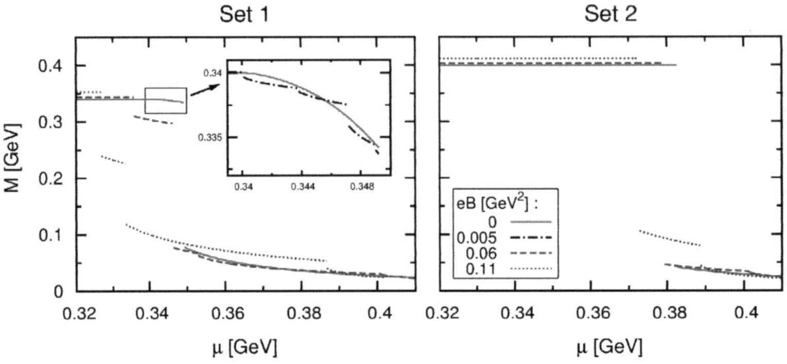

Figure 2. (Color online) Dressed mass as function of chemical potential for different values of magnetic field in the case of symmetric matter with maximum flavor mixing and no vector interactions, for both parameter sets.

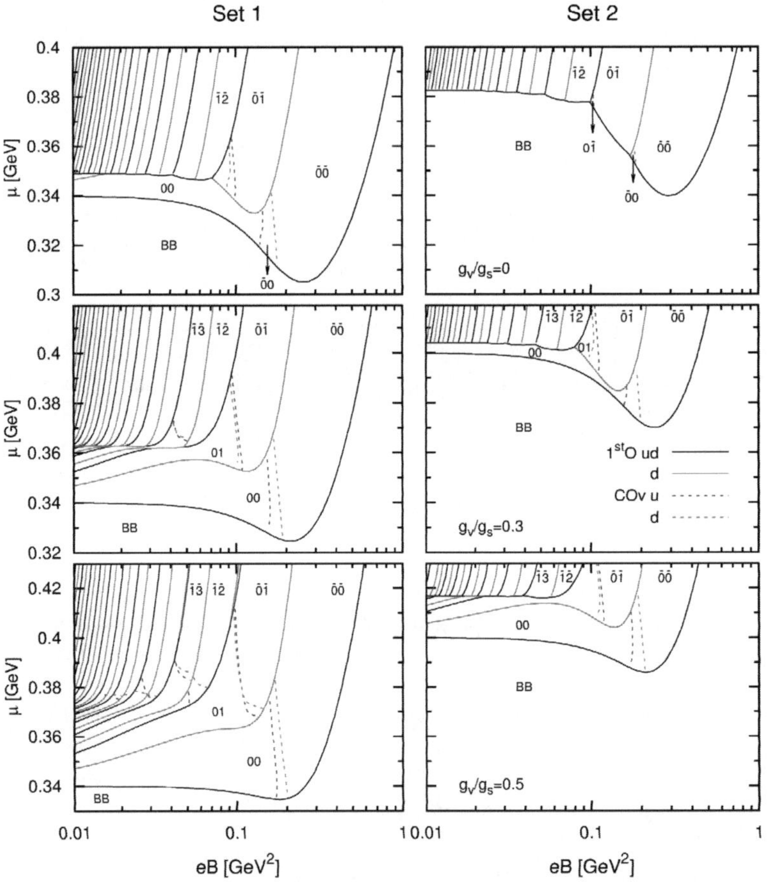

Figure 3. (Color online) Phase diagrams in the $eB-\mu$ plane for symmetric matter with $c_s = 0.2$ and different values of g_v/g_s. Phases are denoted as explained in Fig. 1.

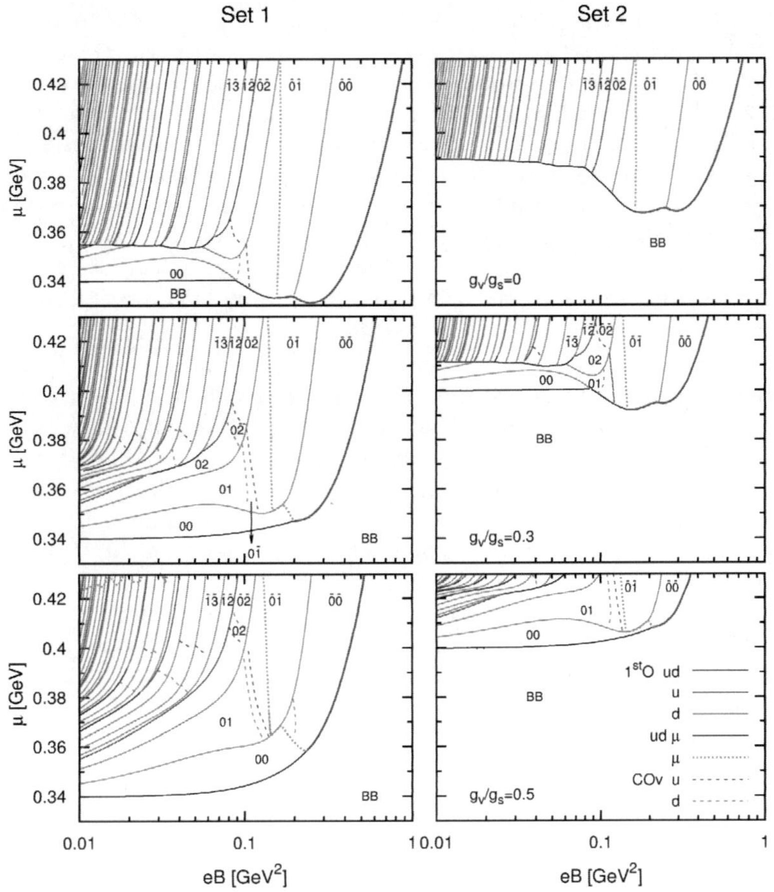

Figure 4. (Color online) Phase diagrams in the $eB - \mu$ plane for neutral matter under β-equilibrium with $c_s = 0.2$ and different values of g_v/g_s. Phases are denoted as explained in Fig. 1. The pink dotted line represents muon transition from vacuum to LLL.

$c_s = 0.2$ was close enough to the full mixing case according to what was established in previous sections, charge neutrality will cause the flavors to behave differently among themselves.

The phase diagrams in the $eB - \mu$ plane for both parameter sets and for increasing values of g_v are plotted in Fig. 4. The chemical potential μ on the horizontal axis is now the quark number chemical potential, which is related to flavor chemical potentials through $\mu_f = \mu - q_f\mu_e$. The introduction of charge neutrality has a few effects similar to those of vector interaction, in the sense that diagrams become similar to the ones corresponding to lower M_0 sets: main transitions separate for low eB and several transitions appear in the region between them. The magnetic anticatalysis effect is reduced also: The depth of the anticatalysis well, as defined in the previous section, is reduced from 35 MeV to 9 MeV in set 1 and from 43 MeV to 22 MeV in set 2. As expected, the introduction of vector interactions will further enhance these effects, so that anticatalysis has completely disappeared for $g_v = 0.5$.

The "Van Alphen - de Haas" (VA-dH) transitions acquire relatively independent behaviors when charge neutrality is introduced. In particular, for intermediate $eB\,(\simeq 0.1\ \text{GeV}^2)$ the transition between $k_u = 0$ and $k_u = 1$ is well separated from the down transition. For symmetric matter, the VA-dH transitions occurred to a very good level of approximation for a critical chemical potential $\mu_c = \sqrt{2kq_f B}$ (being this condition exact for the chiral case), so the relation between the charges produced one up transition every two down transitions. Although this relation is still satisfied for each flavor chemical potential, these are now different between themselves and related through $\mu_u - \mu_d + \mu_e = 0$. As a result of this, in Fig. 4 we roughly have one up transition every four down transitions. With respect to the leptons, it is seen that the electron LLL becomes populated simultaneously with quarks, while the muon transition presents a more complicated dependence with the magnetic field. Namely, the muon LLL requires a very high chemical potential to become populated at low eB, however, the critical μ drops suddenly at intermediate magnetic fields and joins the main quark transition together with electrons.

5. Final Comments

We have investigated how the phase diagram of cold strongly interacting matter is modified in the presence of intense magnetic fields in the context of NJL-type models which include flavor mixing and vector interaction. The whole range of possible flavor mixing values was swept through, going from the situation in which the two flavors are completely decoupled, to the one in which they are fully mixed. For low mixing values, a complex phase diagram is generated, but already for $c_s \simeq 0.1$, phase diagrams display a behavior that is qualitatively very similar to the full mixing case. Since SU(3) estimates suggest a value of $c_s \simeq 0.2$, it can be concluded that the realistic flavor mixing range is similar in behavior to the full mixing case. In what followed, vector interactions and the conditions of β-equilibrium and charge neutrality were introduced. The most notable effect observed is that they attenuate the IMC phenomenon, to the point that it completely disappears when both effects are jointly taken into account. It was also found that introducing vector interaction causes the two main transitions to separate and additional phases to appear between these two, effects that are similar to those occurring when changing to a parameter set that fits to a smaller dressed mass.

Acknowledgments

This work was partially supported by CONICET (Argentina) under grant PIP 00682, by ANPCyT (Argentina) under grant PICT-2011-0113 and by the Mineco (Spain) under contract FPA2013-47443-C2-1-P and CPAN(CSD- 00042) and by Generalitat Valenciana (Spain) under contract PrometeoII/2014/066.

References

[1] Kharzeev D E, Landsteiner K, Schmitt A and Yee H-U 2013 *Lect. Notes Phys.* **871**, 1.
[2] Kharzeev D E , McLerran L D and Warringa H J 2008 *Nucl. Phys.* **A803**, 227.
[3] Duncan R C and Thompson C 1992 Astrophys J. 392, L9.
[4] Karsch F, Laermann E, 2004 *Quark Gluon Plasma 3*, ed.by R.C.Hwa et al. (World Scientific Singapore), p.1.
[5] Fraga E S 2013 *Lect. Notes Phys.* **871**, 121; Gatto R and Ruggieri M, 2013 *Lect. Notes Phys.* **871**, 87; Ferrer E J and de la Incera V 2013 *Lect. Notes Phys.* **871**, 399.
[6] Nambu Y and Jona-Lasinio G 1961 *Phys. Rev.* **122**, 345; 1961 **124**, 246.
[7] Vogl U and Weise W 1991 *Prog. Part. Nucl. Phys.* **27**, 195; Klevansky S 1992 *Rev. Mod. Phys.* **64**, 649; Hatsuda T and Kunihiro T 1994 *Phys. Rep.* **247** 221.
[8] Boomsma J K and Boer D 2010 *Phys. Rev.* D **81**, 074005.
[9] Denke R Z and Pinto M B 2013 *Phys. Rev.* D **88**, 056008 ;
Menezes D P, Pinto M B, Castro L B, Costa P and Providência C 2014 *Phys. Rev.* C **89**, 055207.
[10] Menezes D P, Pinto M B, Avancini S S, Pérez Martínez A and Providência C 2009 *Phys. Rev.* **C 79** 035807; Menezes D P, Pinto M B, Avancini S S and Providência C 2009 *Phys. Rev.* **C 80** 065805; Avancini S S, Menezes D P and Providência C 2011 *Phys. Rev.* **C 83**, 065805.
[11] Kunihiro T 1989 *Phys. Lett.* B **219**, 363 1990 [*Erratum-ibid.* B **245**, 687].
[12] Frank M, Buballa M and Oertel M 2003 *Phys. Lett.* B **562**, 221.
[13] Allen P G and Scoccola N N Phys. Rev. D **88**, 094005 (2013).

Journal of Physics: Conference Series **706** (2016) 052027 doi:10.1088/1742-6596/706/5/052027

Updating an empirical analysis on the proton's central opacity and asymptotia

D A Fagundes[1,2], M J Menon[3] and P V R G Silva[3]

[1] Universidade Federal de Santa Catarina - Campus Blumenau Rua Pomerode, 710, Salto do Norte, 89065-300 Blumenau, SC, Brazil

[2] Instituto de Física Teórica - UNESP Rua Dr. Bento T. Ferraz, 271, Bloco II, 01140-070 São Paulo, SP, Brazil

[3] Instituto de Física Gleb Wataghin, Universidade Estadual de Campinas, 13083-859 Campinas, SP, Brazil

E-mail: daniel.fagundes@ufsc.br, menon@ifi.unicamp.br, precchia@ifi.unicamp.br

Abstract. We present an updated empirical analysis on the ratio of the elastic (integrated) to the total cross section in the c.m. energy interval from 5 GeV to 8 TeV. As in a previous work, we use a suitable analytical parametrization for that ratio (depending on only four free fit parameters) and investigate three asymptotic scenarios: either the black disk limit or scenarios above or below that limit. The dataset includes now the datum at 7 TeV, recently reported by the ATLAS Collaboration. Our analysis favors, once more, a scenario below the black disk, providing an asymptotic ratio consistent with the rational value 1/3, namely a gray disk limit. Upper bounds for the ratio of the diffractive (dissociative) to the inelastic cross section are also presented.

1. Introduction

The total cross section (σ_{tot}) and the elastic cross section (σ_{el}) are fundamental quantities in the study of elastic hadron scattering. Their ratio, as a function of the c.m. energy squared,

$$X(s) \equiv \frac{\sigma_{\text{el}}}{\sigma_{\text{tot}}}(s), \tag{1}$$

plays also an important role [1, 2]: it is related to the ratio of the inelastic cross section (σ_{inel}) to the total cross section via unitarity; it is connected to the central opacity in hadron-hadron collisions (impact parameter representation); it can be related to the ratio between σ_{tot} and the elastic slope B, which is an important quantity in studies on extensive air showers (cosmic-ray experiments).

However, $X(s)$ connects two quantities associated with soft strong scattering states, which, presently, are not directly and unequivocally described from the first principles of QCD. In this situation, a possible strategy of investigation is to obtain phenomenological and/or empirical descriptions of these quantities which may contribute with the search of suitable calculational schemes in the nonpertubative sector of QCD.

In this communication, we present an empirical approach for studying $X(s)$ in the case of proton-proton (pp) and antiproton-proton ($\bar{p}p$) scattering. This analysis follows previous studies by Fagundes and Menon [3, 4] and Fagundes, Menon and Silva [5]. Here, besides the

Journal of Physics: Conference Series **706** (2016) 052027 doi:10.1088/1742-6596/706/5/052027

new experimental data on σ_{tot} and σ_{el} obtained at the CERN-LHC at 7 and 8 TeV by TOTEM Collaboration [6, 7], we also include the recent ATLAS Collaboration results at 7 TeV [8]. By means of a suitable empirical parametrization and fits to the experimental data, we investigate different asymptotic scenarios for $X(s)$. Our analysis favors a scenario below the black-disk limit, consistent with the rational value 1/3. We discuss some preliminary tests on the selection of free parameters and infer that 1/4 may constitute an upper bound for the ratio between the soft diffractive cross section (single/double dissociation) and the inelastic cross section.

2. Empirical parametrization and asymptotic scenarios

Our dataset comprises measurements of σ_{tot} and σ_{el} from pp and $\bar{p}p$ scattering obtained in accelerator experiments, from $\sqrt{s_{\text{min}}} = 5$ GeV up to $\sqrt{s_{\text{max}}} = 8$ TeV [9]. The statistical and systematic uncertainties have been added in quadrature. The experimental points for X are shown in figure 1. In this figure we also display the point at 57 TeV obtained via unitarity from the estimates of σ_{tot} and σ_{inel} by the Pierre Auger Collaboration [10]; this point did not take part in the data reductions: it is shown only as an illustration of cosmic-ray information presently available.

2.1. Empirical parametrization

On the one hand, from the empirical behavior of X (figure 1), we see that this quantity decreases until $\sqrt{s} \sim 100$ GeV and then begins to rise as the energy increases and with a positive curvature. On the other hand it is expected that the ratio $X(s)$ becomes constant when $s \to \infty$, i.e. $X(s)$ saturates. The value of this constant can change from model to model (see discussion below). Since the saturation implies in a negative curvature, it is expected a change of curvature in $X(s)$ at some finite energy. Based on these facts and looking for an economical number of free parameters, we consider the following analytical parametrization:

$$X(s) = A \tanh\{g(s)\}, \tag{2}$$

with

$$g(s) = \alpha + \beta \ln^{\delta}(s/s_0) + \gamma \ln(s/s_0), \tag{3}$$

where A, α, β, γ and δ are dimensionless parameters, in principle, free fit parameters and s_0 is a fixed energy scale. As constructed, the parameter A corresponds to the asymptotic value of $X(s)$, $\lim_{s\to\infty} X(s) = A$ and the hyperbolic tangent takes account of the change of curvature. Therefore, in data reductions, by fixing A we can define an asymptotic scenario or by letting it free we can select a scenario.

2.2. Asymptotic scenarios

The standard scenario in the phenomenological context, typical of eikonal models (for example, [11]), concerns the black disk limit, which predicts $A = 0.5$ [12]. However, theoretical, phenomenological and empirical arguments, allow also to consider scenarios either above or below the black disk, as discussed in what follows.

2.2.1. Scenarios above the black disk From the s-channel unitarity,

$$\frac{\sigma_{\text{el}}}{\sigma_{\text{tot}}} + \frac{\sigma_{\text{inel}}}{\sigma_{\text{tot}}} = 1.$$

Therefore, a maximum value for the ratio X allowed in the formal context reads $A = 1$.

Moreover, two well known formal limits (the Froissart-Lukaszuk-Martin bound for total cross section [13] and a similar bound for the inelastic cross section [14]) predict

$$\sigma_{\mathrm{tot}} \leq \frac{\pi}{m_\pi^2} \ln^2 s \quad \text{and} \quad \sigma_{\mathrm{inel}} \leq \frac{\pi}{4m_\pi^2} \ln^2 s \quad (s \to \infty). \tag{4}$$

Now, assuming that *both limits saturate* when $s \to \infty$, it is possible that $\sigma_{\mathrm{inel}}/\sigma_{\mathrm{tot}} \to 1/4$, and therefore, from unitarity, we have also a "formal" possibility $A = 3/4 = 0.75$ ("formal" result). We recall that a scenario above the black disk is also indicated in the phenomenological context through the U-Matrix unitarity scheme [15].

2.2.2. Scenarios below the black disk Taking into account the TOTEM empirical fit to σ_{el} data [6, 7] and the highest-rank fit result to σ_{tot} data by the COMPETE Collaboration [16], the central values of the free parameters lead to the asymptotic ratio (see [17, 18] for details) $A = 0.436$ (TOTEM/COMPETE).

Moreover, in a recent study on the rise of σ_{tot}, we have considered a parametrization with Regge, Pomeron terms and a leading contribution in the form $\ln^\gamma s$, with γ either fixed to 2 or as a free fit parameter [17, 18, 19]. The parametrization was also extended to fit the σ_{el} data and in all cases investigated we have obtained asymptotic ratios X below the black disk and consistent with the rational limits between 1/3 and 2/5 [19]. As an extreme scenario we consider here the lowest value obtained in the analyses (denoted FMS), which, within the uncertainties from error propagation, reads $A = 0.30$ (minimum FMS).

In what follows we shall consider the asymptotic limits A = 1.0, 0.75, 0.50, 0.436, 0.30, as representative values for investigating the three different scenarios. As already commented, with parametrization (2-3), once A is fixed to each one of the aforementioned values we define an asymptotic scenario and by letting A as a free parameter we can select a scenario.

3. Fits and results

Different fits and procedures have been tested in order to select a best value for the energy scale s_0, which led to the energy cutoff as an optimum value. Therefore, we fix $s_0 = 25$ GeV^2.

In what follows, we first present the fits to only pp data, since, as we shall demonstrate, they allow us to fix (eliminate) one of the free parameters in (2-3). With this particular parametrization the fits are then extended to include $\bar{p}p$ data.

3.1. Preliminary fits to pp data

The dataset consists of 29 points (cutoff at 5 GeV), including those at 7 and 8 TeV by the TOTEM [6, 7] and ATLAS Collaborations [8]. We have fixed A to each value referred to above and letting free the parameters α, β, γ and δ. The results are shown in table 1, together with the statistical information on the fit results: reduced chi-squared ($\chi^2_{\mathrm{red}} = \chi^2/\nu$) and integrated probability $P_{\chi^2} = P(\chi^2, \nu)$, where ν is the number of degrees of freedom.

From table 1 we note that, with the exception of the extreme case $A = 0.3$, all the results for the parameter δ are compatible with the value 0.5 within the uncertainties. That suggested us the possibility to fix δ, reducing, therefore, the number of free parameters. With that assumption the parametrization (2-3) now reads

$$X(s) = A \tanh[\alpha + \beta\sqrt{\ln(s/s_0)} + \gamma \ln(s/s_0)], \tag{5}$$

with α, β, γ free parameters ($s_0 = 25$ GeV^2). We stress the small number of free parameters: 3 in the case of A fixed and 4 if A is let free. This parametrization was introduced in [5] and will be employed in what follows.

Table 1. Fit results to only pp data through parametrization (2-3) (δ as a free parameter, A fixed) and statistical information ($\nu = 25$).

A fixed	α	β	γ	δ	$\chi^2_{\rm red}$	P_{χ^2}
0.3	1.200 ± 0.046	-0.604 ±0.048	0.240 ± 0.018	0.6494 ± 0.0039	0.846	0.684
0.436	0.73 ± 0.16	-0.30 ± 0.11	0.0787 ± 0.050	0.52 ± 0.31	0.822	0.717
0.5	0.62 ± 0.13	-0.245 ±0.094	0.0625 ± 0.038	0.51 ± 0.31	0.820	0.720
0.75	0.392 ± 0.085	-0.149 ± 0.63	0.0362 ± 0.022	0.50 ± 0.31	0.817	0.724
1.0	0.289 ± 0.063	-0.108 ± 0.048	0.020 ± 0.016	0.49 ± 0.32	0.816	0.726

Table 2. Fit results to pp and $\bar{p}p$ data through parametrization (5) with A fixed.

A fixed	α	β	γ	$\chi^2_{\rm red}$	P_{χ^2}
0.3	1.317 ± 0.053	-0.658 ± 0.049	0.168 ± 0.011	0.771	0.846
0.436	0.728 ± 0.024	-0.301 ± 0.021	0.0756 ± 0.0046	0.776	0.841
0.5	0.612 ± 0.019	-0.245 ± 0.017	0.0614 ± 0.0037	0.782	0.832
0.75	0.386 ± 0.011	-0.146 ± 0.010	0.0365 ± 0.0021	0.795	0.816
1.0	0.2846 ± 0.0083	-0.1058 ± 0.0073	0.0264 ± 0.0015	0.794	0.813

3.2. Simultaneous fits to pp and $\bar{p}p$ data

Now our dataset consists of 42 points. As before, we first consider fits with the fixed values for A, using as initial values for the free parameters those obtained in the fits to pp data (table 1).

The fit results and statistical information are shown in table 2 and the corresponding curves together with the experimental data in figure 1. We see that, despite the distinct asymptotic behaviors, all the fit results show consistence with the experimental data presently available. In other words, with parametrization (5) and A fixed, we can not select an asymptotic scenario.

However, a striking result is obtained in the case of A as a free fit parameter. In fact, despite the strong nonlinearity of the fit, using as initial values of the parameters those displayed in table 2, with A free, all the data reductions converge to the same result, given by

$$A = 0.332 \pm 0.049, \quad \alpha = 1.09 \pm 0.28, \quad \beta = -0.51 \pm 0.18, \quad \gamma = 0.129 \pm 0.046 \qquad (6)$$
$$\chi^2/\nu = 0.783, \quad P(\chi^2, \nu) = 0.829$$

The comparison with the experimental data is shown in figure 2, where the uncertainty region has been evaluated through error propagation from the fit parameters (variances and covariances). Also shown in this figure is the prediction for the ratio $\sigma_{\rm inel}/\sigma_{\rm tot}$, obtained via unitarity, together with the experimental data. Numerical predictions for these ratios at some energies of interest are displayed in table 3 (second and third columns).

We conclude that our analysis favors a scenario below that of the black disk, with asymptotic value compatible with the rational limit

$$A = 0.332 \pm 0.049 \rightarrow 1/3.$$

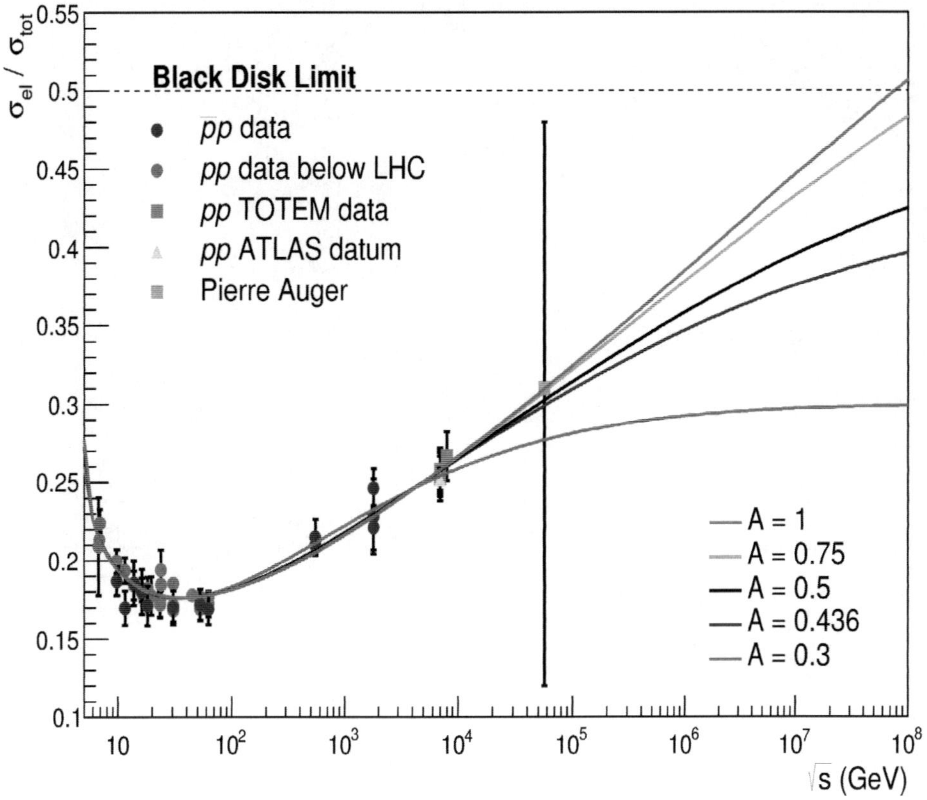

Figure 1. Fit results to pp and $\bar{p}p$ data on $X = \sigma_{\rm el}/\sigma_{\rm tot}$ through parametrization (5), with A fixed (parameters from table 2).

We note that this result is not in conflict with the Pumplin bound [20],

$$\frac{\sigma_{\rm el}}{\sigma_{\rm tot}} + \frac{\sigma_{\rm diff}}{\sigma_{\rm tot}} \leq 1/2,$$

where $\sigma_{\rm diff}$ stands for the soft diffractive cross section (single and double dissociation). If we consider the saturation of the Pumplin bound (equality in the above equation) yet at the LHC energy region [21], we can infer the predictions for the ratio $\sigma_{\rm diff}/\sigma_{\rm tot}$ displayed in table 3 (last column). Moreover, using the Pumplin bound and Unitarity, we can also infer an upper bound for ratio $\sigma_{\rm diff}/\sigma_{\rm inel}$, namely

$$R(s) \equiv \frac{\sigma_{\rm diff}}{\sigma_{\rm inel}} \leq \frac{1/2 - X(s)}{1 - X(s)}.$$

In this case, the maximum asymptotic value for R is also compatible with a rational number:

$$R_{\rm max} = 0.251 \pm 0.055 \;\; \to \;\; 1/4 \quad \text{as} \quad s \to \infty.$$

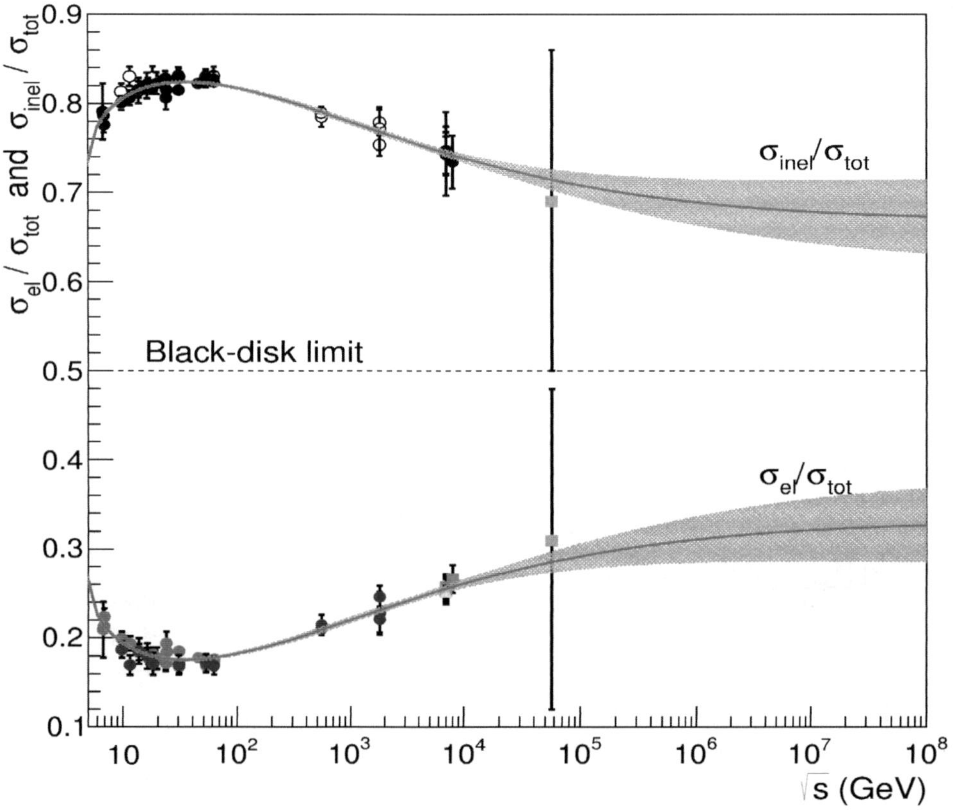

Figure 2. Fit result to $X = \sigma_{\rm el}/\sigma_{\rm tot}$ through parametrization (5) and A free parameter (parameters (6)) and *prediction* for the ratio $\sigma_{\rm inel}/\sigma_{\rm tot}$ (via unitarity), together with their corresponding uncertainty regions.

Table 3. Predictions for different ratios at the LHC energy region.

$\sqrt{s}$ (TeV)	$\sigma_{\rm el}/\sigma_{\rm tot}$	$\sigma_{\rm inel}/\sigma_{\rm tot}$	$\sigma_{\rm diff}/\sigma_{\rm tot}$
2.76	0.2394 ± 0.0036	0.7606 ± 0.0036	0.2606 ± 0.0036
7	0.2556 ± 0.0043	0.7444 ± 0.0043	0.2444 ± 0.0043
8	0.2578 ± 0.0046	0.7422 ± 0.0046	0.2422 ± 0.0046
13	0.2655 ± 0.0058	0.7345 ± 0.0058	0.2345 ± 0.0058
14	0.2666 ± 0.0061	0.7334 ± 0.0061	0.2334 ± 0.0061

4. Conclusions and final remarks

We have presented an empirical analysis of the experimental data presently available on the ratio $X = \sigma_{\rm el}/\sigma_{\rm tot}$ above 5 GeV, including all the recent data at 7 and 8 TeV obtained at the LHC by the TOTEM and ATLAS Collaborations. Our model-independent parametrization, characterized by a small number of free parameters (3 or 4), allows consistent descriptions of all the experimental data analyzed (figures 1 and 2).

By letting free the parameter A, the resulting asymptotic value of X is compatible with a scenario *below* the black-disk limit, namely $0.332 \pm 0.049 \to 1/3$. This value is in agreement with our previous results based in distinct analyses [17, 18, 19]; it is also in plenty agreement with the recent phenomenological approach by Kohara, Ferreira and Kodama [22] which predicts $X \to 1/3$ for $s \to \infty$. Taking into account the Pumplin bound, it was possible to infer an upper bound for the ratio $\sigma_{\rm diff}/\sigma_{\rm inel}$, consistent with the rational value 1/4.

Our next steps concern a review on all the results presented here and in [3, 4, 5], extensions of the analysis treating other choices for $g(s)$ and detailed investigation on the change of curvature in $X(s)$ at the LHC energy region [23].

Acknowledgments

Research supported by FAPESP-CAPES, Contract 2014/00337-8 (DAF) and FAPESP, Contract 2013/27060-3 (PVRGS).

References

[1] Block M M and Cahn R N (1985) *Rev. Mod. Phys.* **57** 563
[2] Dremin I M (2013) *Phys. Usp.* **56** 3 (*Preprint* arXiv: 1206.5474 [hep-ph])
[3] Fagundes D A and Menon M J 2012 *Nucl. Phys.* A **880** 1
[4] Fagundes D A and Menon M J 2013 *AIP Conf. Proc.* **1520** 297 (*Preprint* arXiv:1208.0510 [hep-ph])
[5] Fagundes D A, Menon M J and Silva P V R G 2015 *AIP Conf. Proc.* **1654** 050005 (*Preprint* arxiv:1410.4423 [hep-ph])
[6] Antchev G *et al* (The TOTEM Collaboration) 2013 *Europhys. Lett.* **101** 21004
[7] Antchev G *et al* (The TOTEM Collaboration) 2013 *Phys. Rev. Lett.* **111** 012001
[8] Aad G *et al* (The ATLAS Collaboration) 2014 *Nucl. Phys.* B **889** 486
[9] Olive K A *et al* (Particle Data Group) 2014 *Chin. Phys.* C **38** 090001
[10] Abreu P *et al* (The Pierre Auger Collaboration) 2012 *Phys. Rev. Lett.* **109** 062002
[11] Bourrely C, Soffer J and Wu T T 2011 *Eur. Phys. J.* C **71** 1601
[12] Block M M and Halzen F 2012 *Phys. Rev.* D **86** 051504
[13] Lukaszuk L and Martin A 1967 *Nuovo Cimento* A **52** 122
[14] Martin A 2009 *Phys. Rev.* D **80** 065013
[15] Troshin S M and Tyurin N E 2007 *Int. J. Mod. Phys* A **22** 4437; 1993 *Phys. Lett.* B **316** 175
[16] Cudell J R *et al* (COMPETE Collaboration) 2002 *Phys. Rev. Lett* **89** 201801
[17] Fagundes D A, Menon M J and Silva P V R G 2013 *J. Phys.* G **40** 065005
[18] Menon M J and Silva P V R G 2013 *Int. J. Mod. Phys.* A **28** 1350099
[19] Menon M J and Silva P V R G 2013 *J. Phys.* G **40** 125001
[20] Pumplin J 1973 *Phys Rev.* D **8** 2899
[21] Lipari P and Lusignoli M 2013 *Eur. Phys. J.* C **73** 11
[22] Kohara A K, Ferreira E and Kodama T 2014 *Eur. Phys. J.* C **74** 3175
[23] Fagundes D A, Menon M J and Silva P V R G, Exploring central opacity and asymptotic scenarios in elastic hadron scattering, in preparation.

Multiple photon production in double parton scattering at the LHC

[1]**R. Palota da Silva**[†], [2]**C. Brenner Mariotto**[†] **and** [3]**V. P. Goncalves**[⋆]

[†]Instituto de Matemática, Estatística e Física, Universidade Federal do Rio Grande, Av. Itália, km 8, Campus Carreiros, CEP 96203-900, Rio Grande, RS, Brasil
[⋆]Instituto de Física e Matemática, Universidade Federal de Pelotas, Caixa Postal 354, CEP 96010-900, Pelotas, RS, Brasil

E-mail: [1]`rafaelpalota@furg.br`, [2]`cristianomariotto@furg.br` and [3]`barros@ufpel.edu.br`

Abstract. The high density of gluons in the initial state of hadronic collisions at LHC implies that the probability of multiple parton interactions within one proton-proton collision increases. In particular, the probability of having two or more hard interactions in a collision is not significantly suppressed with respect to the single interaction probability. In this contribution we study for the first time the production of prompt photons in double parton scattering processes. In particular, we estimate the rapidity distribution for the double Compton process, which leads to two photons plus two jets in the final state. Besides, we study the production of three and four photons in the final state, which are backgrounds to physics beyond the Standard Model.

1. Introduction

Multiple photon production in hadronic collisions is a very interesting subject once it provides opportunities to study the predictions of the Standard Model, besides being background to Beyond Standard Model Physics. Recent studies predict the cross sections for 2γ, 3γ and 4γ production in single parton scattering (SPS) processes [1, 2, 3]. Here we estimate the production of the same final states considering double parton scattering (DPS) processes. In particular, the final states are $2\gamma + 2\,jets$, $3\gamma + 1\,jet$ and 4γ.

Multiple parton interactions in one proton-proton colision were already predicted in previous studies [4, 5], but with the energies available in colliders at those times, its contributions were suppressed with respect to the single scattering. However, due to the high energy regime reached at the LHC, the high density of partons in the initial state of hadronic collisions leads to the increase of the multiple parton interaction probability and, for some processes, double parton scattering, which can be described perturbatively, is also significant. This new perspective has motivated the rapid development of the theory of the DPS with several estimates of the cross section for different processes in the recent years [6, 7, 8, 9].

In a double parton scattering one can have, within one hadron-hadron collision, two partons from each colliding hadron interacting via two parton-parton interactions that happen almost independently. Figure 1 shows a general SPS process where one parton from each hadron interacts to form a final state, while in figure (2) one has a double partonic interaction generating a final state. Those final states do not have to be equivalent, but might be competitive. The main goal of the present work is to estimate the rapidity distribution for the production of prompt photons in DPS processes.

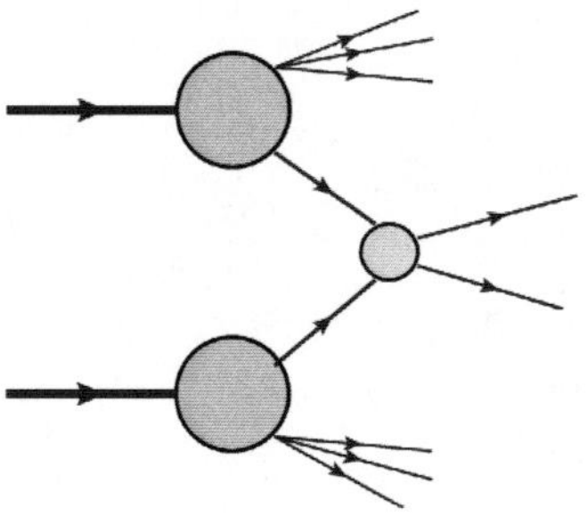

Figure 1. SPS mechanism with one parton from each interacting hadron.

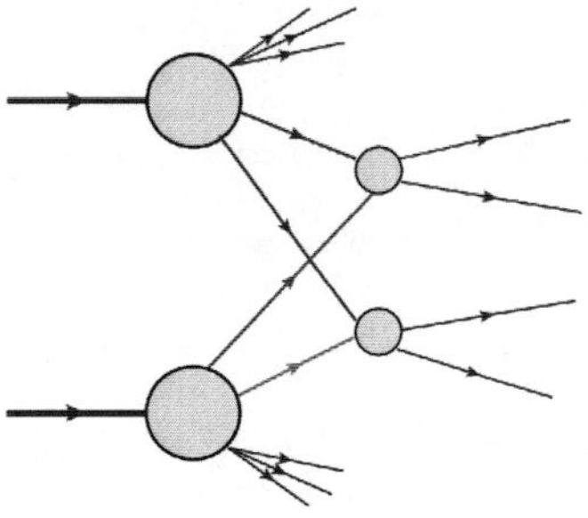

Figure 2. DPS mechanism with two partons from each hadron interacting independently.

2. Theoretical framework

In the case where the two hard scatterings are independent and the same goes for the two incoming partons in each hadron, the DPS cross section can be written as

$$\sigma^{DPS} = \frac{\mathcal{C}}{\sigma_{eff}} \sigma_1^{SPS} \cdot \sigma_2^{SPS} \tag{1}$$

where $\mathcal{C}$ is a combinatorial factor which accounts for distinguishable ($\mathcal{C} = \frac{1}{2}$) and indistinguishable ($\mathcal{C} = 1$) final states, and σ_{eff} is a normalization cross section representing the effective transverse overlap of partonic interactions that produce the DPS process. Its value is estimated through a fit to experimental data [10, 11]. Also, σ_1^{SPS} and σ_2^{SPS} characterize the cross sections for the two hard scattering processes, and can be written (in LO) as usual

$$\sigma^{SPS} = \sum_{a,b,c,d} \int dx_a dx_b f_a^A(x_a, Q^2) f_b^B(x_b, Q^2) \hat{\sigma}(ab \to cd) \tag{2}$$

where $f_i^k(x_i, Q^2)$ are the parton distribution functions (here we use the CTEQ6L parametrization [12]) and $\hat{\sigma}$ are the partonic cross sections for the contributing subprocesses.

Considering the production of at least one photon in each partonic subprocess, and using the same assumptions as in eq. (1), one can write a similar formula for the differential distributions,

$$\frac{d\sigma^{DPS}}{d\eta_1 d\eta_2 dp_{1T}^2 dp_{2T}^2} = \frac{\mathcal{C}}{\sigma_{eff}} \frac{d\sigma_1^{SPS}}{d\eta_1 dp_{1T}^2} \cdot \frac{d\sigma_2^{SPS}}{d\eta_2 dp_{2T}^2}. \tag{3}$$

where η_i and p_{iT} are the rapidity and transverse momentum of one of the photons produced in each SPS process, for which one has

$$\frac{d\sigma^{SPS}}{d\eta dp_T^2} = \sum_{a,b,d} \int_{x_{a\,min}}^{1} dx_a f_a^A(x_a, Q^2) f_b^B(x_b, Q^2) \frac{x_a x_b}{2x_a - x_T e^{\eta}} \frac{d\hat{\sigma}}{dt}(ab \to \gamma d) \tag{4}$$

where $x_T = \frac{2p_T}{\sqrt{s}}$, $x_b = \frac{x_a x_T e^{-\eta}}{2x_a - x_T e^{\eta}}$, $x_{a\,min} = \frac{x_T e^{\eta}}{2 - x_T e^{-\eta}}$. Also, $\frac{d\hat{\sigma}}{dt}$ are the leading-order partonic cross sections for the relevant subprocesses [13]. Integrating equation (3) over the transverse momenta p_{1T} and p_{2T}, one obtains the rapidity distributions

$$\frac{d\sigma^{DPS}}{d\eta_1 d\eta_2} = \frac{\mathcal{C}}{\sigma_{eff}} \frac{d\sigma^{SPS}}{d\eta_1} \cdot \frac{d\sigma^{SPS}}{d\eta_2}. \tag{5}$$

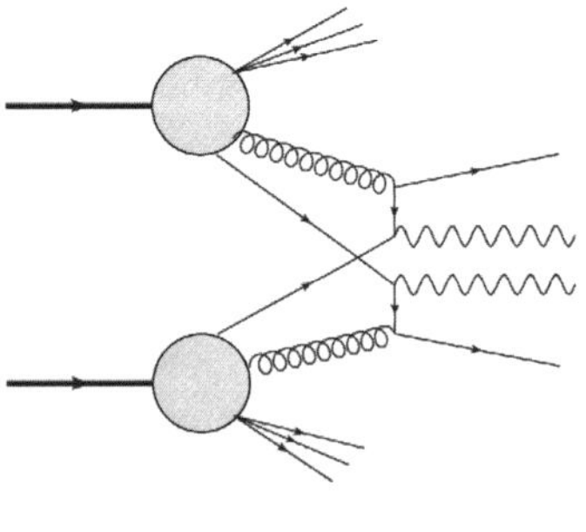

Figure 3. Double Compton scattering for the DPS production of $2\,\gamma + 2\,$jets.

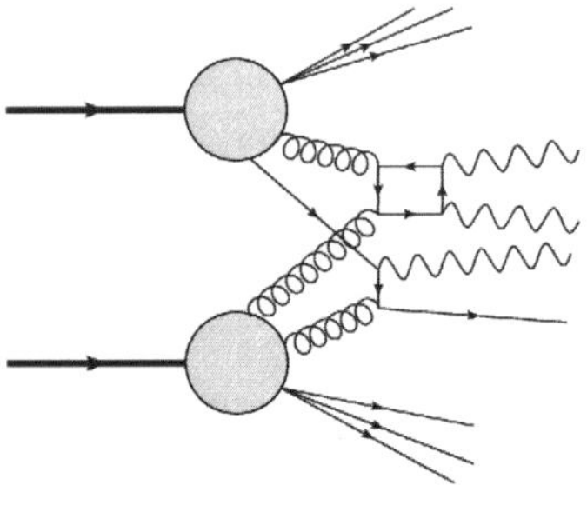

Figure 4. One of the diagrams for DPS production of $3\,\gamma + 1\,$jet.

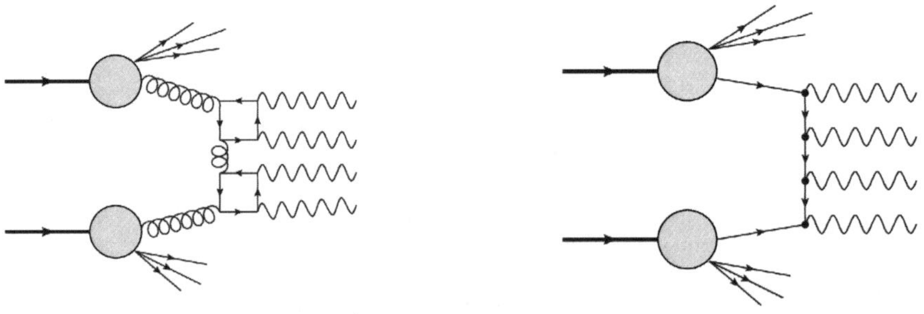

Figure 5. Some typical diagrams for SPS production of $4\,\gamma$.

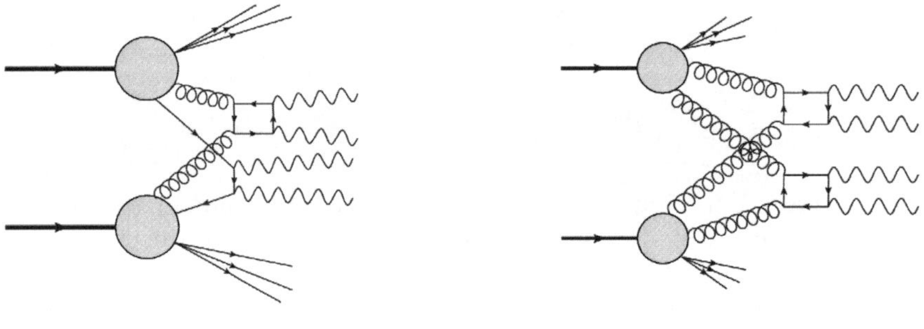

Figure 6. Some typical diagrams for DPS production of $4\,\gamma$.

In what follows we estimate the $2\,\gamma + 2\,jets$ production considering the DPS process, taking into account that in the two SPS processes we have two Compton ($gq \to \gamma q$) scatterings, which is the dominant mechanism [14]. The corresponding diagram is shown in figure 3. In the case of $3\,\gamma + 1\,jet$ production we assume that $\sigma^{DPS}(pp \to 3\gamma X) \propto \sigma^{SPS}(pp \to \gamma X)\sigma^{SPS}(pp \to \gamma\gamma X)$, and one typical diagram is shown in figure 4. Similarly, for $4\,\gamma$ production we assume that $\sigma^{DPS}(pp \to 4\gamma X) \propto \sigma^{SPS}(pp \to 2\gamma X)\sigma^{SPS}(pp \to 2\gamma X)$. For the sake of comparison, we show in figure 5 some typical SPS processes producing $4\,\gamma$, and in figure 6, some typical DPS processes leading to the same final states. In this work we use the aproximation $\sigma_{eff} = 15\,$mb as in Ref. [6]. For the combinatorial factor we use $\mathcal{C} = \frac{1}{2}$ for double Compton and four-photon production, and $\mathcal{C} = 1$ for three-photon production.

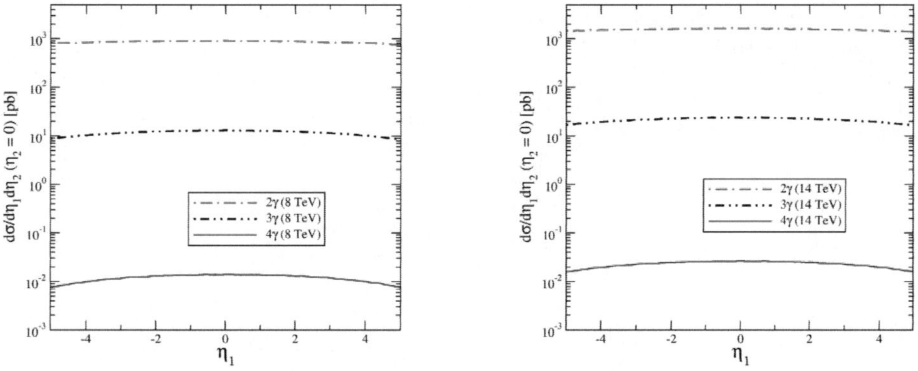

Figure 7. Rapidity distributions for the LHC energies of $\sqrt{s} = 8\,\text{TeV}$ and $14\,\text{TeV}$.

3. Results

Our results for the multiple photon production at LHC energies using the DPS formalism are presented in figure 7, where we show the rapidity distributions for $2\gamma + 2\,jets$ (top lines), $3\gamma + 1\,jet$ (middle lines) and 4γ production (bottom lines) at $\sqrt{s} = 8\,\text{TeV}$ and $\sqrt{s} = 14\,\text{TeV}$. The final state $2\gamma + 2\,jets$ have the larger cross sections, which are somewhat flat in the rapidity region considered and peaks around 10^3 pb. For $3\gamma + 1\,jet$ and for 4γ, the maximum values of the cross sections are above 10^1 pb and 10^{-2} pb, respectivelly. These results indicates that measurements of DPS mechanism for multiple photon production are feasible to be measured at the LHC. In order to do so, it is important, however, that one finds the appropriate regions of phase space where the DPS processes could be distinguished from the same final states coming from SPS processes.

4. Summary

In this contribution we obtain, for the first time, results for multiple photon production in DPS processes. We show predictions for rapidity distributions for $2\gamma + 2\,jets$, $3\gamma + 1\,jet$ and 4γ final states at the LHC energies of $\sqrt{s} = 8\,\text{TeV}$ and $14\,\text{TeV}$. Our results show that all this processes have relevant cross sections, which indicates that they are feasible to be measured once one finds kinematic regions where they can be distinguished from the same final states coming from SPS processes. This is particularly important in experimental analyses of the multiple photon production as an observable to test the Standard Model and probe possible scenarios for Beyond the Standard Model Physics.

Acknowledgments

This work is partially financed by CAPES, CNPq, FAPERGS and by the Programa de Pós-Graduação em Física FURG, Brazil.

References

[1] Gehrmann T, Greiner N and Heinrich G 2013 *Phys. Rev. Lett* **111** 222002
[2] Campbell J M and Williams C 2014 *Phys. Rev.* D **89** 113001
[3] Dennen T and Williams C 2015 *Phys. Rev.* D **91** 054012
[4] Landshoff P V and Polinghorne J C 1978 *Phys. Rev.* D **18** 3344
[5] Goebel C, Scott D M and Halzen F 1980 *Phys. Rev.* D **22** 2789
[6] Luszczak M, Maciula R and Szczurek A 2012 *Phys. Rev.* D **85** 094034
[7] Maciula R and Szczurek A 2013 *Phys. Rev.* D **87** 074039
[8] Cazaroto E R, Goncalves V P and Navarra F S 2013 *Phys. Rev.* D **88** 034005
[9] Maciula R and Szczurek A 2014 *Phys. Rev.* D **90** 014022

[10] Abe F *et al.* (CDF Collaboration) 1997 *Phys. Rev.* D **56** 3811
[11] Abazov V M 2010 *Phys. Rev.* D **81** 052012
[12] Pumplin J *et al.* 2002 *JHEP* **07** 012
[13] Owens J F 1987 *Rev. Mod. Phys.* **59** 465
[14] Brenner Mariotto C and Goncalves V P 2013 *Phys. Rev.* D **88** 074023

Impact of a magnetic field on the thermodynamics of magnetized quark matter

R.L.S. Farias[1,2], V.S. Timóteo[3], S. Avancini[4], M.B. Pinto[4], G. Krein[5]

[1] Departamento de Física, Universidade Federal de Santa Maria, 97105-900, Santa Maria, RS, Brazil
[2] Physics Department, Kent State University, Kent, OH 44242, United States
[3] Grupo de Óptica e Modelagem Numérica - GOMNI, Faculdade de Tecnologia - FT Universidade Estadual de Campinas - UNICAMP, 13484-332, Limeira, SP , Brasil
[4] Departamento de Física, Universidade Federal de Santa Catarina, 88040-900 Florianópolis, Santa Catarina, Brazil
[5] Instituto de Física Teórica, Universidade Estadual Paulista, Rua Dr. Bento Teobaldo Ferraz, 271 - Bloco II, 01140-070 São Paulo, SP, Brazil

E-mail: rfarias@kent.edu

Abstract. We investigate the effect of a magnetic field on the thermodynamics of magnetized quark matter at finite temperature. By using the Nambu Jona–Lasino (NJL) model, we show that the lattice results for the quark consensate can be reproduced when the coupling constant G of the model decreases with the magnetic field and the temperature. Our results show that thermodynamic quantities and quark condensates are very sensitive to the dependence of G with the temperature, even in the absence of a magnetic field.

1. Introduction

Motivated by the fact that strong magnetic fields may be produced in noncentral heavy-ion collisions [1, 2], investigations of the effects of a magnetic field on the phase diagram of strongly interacting matter became a subject of great interest in recent years.

At zero temperature, the great majority of effective models for Quantum Chromodynamics (QCD) are in agreement with respect to the occurrence of Magnetic Catalysis (MC) effect, in that the quark condensate increases with the magnetic field B. On the other hand, finite temperature lattice QCD simulations show that the pseudo-critical temperature T_{pc} for chiral symmetry restoration decreases as the strength of the magnetic field increases, a phenomenon known as Inverse Magnetic Catalysis (IMC) [3]. The great majority of effective models fail to explain this phenomenon — see Ref. [4] for a recent review.

A large body of work has been devoted to the search for possible mechanisms responsible for the IMC. In a recent work by some of us [5], we proposed that the IMC found by lattice simulations can be explained within the NJL model if the coupling constant G decreases with both the magnetic field and the temperature. A similar mechanism was proposed within $SU(3)$ PNJL models, but with G depending only on the magnetic field [6]. Another possible explanation was provided within the framework of holography [7]. In the present work we refine the running of the coupling $G(eB, T)$ quantitatively and reinvestigate the effects of the magnetic field on thermodynamic quantities of the magnetized quark matter within the $SU(2)$ NJL model.

2. The magnetized NJL model

We use the SU(2) version of the NJL model [8], defined by the Lagrangian density

$$\mathcal{L}_{\rm NJL} = \bar{\psi}\left(i\partial\!\!\!/ - m\right)\psi + G\left[(\bar{\psi}\psi)^2 - (\bar{\psi}\gamma_5\vec{\tau}\psi)^2\right] . \tag{1}$$

The field ψ represents a flavor iso-doublet of u and d quark flavors and N_c-plet of quark fields, $\vec{\tau}$ are isospin Pauli matrices. Since the model is non-renormalizable, we need to specify some regularization scheme; here we use a sharp momentum cutoff Λ as an ultraviolet regulator. Therefore, Λ, G and the quark masses $m_f, f = u, d$ are free parameters which are fixed [9, 10] by fitting the values of the pion mass m_π, pion decay constant f_π and quark condensate $\langle\bar{\psi}_f\psi_f\rangle$.

Thermodynamic quantities are computed from the grand thermodynamical potential in the mean field approximation — see e.g. Ref. [10] for details:

$$\Omega^{\rm NJL} = \frac{(M-m)^2}{4G} + \frac{i}{2}\int\frac{d^4p}{(2\pi)^4}\ln[-p^2+M^2] . \tag{2}$$

In order of include the effects of T and B simultaneously in the thermodynamic potential, it is usual to make use of the following transformations [11]:

$$p_0 \rightarrow i(\omega_n - i\mu) , \tag{3}$$

$$p^2 \rightarrow p_z^2 + (2k+1-s)|q_f|B , \tag{4}$$

$$\int_{-\infty}^{+\infty}\frac{d^4p}{(2\pi)^4} \rightarrow i\frac{T|q_f|B}{2\pi}\sum_{n=-\infty}^{\infty}\sum_{k=0}^{\infty}\int_{-\infty}^{+\infty}\frac{dp_z}{2\pi} , \tag{5}$$

with $s = \pm 1$, $k = 0, 1, 2, \cdots$. The Matsubara frequencies for fermions are given by $\omega_n = (2n+1)\pi T$, with $n = 0, \pm 1, \pm 2, \cdots$ and k represents the Landau levels. After these replacements, the grand potential including the effects of the magnetic field can be written as

$$\Omega^{\rm NJL} = \frac{(M-m)^2}{4G} + \Omega^{\rm NJL}_{\rm vac} + \Omega^{\rm NJL}_{\rm mag} + \Omega^{\rm NJL}_{\rm med} , \tag{6}$$

where

$$\Omega^{\rm NJL}_{\rm vac} = -2N_cN_f\int\frac{d^3p}{(2\pi)^3}\sqrt{p^2+M^2} , \tag{7}$$

and

$$\Omega^{\rm NJL}_{\rm mag} = -\sum_{f=u}^{d}\frac{N_c(|q_f|B)^2}{2\pi^2}\left\{\zeta'[-1,x_f] - \frac{1}{2}[x_f^2 - x_f]\ln(x_f) + \frac{x_f^2}{4}\right\} , \tag{8}$$

with $x_f = M^2/(2|q_f|B)$, $\zeta'[-1, x_f] = d\zeta(z, x_f)/dz|_{z=-1}$ and $\zeta(z, x_f)$ representing the Hurwitz zeta function. The last term in Eq. (6), which contains the effects of temperature and magnetic field, is given by

$$\begin{aligned}\Omega^{\rm NJL}_{\rm med} &= -\frac{N_c}{2\pi}\sum_{f=u}^{d}\sum_{k=0}^{\infty}\alpha_k(|q_f|B)\int_{-\infty}^{+\infty}\frac{dp_z}{2\pi} \\ &\times \left\{T\ln[1+e^{-[E_{p,k}(B)+\mu]/T}] + T\ln[1+e^{-[E_{p,k}(B)-\mu]/T}]\right\} ,\end{aligned} \tag{9}$$

where $E_{p,k}(B) = \sqrt{p_z^2 + 2k|q_f|B + M^2}$ and $\alpha_k = 2 - \delta_{0k}$. More details in the derivation of these expressions can be found in Refs. [11, 12]. The effective quark mass M_f is obtained in a self-consistent way by solving the gap equation

$$M_f = m_f - 2G\sum_f\langle\bar{\psi}_f\psi_f\rangle , \tag{10}$$

where $\langle\bar{\psi}_f\psi_f\rangle$ represents the quark condensate of flavour f. The explicit expression for the quark condensate at finite T and B is given by

$$\begin{aligned}\langle\bar{\psi}_f\psi_f\rangle &= -\frac{N_c M}{2\pi^2}\left\{\Lambda\sqrt{\Lambda^2+M^2}-\frac{M^2}{2}\ln\left[\frac{(\Lambda+\sqrt{\Lambda^2+M^2})^2}{M^2}\right]\right\} \\ &\quad -\frac{N_c M}{2\pi^2}|q_f|B\left\{\ln[\Gamma(x_f)]-\frac{1}{2}\ln(2\pi)+x_f-\frac{1}{2}(2x_f-1)\ln(x_f)\right\} \\ &\quad +\frac{N_c M}{2\pi^2}\sum_{k=0}^{\infty}\alpha_k|q_f|B\int_{-\infty}^{\infty}\frac{dp_z}{E_{p,k}(B)}\left\{\frac{1}{e^{[E_{p,k}(B)]/T}+1}\right\}. \qquad (11)\end{aligned}$$

It is important to note that we are working with Gaussian natural units where $1\,\mathrm{GeV}^2 \simeq 1.44\times10^{19}$ G and $|q_f|$ is the absolute value of the quark electric charge. At finite magnetic fields, the quark condensate for the flavors u and d are different due to their different electric charges ($|q_u| = 2e/3$, $|q_d| = e/3$, $e = 1/\sqrt{137}$).

An interesting feature that occurs in the two flavour case is that, in principle, one should have two coupled gap equations for the two distinct flavors, but in this case the different condensates contribute to M_u and M_d in a symmetric way and, since we are using $m_u = m_d = m$, one can write $M_u = M_d = M$ (more details see Ref. [11]).

3. Fitting NJL coupling constant with the lattice results

In this section we describe the fitting procedure used to obtain the thermomagnetic dependence of the NJL coupling constant. We fit the NJL coupling $G(eB,T)$ so that the NJL model reproduces the average of the quark condensates, $(\Sigma_u+\Sigma_d)/2$, obtained by the lattice simulation of Ref. [3]. The lattice results of Ref. [3] are for zero temperature and high temperatures, with no data points between $T = 0$ and $T = 113$ MeV. Given the fact that the discrepancies between lattice results and effective models appear in the region where chiral symmetry is partially restored (crossover), we fit the NJL coupling constant with the lattice data in this regime and then extrapolate our results to zero temperature.

We can get a good fit to the lattice data for the average of the quark condensates, $(\Sigma_u+\Sigma_d)/2$, using the following equation for NJL coupling constant:

$$G(eB,T) = c\left[1-\frac{1}{1+\exp\left[d\,(T_a-T)\right]}\right]+s, \qquad (12)$$

where the parameters c, s, β and T_a depend on eB. For zero magnetic field, the values of the fitting parameters are: $c = 0.9\ \mathrm{GeV}^{-2}$, $s = 3.7311\ \mathrm{GeV}^{-2}$, $d = 0.4\ \mathrm{MeV}^{-1}$, and $T_a = 168$ MeV. The extrapolation of our fit to $T = 0$ gives $G(0,0) = 4.6311\ \mathrm{GeV}^{-2}$. Using $\Lambda = 0.650$ GeV and $m_0 = 5.5$ MeV, we obtain $f_\pi = 84.4498\,\mathrm{MeV}$, $m_\pi = 142.177\,\mathrm{MeV}$ and $\langle\bar{\psi}_f\psi_f\rangle^{1/3} = -236.374\,\mathrm{MeV}$ in the vacuum.

In panel (a) of Fig. 1 we compare the temperature dependence of average of the condensates $(\Sigma_u+\Sigma_d)/2$ at zero magnetic field obtained with the NJL model with a coupling G temperature independent (solid black line) and with $G(0,T)$ (red dashed line) given by Eq. (12). In the transition region, around $T \simeq 170$ MeV, one sees that the difference between the curves is quite substantial. In panel (b) of the same figure, we show the running of the G as given by Eq. (12) for zero magnetic field.

In the next section, we investigate the impact on thermodynamic quantities calculated with the NJL model when using a running $G(eB,T)$ as given in Eq. (12).

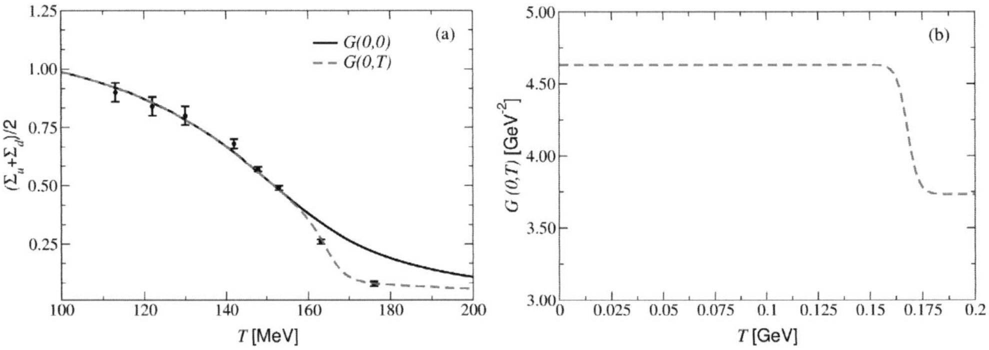

Figure 1. Panel (a): Condensate average as a function of the temperature at zero magnetic field, where data points are from the lattice simulations of Ref. [3]. Panel (b): NJL coupling constant $G(eB, T)$ at zero magnetic field.

4. Numerical Results

We start examining the effects of the running G with the temperature in the quark mass M and in quark condensate $\langle\bar{\psi}_f\psi_f\rangle$ — for zero magnetic field, the u and d condensates are equal (recall we are using $m_u = m_d$). We can observe in Fig. 2 that in the critical region these quantities are very sensitive to the running with T and at high temperatures the model displays a crossover (where chiral symmetry is partially restored) similar to the case with no running G. The difference is that the drop in the mass and in the condensate is much sharper when we let the coupling run with the temperature.

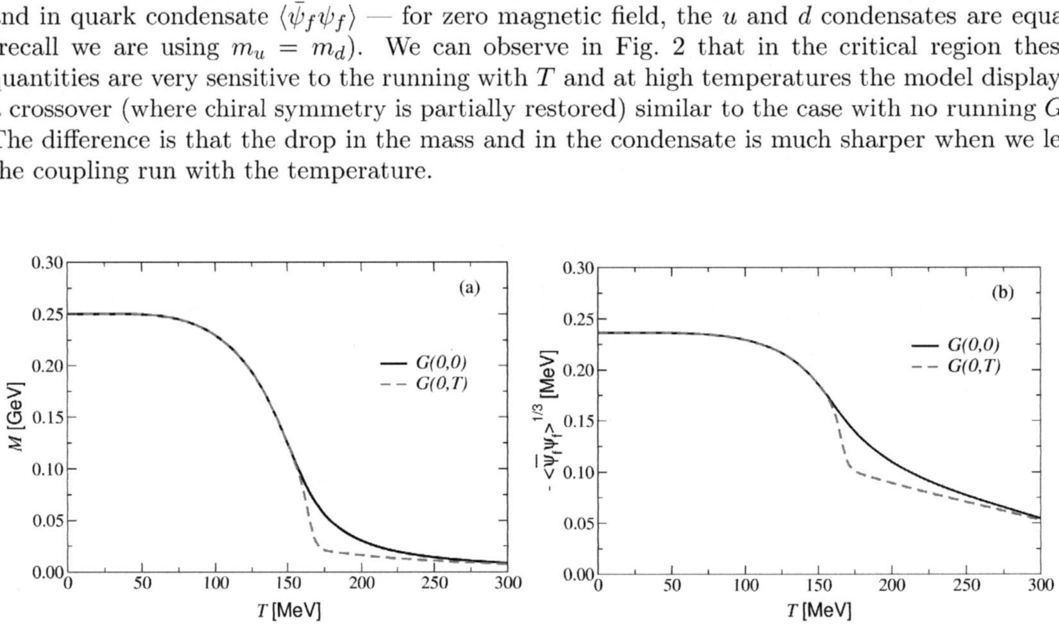

Figure 2. Temperature dependence of the constituent quark mass M (panel (a)) and the quark condensate (panel (b)).

The model displays a crossover and the pseudo-critical temperature depends on the observable used to locate the transition. Here we use the peak of the thermal susceptibility, which is given by

$$\chi_T = -m_\pi \frac{\partial \sigma}{\partial T}, \quad \text{where} \quad \sigma = \frac{\langle\bar{\psi}_u\psi_u\rangle(B,T) + \langle\bar{\psi}_d\psi_d\rangle(B,T)}{\langle\bar{\psi}_u\psi_u\rangle(B,0) + \langle\bar{\psi}_d\psi_d\rangle(B,0)} \ . \tag{13}$$

Other quantities of interest are pressure, the negative of the free energy, $p = -\Omega(M, eB, T)$, the entropy density, obtained from $s = dp/dT$, and the energy density $\varepsilon = -p + T\, s + \mu\, \rho$. We normalize the pressure by subtracting the pressure at zero temperature, $p(eB, 0)$ so that the normalized pressure, p_N, is given by

$$p_N(eB, T) = p(eB, T) - p(eB, 0)\ . \tag{14}$$

In addition, we also consider the interaction measure, speed of sound, and specific heat, given respectively by

$$\Delta = \frac{\varepsilon - 3p_N}{T^4}, \qquad c_s^2 = \left(\frac{\partial p_N}{\partial \varepsilon}\right)_v, \qquad c_v = \left(\frac{\partial \varepsilon}{\partial T}\right)_v. \tag{15}$$

In panel (a) of Fig. 3, we show the thermal susceptibility defined by Eq. (13) as a function of the temperature at zero magnetic field. The result indicates that the running of coupling with the temperature change the position of the χ_T peak and T_{pc} is slightly larger if we compare with the case with no running. The interaction measure Δ is shown in panel (b) of the figure.

We next examine the effect of the running of $G(eB = 0, T)$ with the temperature in all relevant thermodynamic quantities.

In Fig. 4 we show our results for the scaled pressure p_N/T^4, scaled energy density ε/T^4, the equation of state (eos) p_N/ε and scaled entropy density s/T^3. In panel (a) we see that the running $G(0, T)$ does not affect the behaviour of the scaled pressure and in panel (b) we observe the influence of the $G(0, T)$ in the scaled energy density that also will appear in the critical region of the eos shown in panel (c). In panel (d) we show that the scaled entropy is not affected by the running.

Fig. 5 displays the temperature dependence of the speed of sound squared c_s^2, scaled specific heat c_v/T^3. Note that c_s^2 and c_v display different behaviours compared to the case with no running. With the running, c_s^2 shows a downward cusp in T_{pc} while c_v/T^3 shows a upward cusp in T_{pc} as expected.

In figure 6 we present our results for the condensate average as a function of temperature for different values of the magnetic field. More details on the fit of $G(eB, T)$, the parametrization and their effect on the thermodynamic quantities of magnetized quark matter will be discussed in a forthcoming work [13].

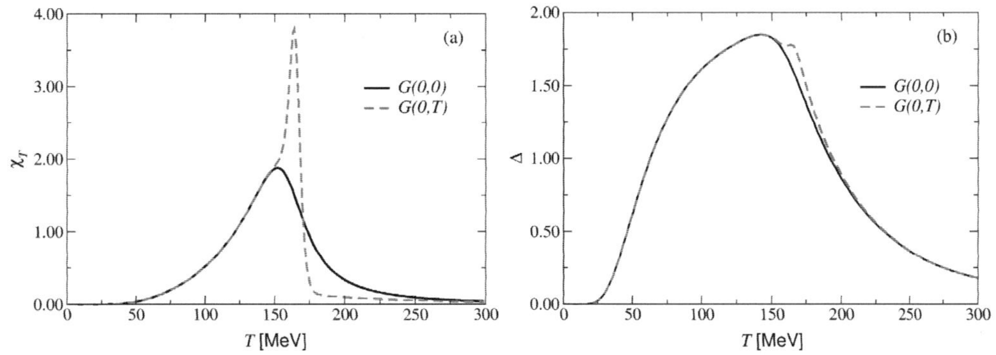

Figure 3. Temperature dependence of the normalized thermal susceptibility χ_T (panel (a)) and interaction measure (panel(b)) for zero magnetic field.

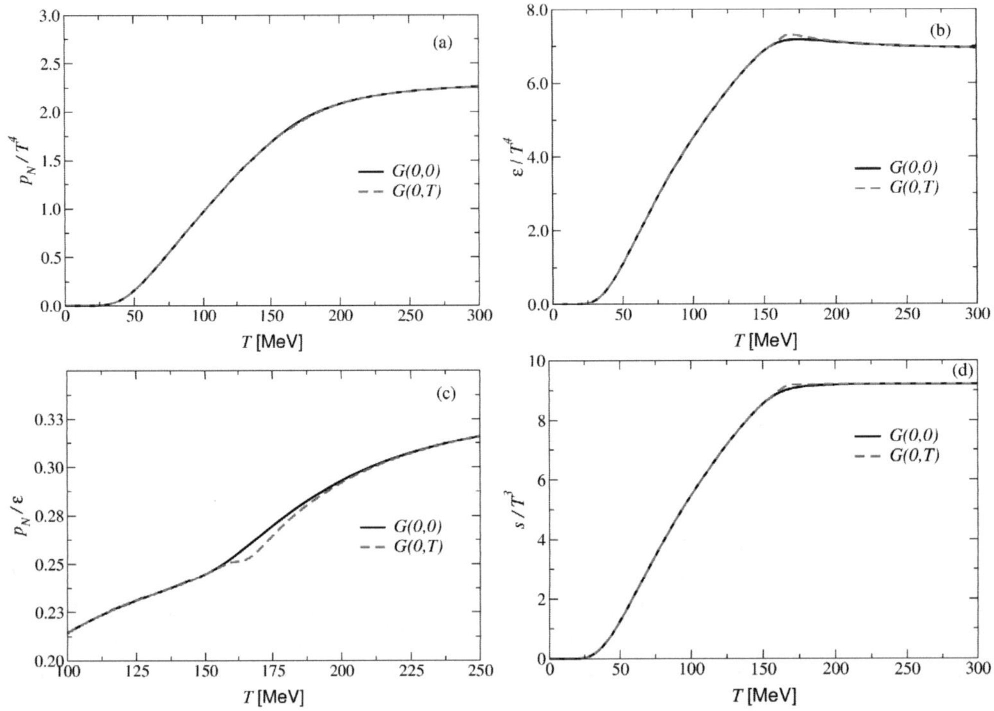

Figure 4. Panel (a): Scaled pressure p_N/T^4. Panel (b): Scaled energy density ε/T^4. Panel (c): Equation of state p_N/ε. Panel (d): Scaled entropy density $s/T^3(s = dP/dT)$.

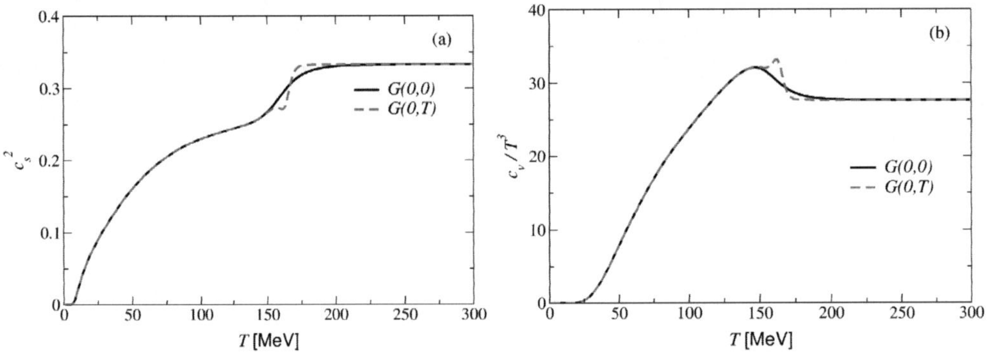

Figure 5. Temperature dependence of speed of sound squared (panel (a)) and scaled specific heat (panel (b)) for zero magnetic field.

5. Final Remarks

In this work we have considered the two flavor NJL model for hot and magnetized quark matter within the mean field approximation. At vanishing magnetic fields, we have shown that lattice QCD results can be reproduced in the NJL model if the coupling constant has a temperature dependence. We have analysed the effects of the running $G = G(0,T)$ on the quark masses, quark condensates as well as on thermodynamic quantities. We have improved a previous attempt in

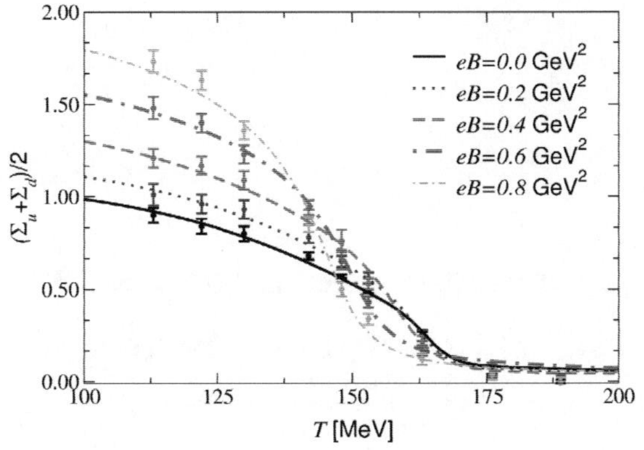

Figure 6. The condensate average as a function of the temperature. Data points are from the lattice simulations of Ref. [3]. The curves were obtained using our fit of $G(eB, T)$[13].

Ref. [5] and fitted the NJL coupling constant so that the model reproduces the average of quark condensates computed on the lattice [3].

Our numerical results show that this running with T introduces modifications in the behaviour of the quark condensate, quark mass and in the thermodynamic properties of the system. The equation of state and thermodynamic quantities like the sound velocity squared and the specific heat are very sensitive to this running in the region around T_{pc}. At finite magnetic fields and temperatures we have presented results for the average of quark condensates that are in good agreement with lattice QCD simulations of Ref. [3]. A detailed investigation of the effect of the running with both T and B on the thermodynamics of magnetized quark matter is in progress [13].

Acknowledgments
We thank G. Endrodi for helpful discussions and for providing the lattice data for u and d quark condensates. We also thank FAEPEX, FAPESP, FAPESC and CNPq for financial support.

References
[1] Fukushima K, Kharzeev D E and Warringa H J 2008 *Phys. Rev.* D **78** 074033.
[2] Kharzeev D E and Warringa H J 2009 *Phys. Rev.* D **80** 0304028.
[3] Bali G S, Bruckmann F, Endrodi G, Fodor Z, Katz S D and Schafer A 2012 *Phys. Rev.* D **86** 071502(R), Bali G S, Bruckmann F, Endrodi G, Fodor Z, Katz S D, Krieg S, Schafer A and Szabo K K 2012 *J. High Energy Phys.* 02 **044**.
[4] Andersen J O, Naylor W R and Tranberg A arXiv:14117176 [hep-ph] *Phase diagram of QCD in a magnetic field: A review.*
[5] Farias R L S, Gomes K P, Krein G and Pinto M B 2014 *Phys. Rev.* C **90** 025203.
[6] Ferreira M, Costa P, Lourenço O, Frederico T and Providência C 2014 *Phys. Rev.* D **89** 116011.
[7] Rougemont R, Critelli R and Noronha J arXiv: 150507894 [hep-th] *Holographic calculation of the QCD crossover temperature in a magnetic field.*
[8] Nambu Y and Jona-Lasinio G 1961 *Phys. Rev.* **122** 345.
[9] Klevansky S P 1992 *Rev. Mod. Phys.* **64** 649.
[10] Buballa M 2005 *Phys. Rep.* **407** 205.
[11] Ferrari G N, Garcia A F and Pinto M B 2012 *Phys. Rev.* D **86** 096005.
[12] Menezes D P, Pinto M B, Avancini S S, Martínez A P and Providência C 2009 *Phys. Rev.* C **79** 035807.
[13] Farias R L S, Timoteo V S, Avancini S S, Pinto M B and Krein G, work in progress.

Journal of Physics: Conference Series **706** (2016) 052030 doi:10.1088/1742-6596/706/5/052030

Information Theoretical Methods as Discerning Quantifiers of the Equations of State of Neutron Stars

Rodrigo Alvares de Souza[⋆], **Marcio G B de Avellar**[⋆], **Jorge E Horvath**[⋆], **D. M. Paret**[⋆⋆]

[⋆] Instituto de Astronomia, Geofísica e Ciências Atmosféricas, Universidade de São Paulo, São Paulo, Brazil

[⋆⋆]Facultad de Física, Universidad de la Habana, San Lázaro y L, Vedado La Habana, 10400, Cuba.

E-mail: rodrigo.souza@usp.br

Abstract. In this work we use the statistical measures of information entropy, disequilibrium and complexity to discriminate different approaches and parametrizations for different equations of state for quark stars. We confirm the usefulness of such quantities to quantify the role of interactions in such stars. We find that within this approach, a quark matter equation of state such as SU(2) NJL with vectorial coupling and phase transition is slightly favoured and deserves deeper studies.

1. Introduction

The basic concept of information theory is the Shannon Information, also known as Shannon Entropy or Information Entropy. Shannon defined, in 1948, an expression that measures the information (or randomness, or uncertainty, or ignorance) about a system. Obeying a set of mathematical properties defined by him, the information content of a system in terms of probabilities of a event to occur is:

$$H = -K \sum_i p_i log_b[p_i] \quad or \quad H = -K \int p(x) log_b[p(x)] dx, \tag{1}$$

respectively for the discrete and continuous cases.

From this, people started thinking about how complex a system can be and how to measure this complexity calculating it from a mathematical definition. Thus, the statistical measure of complexity introduced by Lopez-Ruiz, Mancini and Calbet [2] relates the complexity of a system to the information stored in it and the distance to a situation in which all possible states of the system are equiprobable. This definition encodes the concepts of order and disorder of a given arrangement of the system.

It was assumed in the literature that the probability distribution is proportional to the energy density (or the mass density in the non-relativistic case) profile of the star, $\epsilon(r)$, obtained when solving the equations of structure of the star (in the case of neutron stars, the TolmanOppenheimerVolkoff equations, the relativistic version of hydrostatic equilibrium). This was justified by the statement that the energy/mass density is related to the probability of finding some particles at a given defined location inside the star. Even though it may be criticized, it proved difficult to suggest a better alternative from the available physical quantities.

To illustrate these concepts we may think about two ideal systems frequently used in physics: the ideal gas and the perfect crystal, extremes in all aspects and opposites as well. Because they both are idealized systems, they should be thought as minimally complex systems. However, while the latter is totally ordered, the former is totally disordered; i.e. while for the perfect crystal one state is more probable than the others, in the ideal gas all states are equiprobable. Summarizing this intuitive view:

Perfect crystal: zero complexity by definition; strict symmetry rules $\Rightarrow$ probability density centered around the prevailing state of perfect symmetry $\Rightarrow$ minimal information. Completely ordered.

Ideal gas: zero complexity by definition; accessible states are equiprobable $\Rightarrow$ maximal information. Totally disordered.

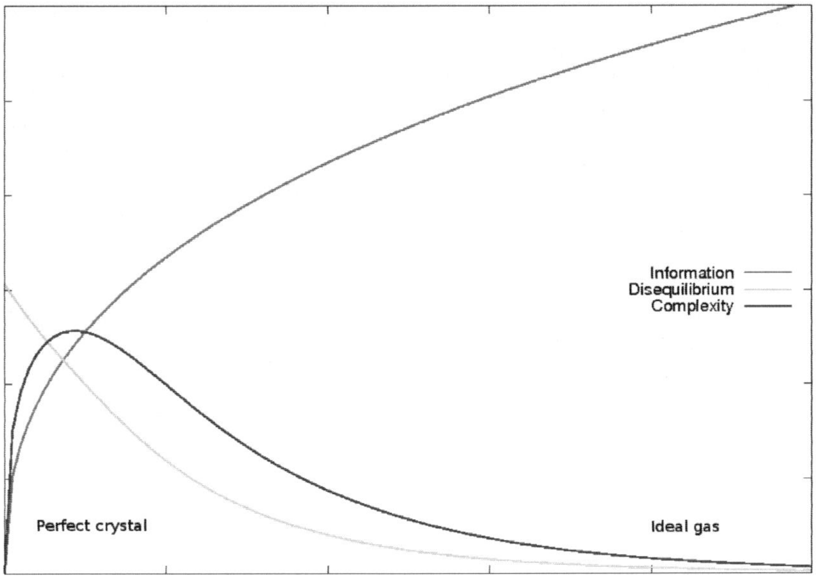

Figure 1. Intuition of what should be complexity, at least asymptotically.

At this stage an suitable expression for complexity arises:

$$C \equiv H \times D \quad or \quad C \equiv e^H \times D, \tag{2}$$

where

$$D = \sum_i \left[p_i - \frac{1}{N}\right]^2 \quad or \quad D = \int p^2(x)dx$$

is the distance to the (equilibrium) equiprobability of states.

If we measure the complexity of some system, and allow its proximity to a crystal or a gas to encode the degree of order of the system under consideration, then one can wonder what state of the system is preferable for Nature to realize. In particular, by calculating the information entropy of a star allowed by a chosen equation of state, we find a quite univocal behaviour of this quantity with the mass and radius of the stars in the stellar sequence.

2. The equations of state

To clarify the issue of the composition using these concepts, we have compared in Ref. [1] the information entropy, the disequilibrium and the complexity for two different sample cases: the quark stars constructed from a MIT Bag model equation of state (an admixture of quarks up, down and strange in approximately equal amounts and hadronic neutron stars from SLy4 equation of state (neutron rich in the core) [2]. We found that the information entropy trend for both cases is pretty much the same when viewed as a function of the stellar mass, but differs a lot when plotted as a function of the stellar radius.

If higher D implies higher distance to the equiprobability and a trend to be more crystalline (and ordered), and additionally we assume that order has a cost, then for each EoS separately Nature prefers to form less massive objects, albeit more complex ones. This trend makes sense since ideal systems (in this case the perfect crystal and the ideal gas) have complexity zero by construction and should not form in Nature (because they are just idealizations). In other words, the complexity of any real system have to be greater than zero.

Here we extended that work by comparing a larger set of equations of state (Ref.[3]), enumerated and briefly described below:

(i) Original MIT Bag Model [4]: Neutron stars can be actually strange stars, stars in which the nuclei are subject to so high pressures that the nuclear boundaries dissolve, resulting in a soup of roughly equal numbers of quarks up, down and strange plus electrons to keep the star neutral.

(ii) Alford-Han-Reddy (AHR) [5]: The statement here is that at sufficiently low surface tension, the same strange quark matter of the above items will not be self-bound and at low pressure and temperature the matter will form a crystal of positively charged strangelets in a neutralizing background of electrons.

(iii) Dexheimer-Schramm (DS1 and DS2) [6]: the equation of state used here is actually a new parametrization of the equation of state presented in [6].

(iv) Coelho et al. [7, 8]: This equation of state is based on the SU(2) NJL model with vectorial coupling and a phase transition to a hadron phase described by the GM1 parametrization for different values of the ratio g_v/g_s for the strength of the vectorial coupling.

(v) Douchin-Haensel (SLy4) [2]: Based on the effective nuclear interaction SLy of the Skyrme type, which is particularly suitable for the application to the calculation of the properties of very neutron rich matter.

3. Results and Discussion

We have compared our statistical measures for all the EoSs. Notice that the quantity which we have to compare, in terms of what Nature would want to form, is the complexity. Recalling that complexity is a composite quantity that encodes a "distance" of the real object to the two extremes and opposites ideal systems, the optimal complexity (the preferred object) is the one that at the same time maximizes the information entropy while keeping the disequilibrium at a safe distance of the extremes.

As an example, we show here the analysis of the information entropy, disequilibrium and complexity for the Sly4 EoS against the SU(2) NJL EoS. In Figure 2 we note that from the complexity measures we can also have different regimes: for masses $0.109M_{\odot} \leq M \leq 0.74M_{\odot}$, a slight preference for the SLy4 EoS; for masses $\geq 0.74M_{\odot}$ Nature would prefer to make star with the SU(2) NJL EoS.

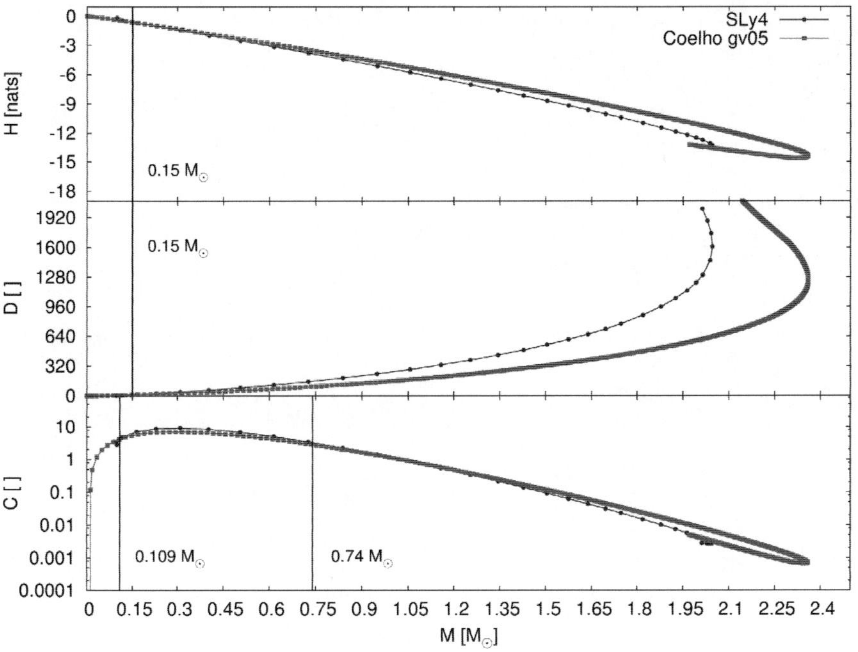

Figure 2. From the complexity measures, we see that we can also have different regimes: for masses $0.109M_{\odot} \leq M \leq 0.74M_{\odot}$, a slight preference for the SLy4 EoS; for masses $\geq 0.74M_{\odot}$ Nature would prefer to make star with the SU(2) NJL EoS.

From the point of view of *complexity*, as stated earlier, although all sequences of neutron stars in our study are similar, there is a noticeable preference for the SU(2) NJL equation of state with phase transition and vectorial coupling for masses $\geq 0.75M_{\odot}$.

In Figure 3 we present the complexity versus information entropy for all of our equations of state. Allowing then the possibility of a phase transition hadrons $\rightarrow$ quarks and considering that the information entropy evolution is a good indicator of the time evolution of the system, then we see that for a given mass, such 1.73 $M_{\odot}$, for example, the object will start as a SLy4 star and end up as a SU(2) NJL star (see the arrow). The higher the mass, more attractive and pronounced is the transition.

Finally, from Figure 3 we see that the SU(2) NJL would be the final state of a neutron star that is born as a SLy4 star. Actually, all neutron stars would be SU(2) NJL stars. On the other hand, the threshold of 1.06 $M_{\odot}$ also marks the possibility of coexistence of two families of neutron stars: below this threshold down to ~ 0.85 $M_{\odot}$, Sly4 stars; above it AHS or DS1 stars. Besides, SU(2) NJL is almost in all circumstances preferred over the other quark stars.

This does not mean, however, that the SU(2) NJL equation of state with phase transition and vectorial coupling is the "true" equation of state of neutron stars. The real meaning is that our results suggest a direction where our studies should go to, they suggest that this kind of equation of state deserves more studies. Another outcome is this kind of quantifiers is that they suggest also that there could be two or more families of neutron stars in Nature depending on the masses.

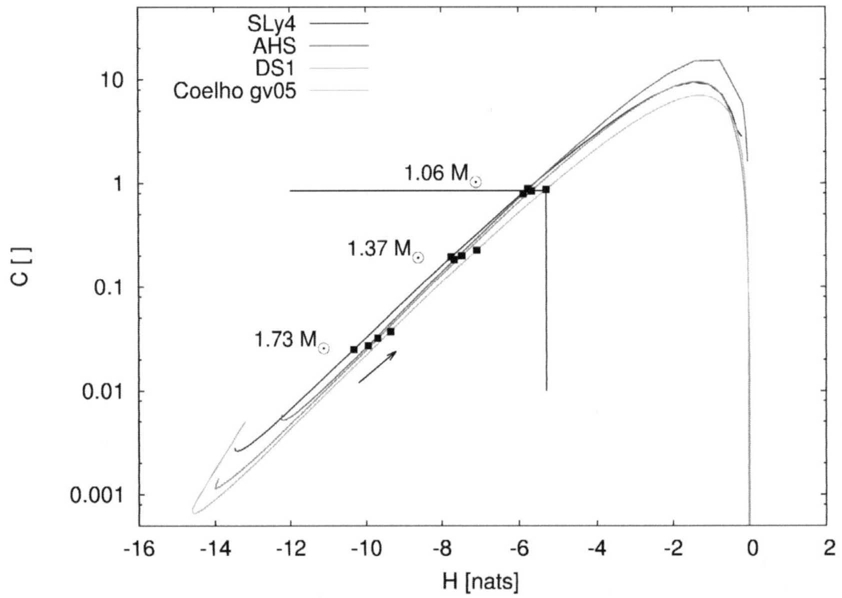

Figure 3. Complexity versus information entropy for all EoSs used in this work. The straight lines delimit the region of observed masses; the black squares represent the posotion of the given mass for each EoS. The arrow shows the trend of evolution in time for a given mass.

4. Conclusions

We have used the definitions of information content, disequilibrium and complexity to discriminate among the equations of state for white dwarfs and, more interestingly, neutron stars (Reference [3]). However, while this statement (jump from a low density WD to a high density NS) seems to be valid from the information theoretic point of view, we cannot neglect the fact that an analogous energy barrier exist when considering the formation of neutron stars with quarks. The pressure of the collapse must be high enough to attain central densities at least few times $10^{14} g/cm^3$ in order to be entropically favourable to form quarks, at least in principle. Unfortunately, the specific mechanism(s) of formation of neutron stars are still poorly understood to answer the issue of the presence of quarks. For example, the mass of the core that will become a neutron star is a major source of uncertainty the stellar evolution theory for progenitor masses between $7.5 - 11 M_{\odot}$, and with it the explosion mechanism itself for progenitor masses between $11 - 25 M_{\odot}$ [9]. It may happen that the energy (and entropy) barrier to form quarks is never attained, or rather that it is a trivial fact for all explosive formation scenarios [10].

References

[1] M. G. B. de Avellar, J. E. Horvath, Physics Letters A 376 (2012) 1085.
[2] F. Douchin, P. Haensel, Astronomy and Astrophysics 380 (2001) 151.
[3] M. G. B. de Avellar, R.A. de Souza, J. E. Horvath, D.M. Paret, Physics Letters A 378 (2014) 47.
[4] E. Witten, Physical Review D 30 (1984) 272.
[5] M. Alford, S. Han, S. J. Reddy, Phys. G: Nucl. Part. Phys. 39 (2012) 065201.
[6] J. Steinheimer, S. Schramm, H. J. Stöcker, Phys. G: Nucl. Part. Phys. 38 (2011) 035001.
[7] J. Coelho, C. H. Lenzia, M. Malheiro, R. M. Marinho Jr., C. Providência, M. Fiolhais, Nuclear Physics B (Proc. Suppl.) 199 (2010) 325.
[8] J. Coelho, Ph.D. Thesis, Instituto de Tecnologia Aeronáutica (ITA), São José dos Campos/SP, Brazil, 2009.

[9] C. L. Fryer, K. Belczynski, G. Wiktorowicz, M. Dominik, V. Kalogera, D. E. Holz, The Astrophysical Journal 749 (2012) 91.
[10] O. G. Benvenuto, J. E. Horvath, Physical Review Letters 63 (1989) 716.

Acknowledgments
MGBA wishes to acknowledge the FAPESP agency for the financial support. JEH wishes to acknowledge the financial support of the CNPq Agency (Brazil) through a research scholarship. DMP acknowledges the fellowship CLAF-ICTP and also thanks to IAG-USP for the hospitality.

Double Parton Scattering at the LHC

Rodrigo Konrath

Department of Physics, Federal University of Santa Catarina, Florianópolis, SC, Brazil

E-mail: rodrigo.konrath@posgrad.ufsc.br

Abstract. In this short presentation, we will briefly discuss the status of Double Parton Scattering (DPS) at the Large Hadron Collider (LHC). Often DPS, which can be described as perturbative process, can occur when two pairs of partons interact in a same and single p-p (proton-proton) collision. Until the advent of LHC, this kind of process was not very well understood, because collisions at very high energies are necessary for a more visible double scattering contribution. We start our presentation speaking about what is the DPS phenomena. Next, we discuss the cross section calculation for the DPS process with its details. Naively, the DPS process can be calculated by the multiplication of two independent parton cross sections and dividing the result by an effective cross section. We review this possibility. And finally, we explain the concept of effective cross section and show some experimental results measured by different experiments.

1. Introduction

Because of asymptotic freedom, the collisions between hadrons can be calculated perturbatively in the high energy limit as a parton collision. In high energies, the partons can be studied through the perturbative QCD, while the hadrons are studied by non-perturbative QCD. This separation is due the factorization, which define parton distributions inside each hadron. It should be remembered that a perturbative process is a process where the hard scale is well defined. A example of hard process is the production of heavy objects.

Fig. 1 shows the values for the cross section for various processes. As we can see, the cross section for a p-p scattering is dominated by soft-QCD [1], which gives interactions with low momentum transfer. In comparision, the total cross section does not change very much with the increase in energy. However, the Higgs mechanism and the production of W and Z bosons are highly dependent on energy.

Multi Parton Interactions (MPI) occurs when more than one parton takes part in the same hadron - hadron collision. With the advent of the LHC in the last decade, this kind of physical process can be more detailed studied. A typical MPI interaction at lower scales is the double parton scattering (DPS), that occurs when two partons from two different protons interact, producing particles with large mass or transverse momentum. Fig. 2 illustrates this phenomenon, which is the simplest MPI process.

The occurrence of a DPS process in proton collisions depends strongly on the density of partons inside the colliding hadrons. With the LHC (which operates at 13 TeV) smaller momentum fractions are probed, instead other colliders. Consequently, more smaller momentum fractions leads to a increase in parton density. Therefore, a increase in parton density means that there a more probability to occur DPS.

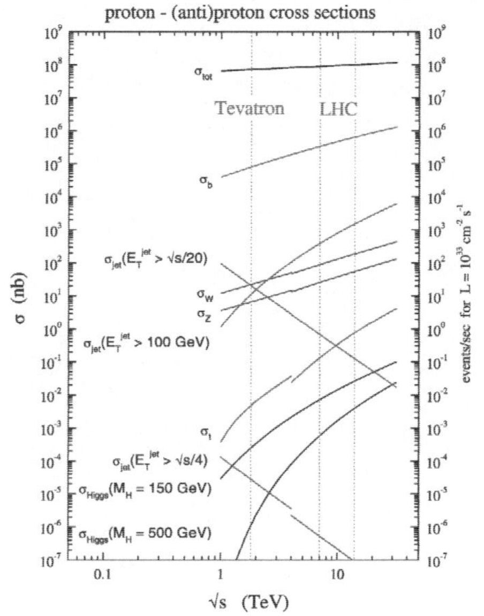

Figure 1. Cross Section values for many processes. The discontinuity in some lines shows the change from $p\bar{p}$ collisions at the Tevatron to pp collisions at the LHC. Processes driven by strong interactions have cross sections larger than processes governed by electroweak interactions. This Figure was obtained from [2].

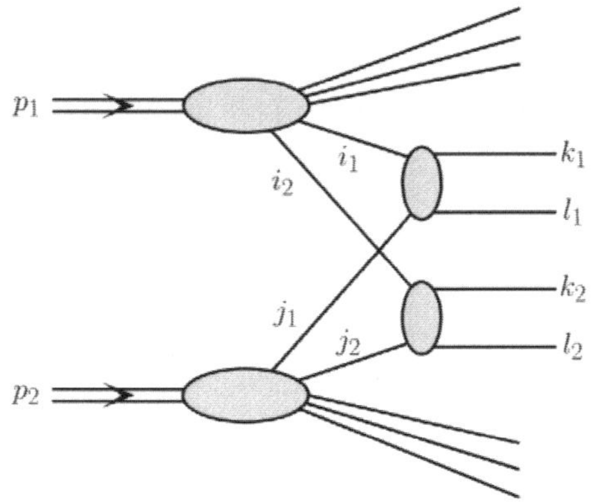

Figure 2. Mechanism of Double Parton Scattering in which the partons of the first proton are labelled i_1 and i_2 and of the second proton j_1 and j_2. The two hard scattering subprocess are $A(i_1 j_1 \to k_1 l_1)$ and $B(i_2 j_2 \to k_2 l_2)$. This Figure was adapted from [3].

The aim of this study is to review the DPS process. The simplest calculation of the cross section for this process is given by the multiplication of the two SPS (Single Parton Scattering) cross sections from the two resulting hard scattering subprocess. However, is required a normalization of the result through a factor called the *effective cross section*.

Journal of Physics: Conference Series **706** (2016) 052031 doi:10.1088/1742-6596/706/5/052031

2. Cross Section for the DPS process

To calculate the cross section for the DPS process illustrated above, it is required calculate the SPS cross section for the independent processes. The differential cross section for the SPS process are:

$$d\sigma_1^{\rm SPS} = \sum_{i_1,j_1} \int f_{p_1}^{i_1}(x_1,\mu_1) f_{p_2}^{j_1}(x_1^{'},\mu_1) d\hat{\sigma}_{(i_1 j_1 \to k_1 l_1)}(x_1, x_1^{'}, \mu_1) dx_1 dx_1^{'} \tag{1}$$

$$d\sigma_2^{\rm SPS} = \sum_{i_2,j_2} \int f_{p_1}^{i_2}(x_2,\mu_2) f_{p_2}^{j_2}(x_2^{'},\mu_2) d\hat{\sigma}_{(i_2 j_2 \to k_2 l_2)}(x_2, x_2^{'}, \mu_2) dx_2 dx_2^{'} \tag{2}$$

In the above equations, $f(x_i, \mu_i)$ are the parton distribution functions (PDF's) and $d\sigma$ are the perturbatively-calculable partonic cross section. The differential DPS cross section is then calculated by the equation (3):

$$d\sigma^{\rm DPS} = \frac{m}{2} \frac{d\sigma_1^{\rm SPS} d\sigma_2^{\rm SPS}}{\sigma_{\rm eff}} \tag{3}$$

The factor m that appears in (3) determine the distinctness from the final states created after the collisions. If $m = 1$, the final states are indistinguishable and if $m = 2$, the final states are distinguishable [4]. The effective cross section $\sigma_{\rm eff}$ makes a quantitative prediction of the transverse distribution of partons in a hadron collision [5].

2.1. Factorization

As a first approximation, the DPS process basically depends of the longitudinal momentum fractions and their corresponding factorization scales. However, recent studies [6] and [7], show that these are not enough to describe the p-p scattering; we need the transverse distance **b** between the two partons, besides two longitudinal momentum fractions from the two partons x_1 e x_2. Thus, we need a double parton distribution designated $F(x_1, x_2, \mathbf{b})$.

It can be assumed that the density of the two partons factorizes as [8]:

$$F(x_1, x_2, \mathbf{b}) = f(x_1) f(x_2) G(\mathbf{b}) \tag{4}$$

The double PDF can be decomposed into longitudinal and transverse components, which brings us two assumptions [9]:

i) the transverse component is expressed in terms of the overlap function

$$G(\mathbf{b}) = \int F_\perp(\mathbf{b}_1) F_\perp(\mathbf{b}_1 - \mathbf{b}) d^2\mathbf{b}_1 \tag{5}$$

which represents the effective transverse overlap area of partonic interactions that produce the characteristic phenomena of the DPS process. The parton function $F_\perp(\mathbf{b})$ gives us the partons density inside the colliding protons

ii)the longitudinal component can be calculated as the "diagonal" product of the two independent single-PDF.

In some cases (4) obviously is not valid, for example, when $x_1 = 0.8$ and $x_2 = 0.7$, by momentum conservation. However, when $x_i \to 0$, this is a reasonable approximation.

2.2. Efective Cross Section

The effective cross section $\sigma_{\rm eff}$ can be written as:

$$\frac{1}{\sigma_{\rm eff}} = \int d^2 b G^2(\mathbf{b}) = \int \frac{d^2\mathbf{r}}{(2\pi)^2} [F(\mathbf{r})]^4 \tag{6}$$

The expression after the second equality in (6) was recently found by [10] and uses the Fourier transform $F(\mathbf{r})$ of $G(\mathbf{b})$, which is dependent only on $\mathbf{r}^2$.

3. Effective Cross Sections Measurements

Fig. 3 shows the effective cross sections measured by different experiments and energies. At first sight, it looks like it is growing with the energy, but this one has to take into account the large error bars. The average effective cross section, according to the more recent experiments, is approximately 15 mb.

The effective cross section will be measured again this year 2015, when the LHC will resume its activities with an energy around 13 TeV.

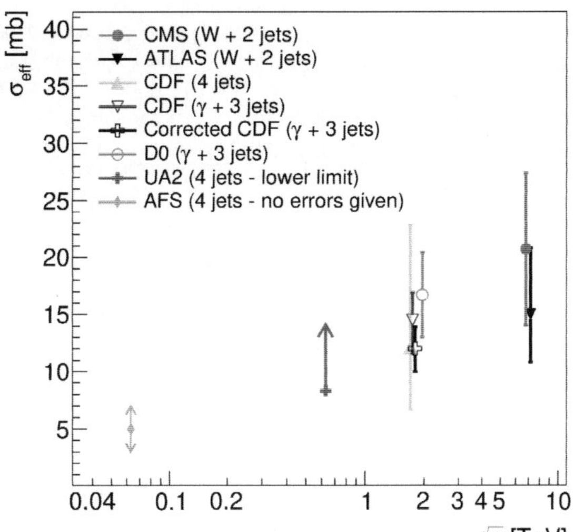

Figure 3. $\sigma_{\rm eff}$ measured by different experiments using different processes [11]. The "Corrected CDF" data point indicates a $\sigma_{\rm eff}$ value corrected [12]. This correction was performed, taking into account a normalized overlap function $G(\mathbf{b})$, which leads us a $\sigma_{\rm eff}$ independent of the individuals SPS cross sections. This Figure was obtained from [5].

4. Conclusion

We have reviewed the nature of the DPS process, showing the correct form of the simplest cross section calculation. Also was shown that this calculation, requires a normalization factor, the effective cross section, which makes a quantitative prediction of the transverse distribution of partons in a hadron collision. We also conclude, according to the more recent experiments, that the average effective cross section is approximately 15 mb. So, we can affirm that the Double Parton Scattering is a promising line of research in physics with very good work perspectives for the future, that is seen in the quality and quantity of published articles in last years. Indeed, this area offers opportunities to develop and test new phenomenological models.

Acknowledgements

We thank the organizers of the XIII International Workshop on Hadron Physics 2015 for the opportunity to present our work. We also thank to the Federal University of Santa Catarina by the support provided for this study. This work has been supported by CAPES and CNPq.

References

[1] N. Cartiglia, arXiv:1305.6131 [hep-ex].
[2] K. Jakobs, arXiv:1206.7024 [hep-ex].
[3] R. Maciula and A. Szczurek, Phys. Rev. D **90**, no. 1, 014022 (2014) [arXiv:1403.2595 [hep-ph]].
[4] G. Calucci and D. Treleani, Phys. Rev. D **83**, 016012 (2011) [arXiv:1009.5881 [hep-ph]].
[5] S. Chatrchyan *et al.* [CMS Collaboration], JHEP **1403**, 032 (2014) [arXiv:1312.5729 [hep-ex]].
[6] J. R. Gaunt, R. Maciula, A. Szczurek, arXiv:1407.5821 [hep-ph].
[7] A. V. Manohar, W. J. Waalewijn, Phys. Rev. D **85**, 114009 (2012) [arXiv:1202.3794 [hep-ph]].
[8] M. Diehl, PoS DIS **2013**, 074 (2013) [arXiv:1306.6480 [hep-ph]].
[9] D. d'Enterria, A. M. Snigirev, arXiv:1403.2335 [hep-ph].
[10] B. Blok, Y. Dokshitser, L. Frankfurt and M. Strikman, Eur. Phys. J. C **72**, 1963 (2012) [arXiv:1106.5533 [hep-ph]].
[11] UA2 Collaboration, Phys. Lett. B 268 (1991) 145, [doi:10.1016/0370-2693(91)90937-L]; AFS Collaboration, Z. Phys. C 34 (1987) 163, [doi:10.1007/BF01566757]; F. Abe *et al.* [CDF Collaboration], Phys. Rev. D **47**, 4857 (1993); CDF Collaboration, Phys. Rev. D 56 (1997) 3811, [doi:10.1103/PhysRevD.56.3811]; V. M. Abazov *et al.* [D0 Collaboration], Phys. Rev. D **81**, 052012 (2010) [arXiv:0912.5104 [hep-ex]]; G. Aad *et al.* [ATLAS Collaboration], New J. Phys. **15**, 033038 (2013) [arXiv:1301.6872 [hep-ex]].
[12] M. Bahr, M. Myska, M. H. Seymour and A. Siodmok, JHEP **1303**, 129 (2013) [arXiv:1302.4325 [hep-ph]].

Strong magnetic fields and SGRs/AXPs as white dwarf pulsar: a source of ultra-high energy cosmic rays

Ronaldo V. Lobato and M. Malheiro

Departamento de Física, Instituto Tecnológico de Aeronáutica, ITA - DCTA, Vila das Acácias, São José dos Campos, 12228-900 SP, Brazil

E-mail: rvlobato@ita.br, malheiro@ita.br

Abstract. The origin of highest energy cosmic rays still remains a mystery in Astrophysics. In this work we consider the Soft Gamma Repeaters (SGRs) and Anomalous X-ray Pulsars (AXPs) as possible sources of ultra-high cosmic rays. These stars described as white dwarfs pulsars can achieved large electric potential differences in their surface and accelerate particles up to Lorentz factors $\gamma \sim 10^{10}$. Pulsars offer favorable sites for the injection of electrons and heavy nuclei, and accelerate them to ultrahigh energies. Once accelerated in the pulsar this particles can escape from the magnetosphere and produce the radiation observed. Here, we discuss the possibility of SGRs/AXPs as white dwarf pulsars to be possible sources of ultra-high energetic photons with $E \sim 10^{21}$eV.

1. Introduction

Following and updating ours previous [1] work, we study the influence of large magnetic fields in white dwarfs (WD), their role at the magnetosphere, where the electron-positron pair and electromagnetic radiation occurs. For neutron stars there are many evidences of photons from radio to γ-rays energies, produced by electric potentials in neutron star magnetosphere that are able to accelerate particles to high energies. To explain it, a number of models were proposed: polar cap model [2, 3], slot gap model and outer gap model [4, 5]. There are some works exploring the possibility that the enhance of positron fraction observed by PAMELA could be explained by local pulsars [6, 7] and others considering white dwarfs pulsars [8]. We extend this idea considering the possibility of SGRs/AXPs being white dwarfs. We will see that fast and highly magnetized white dwarfs are also able to have large electric fields in their surface induced by their magnetic field and accelerate particles producing ultra-high energy cosmic rays.

White dwarfs with large magnetic fields of 10^6 to 10^9G and very massive have been discovered in the last years [9]. This fact opens the possibility to have white dwarfs pulsars [10], the theoretical possibility already was considered in [11] and it is shown that generation of hard X-ray and gamma rays are possible by fast rotating magnetic white dwarfs. The first fast white dwarf pulsar founded in the observations was AE Aquarii with a short period $P = 33.08$s [12]. It shows high energy emission in X that can be explained in terms of a rotation powered pulsar. Furthermore, 10 years ago already a candidate to a white dwarf pulsar has been proposed in [13] where high energy has been observed in a source in the Galactic center with a period $P = 77.13$ minutes.

1.1. Soft Gamma Repeaters and Anomalous X-ray Pulsars SGRs/AXPs
Soft Gamma Repeaters (SGRs) and Anomalus X-Ray Pulsars (AXPs) are very slow rotating pulsars $P \sim (2-12)$s, have huge magnetic fields (100-1000) times larger than radio pulsars, and are understood as neutron star (known as Magnetars [14]). Their spin-down rates $\dot{P} \sim (10^{-13} - 10^{-10})$s/s are larger than the ones of normal pulsars $\dot{P} \sim (10^{-15} - 10^{-14})$s/s [15]. To explain their luminosity is considered a reservoir of high magnetic energy liberated by the decay of their huge magnetic field. They cannot be understood as neutron star rotating-powered pulsars, since their rotational energy is smaller than their luminosity in X-rays. Currently there are 23 SRGs/AXPs [16] that are classified as magnetars and 5 until now only candidates.

2. SGRs/AXPs as White Dwarf Pulsars
In the last years, white dwarfs (WD) very massive, fast and with large magnetic field ($10^6 - 10^9$G) observed for this type of star (10^4G) were discovered [17, 9]. They share some characteristics with low B SGRs/AXPs recently detected, when understood as white dwarf pulsars [10]. Two more known fast WD are AE Aquarii [18, 19, 20, 21] with a period $P = 33.078$ and RXJ 0648.0-4418 with a period $P = 13.2$ [22, 23, 24].
An alternative model has been proposed considering SGRs/AXPs as white dwarfs pulsars [25, 10]. The magnetars model to SGRs/AXPs cannot explain some observation (for more details see [10]). As show in [26], the process of release energy by dipole radiation in a white dwarf can be explained in terms of a canonical spin-powered pulsars model, because in several aspects they are similar.
For example, if we consider a star with $M = 1.4M_\odot$ and $R = 10^6$ cm, the magnetic field at poles is:

$$B_{\rm p}^{\rm NS} = 3.2 \times 10^{19}(P\dot{P})^{1/2}{\rm G}, \tag{1}$$

In the case of a white dwarf with $M = 1.4M_\odot$ and $R = 3 \times 10^8$ cm, there is a new scale for the magnetic field at poles:

$$B_{\rm p}^{\rm WD} = 4.21 \times 10^{14}(P\dot{P})^{1/2}{\rm G}. \tag{2}$$

Following the last consideration there are new values for the mass density, moment of inertia, dipole moment and rotation energy. It explains a large class of problems when SGRs/AXPs are considered as neutron stars.

2.1. White dwarfs under presence of a high magnetic field
Magnetic fields has a important role in the structure of the WD, because the pressure not depend only on ρ but on magnetic field also [27],

$$\frac{\partial P}{\partial \rho}\bigg|_B \nabla\rho + \frac{\partial P}{\partial B}\bigg|_\rho \nabla B = -\rho\nabla\Phi + \frac{1}{4\pi}\nabla \times B \times B. \tag{3}$$

Several works explore this importance [28, 29, 30, 31] showing that white dwarfs with strong interior magnetic fields are more massive [29, 32]. In [33, 34, 35] self-consistent computations shown that the minimum period for white dwarfs are for example ~ 0.3s to star made of ^{4}He. To white dwarfs this is a high frequency, near of the Kepler frequency where the balance between gravity and centrifugal forces is equal. A star in this frequency losses mass, and become unstable.

3. Rotation Powered Pulsars and emission mechanism

There is a class of pulsars known as rotation-powered pulsars (RPPs), where we have rotating magnets, that emit dipole radiation with a rate

$$\dot{E}_{\rm rot} = \frac{2\mu^2\Omega^4\sin^2\alpha}{3c^3}, \tag{4}$$

where c is the speed of light. A dominant effect as to increase of their period P, by the breaking in the rotation star frequency by the conversion of rotational energy

$$\dot{E}_{\rm rot} = I\Omega\dot{\Omega} + \frac{1}{2}\dot{I}\Omega^2 \approx I\Omega\dot{\Omega}. \tag{5}$$

into electromagnetic energy [27].
A conducting magnetized sphere rotating in vacuum will induce in the surrounded space an electric field $\vec{E}$ with a component along B,

$$E_{\parallel} = \vec{E}\cdot\vec{B}/|\vec{B}| \neq 0. \tag{6}$$

Near the surface this electric field produce a force that surpass the gravitational attraction and put off particles from the surface and eject into magnetosphere. This particle will co-rotate with the star until the distance where the linear velocity reaches the speed of light, $v = c$. This region is the light cylinder with radius

$$R_{\rm L} = c/\Omega \tag{7}$$

and much more particles achieve this region more relativistic they are. Magnetic field lines beyond the light cylinder are open and delimit a region at surface called polar cap.
The plasma in the magnetosphere is governed by electric forces and flowed. If we assume the electric conductivity as infinity, the version of Ohm's law $J = \sigma(E + vB/c)$ implies [36],

$$\vec{E} + \frac{1}{c}(\vec{\Omega}\times\vec{r})\times\vec{B} = 0. \tag{8}$$

This parallel electric field determines a charge distribution, known as Goldreich-Julian charge density [37]

$$\rho_{\rm c} = \frac{1}{4\pi}\vec{\nabla}\cdot\vec{E} = -\frac{1}{2\pi c}\vec{\Omega}\cdot\vec{B}. \tag{9}$$

For a magnetic dipole field pointing along $\theta = 0$,

$$B(r) = \frac{B_0R^3}{r^3}(2e_r\cos\theta + e_\theta\sin\theta), \tag{10}$$

the typical length where the particle are accelerated is essentially the same of the radius of polar cap, so the potential difference follow as [38]

$$\nabla\Phi \sim \frac{\Omega^2R^3B_0}{2c^2}. \tag{11}$$

This potential would be available to accelerate particles from the surface of the star along the open magnetic fields lines that leave the magnetosphere with high energy.
There are some sites in the magnetosphere where occurs the particle acceleration, one of this is the polar cap region, just above the magnetic pole. Electrons and ions emerge with ultra-relativistic energies, moving along the magnetic field lines. The perpendicular energy is rapidly

dissipated by synchrotron radiation, but the longitudinal energy acquired could radiates far way of the star by curvature radiation,

$$E_\gamma = \frac{3}{2}\frac{\hbar\gamma^3 c}{r_c} \tag{12}$$

where

$$r_c \sim (Rc/\Omega)^{1/2}, \tag{13}$$

is the curvature radius of the field lines. This curvature radiation interact with the magnetic field and produce secondary $e^{\pm}$ pairs, $\gamma + B \rightarrow e^- + e^+$ leading to a pair creation avalanche. The value of γ is given by the potential difference in the polar cap region,

$$\gamma = \frac{e\Delta V}{mc^2}, \tag{14}$$

that in neutron stars pulsars is able to produce electrons that achieve Lorentz factors $\gamma \gtrsim 10^7$. In the polar cap model, there is a gap h [2] above the polar region, and the potential difference becomes

$$\Delta V = \frac{B_{\rm p}\Omega h^2}{2c}, \tag{15}$$

where h is given by

$$h \approx \left(\frac{R^3\Omega}{c}\right)^{1/2}. \tag{16}$$

4. Discussion

As we discuss the previous sections, the magnetic field of a white dwarfs can be calculated by the dipole expression. Thus, for SGRs/AXPs with P and $\dot{P}$ from [16], the magnetic fields are

$$B_{\rm p}^{\rm WD} \sim 10^9\text{G}. \tag{17}$$

This allow us to estimate the charge density in the magnetosphere, i.e., the Goldreich-Julian density given by equation (9)

$$\rho_c^{\rm WD} \sim 10^{-2}\text{cm}^{-3}, \tag{18}$$

this value is 10^5 smaller than the one when SGRs/AXPs are considered as neutron star which has a magnetic field $B \sim 10^{14}$G. So, in the magnetosphere of SGRs/AXPs as white dwarfs there are less charge particles than in the neutron star magnetosphere.

We can calculate the maximum potential difference achieved on the surface of SGRs/AXPs as white dwarfs by equation (15) with $\Omega \sim 1$Hz, $R \sim 3 \times 10^3$km, and obtain,

$$\Delta V \sim 10^{16}\text{V}. \tag{19}$$

This difference is achieved for length $h \sim 10^3$cm and allows the particles accelerated to reach ultra-relativistic energies more than neutron stars pulsars, i.e.. White dwarfs can accelerate electrons, following equation (14), with a Lorentz factor at least $\gamma \sim 10^{10}$, one thousand large than neutron star pulsars, which can produce curvature photons with an energy up to $\sim 10^{21}$eV (ions iron can be also accelerated up to high energy). Since the photon is emitted in a cone with a angular aperture $1/\gamma$ and there is a low charged particle density, some photons or another low mass particle produced by $\gamma + \gamma$ with this energy can escape far away from the star and through the space to the Earth. SGRs/AXPs are within the GZK limit [39] ($\approx$ 10Mpc for protons with a energy of 10^{19}eV) and the particles accelerated in this sources could be a fraction of ultra-high cosmic ray observed in our planet.

5. Acknowledgments

This work was financially supported by CAPES (Coordenação de Aperfeiçoamento de Pessoal de Nível Superior), CNPq (Conselho Nacional de Desenvolvimento Científico e Tecnológico) and FAPESP thematic project 2015/26258-4. The authors acknowledgements also to McGill Pulsar Group.

References

[1] Lobato R V, Coelho J and Malheiro M 2015 *Journal of Physics: Conference Series* **630** 012015 ISSN 1742-6596
[2] Ruderman M A and Sutherland P G 1975 *The Astrophysical Journal* **196** 51–72 ISSN 0004-637X
[3] Arons J 1983 Pair creation above pulsar polar caps - Geometrical structure and energetics of slot gaps
[4] Cheng K S, Ho C and Ruderman M 1986 *The Astrophysical Journal* **300** 500–539 ISSN 0004-637X
[5] Cheng K S, Ho C and Ruderman M 1986 *The Astrophysical Journal* **300** 522 ISSN 0004-637X
[6] Aharonian F, Atoyan A and Voelk H 1995 *Astronomy and Astrophysics* **294** L41–L44 ISSN 0004-6361
[7] Hooper D, Blasi P and Serpico P D 2008 *Journal of Cosmology and Astroparticle Physics* **2009** 10 ISSN 1475-7516 (*Preprint* 0810.1527)
[8] Kashiyama K, Ioka K and Kawanaka N 2011 *Physical Review D - Particles, Fields, Gravitation and Cosmology* **83** 1–17 ISSN 15507998 (*Preprint* 1009.1141)
[9] Kepler S O, Pelisoli I, Jordan S, Kleinman S J, Koester D, K B, P V, Castanheira B G, Nitta A, Costa J E S, Winget D E, Kanaan A and Fraga L 2013 *Monthly Notices of the Royal Astronomical Society* **429** 2934–2944 ISSN 0035-8711, 1365-2966
[10] Coelho J G and Malheiro M 2014 *Publications of the Astronomical Society of Japan* **66** 14–14 ISSN 0004-6264, 2053-051X
[11] Usov V V 1988 *Soviet Astronomy Letters, Vol.14, NO. 4, P. 258, 1988* **14** 258
[12] Terada Y, Hayashi T, Ishida M, Mukai K, Dotani T, Okada S, Nakamura R, Naik S, Bamba A and Makishima K 2008 *Publications of the Astronomical Society of Japan* **60** 387–397 ISSN 0004-6264, 2053-051X
[13] Zhang B and Gil J 2005 *The Astrophysical Journal Letters* **631** L143 ISSN 1538-4357
[14] Duncan R C and Thompson C 1992 *The Astrophysical Journal Letters* **392** L9–L13 ISSN 0004-637X
[15] Turolla R, Zane S and Watts A L 2015 *arXiv:1507.02924 [astro-ph]* (*Preprint* 1507.0292)
[16] Olausen S A and Kaspi V M 2014 *The Astrophysical Journal Supplement Series* **212** 6 ISSN 0067-0049, 1538-4365
[17] Castanheira B G, Kepler S O, Kleinman S J, Nitta A and Fraga L 2013 *Monthly Notices of the Royal Astronomical Society* **430** 50–59 ISSN 0035-8711, 1365-2966
[18] Crawford J A and Kraft R P 1956 *The Astrophysical Journal* **123** 44 ISSN 0004-637X
[19] Patterson J 1979 *The Astrophysical Journal* **234** 978–992 ISSN 0004-637X
[20] Meintjes P J, Raubenheimer B C, de Jager O C, Brink C, Nel H I, North A R, van Urk G and Visser B 1992 *The Astrophysical Journal* **401** 325–336 ISSN 0004-637X
[21] Beskrovnaya N G and Ikhsanov N R 2014 *Advances in Space Research* **55** 8 ISSN 18791948 (*Preprint* 1405.2182)
[22] Mereghetti S, La Palombara N, Tiengo A, Pizzolato F, Esposito P, Woudt P a, Israel G L and Stella L 2011 *The Astrophysical Journal* **737** 51 ISSN 0004-637X
[23] Palombara N L, Mereghetti S, Tiengo A and Esposito P 2012 *The Astrophysical Journal Letters* **750** L34 ISSN 2041-8205
[24] Mereghetti S, Palombara N L, Tiengo A, Sartore N, Esposito P, Israel G L and Stella L 2013 *Astronomy & Astrophysics* **46** 1–5
[25] Malheiro M, Rueda J A and Ruffini R 2012 *Publications of the Astronomical Society of Japan* **64** 56 ISSN 0004-6264, 2053-051X
[26] Usov V 1988 *Pisma v Astronomicheskii Zhurnal* **14** 606–609
[27] Shapiro S L and Teukolsky S A 2008 *Black Holes, White Dwarfs and Neutron Stars: The Physics of Compact Objects* (John Wiley & Sons) ISBN 9783527617678
[28] Suh I S and Mathews G J 1999 *The Astrophysical Journal* **530** 7 ISSN 0004-637X (*Preprint* 9906239)
[29] Coelho J G, Marinho R M, Malheiro M, Negreiros R, Cáceres D L, Rueda J A and Ruffini R 2014 *The Astrophysical Journal* **794** 86 ISSN 0004-637X
[30] Chamel N, Molter E, Fantina A F and Arteaga D P n 2014 *Physical Review D - Particles, Fields, Gravitation and Cosmology* **90** 043002 ISSN 15502368
[31] Paret D M, Martinez A P and Horvath J E 2015 *arXiv:1501.04619 [astro-ph]*
[32] Das U and Mukhopadhyay B 2014 *Journal of Cosmology and Astroparticle Physics* **2014** 050 ISSN 1475-7516

[33] Boshkayev K, Rueda J and Ruffini R 2013 *International Journal of Modern Physics: Conference Series* **23** 193–197
[34] Boshkayev K, Rueda J A, Ruffini R and Siutsou I 2013 *The Astrophysical Journal* **762** 117 ISSN 0004-637X
[35] Boshkayev K, Rueda J A, Ruffini R and Siutsou I 2014 *Journal of the Korean Physical Society* **65** 855–860 ISSN 0374-4884, 1976-8524
[36] Krall N A, Trivelpiece A W and Gross R A 1973 *American Journal of Physics* **41** 1380–1381 ISSN 0002-9505, 1943-2909
[37] Goldreich P and Julian W H 1969 *The Astrophysical Journal* **157** 869 ISSN 0004-637X
[38] Sturrock P A 1971 *The Astrophysical Journal* **164** 529 ISSN 0004-637X
[39] Greisen K 1966 *Physical Review Letters* **16** 748–750 ISSN 00319007

Two-Component Theory of Classical Proca Fields in Curved Spacetimes with Torsionless Affinities

S. I. Santos Júnior
Department of Physics
Centre for Technological Sciences-UDESC
Joinville 89223-100, Santa Catarina, Brazil.
e-mail: samuel.isidoro.santos@gmail.com

J. G. Cardoso
Department of Mathematics
Centre for Technological Sciences-UDESC
e-mail: jorge.cardoso@udesc.br

Abstract

The world formulation of the full theory of classical Proca fields in generally relativistic spacetimes is reviewed. Subsequently the entire set of field equations is transcribed in a straightforward way into the framework of one of the Infeld-van der Waerden formalisms. Some well-known calculational techniques are then utilized for deriving the wave equations that control the propagation of the fields allowed for. It appears that no interaction couplings between such fields and electromagnetic curvatures are ultimately carried by the wave equations at issue. What results is, in effect, that the only interactions which occur in the theoretical context under consideration involve strictly Proca fields and wave functions for gravitons.

1 Introduction

Traditionally, the Infeld-van der Waerden $\gamma\varepsilon$-formalisms [1] constitute the classical two-component spinor framework for general relativity. The construction of these formalisms was primarily aimed at exhibiting an elementary description of the dynamics of Dirac fields in generally relativistic spacetimes. Such formalisms had been designed originally much earlier than the achievement of the definitive conditions for a curved space to admit spinor structures locally [2-4]. The legitimacy of the procedures for building up them relies crucially upon the existence of sets of Hermitian connecting objects at non-singular spacetime points. Their affine prescriptions were formally shaped upon the ones that occur in the realm of general relativity. Thus, the generalized Weyl gauge group [5] is taken to operate on spin spaces set up locally in a way that does not depend at all upon the action of manifold mapping groups. Loosely speaking, all curvature spinors arise from the decomposition of mixed world-spin quantities that result from the action of torsionless covariant derivative commutators on arbitrary spin vectors [1, 6]. The $\gamma\varepsilon$-formalisms were extensively utilized over the years by many authors in several different ways [6-29], noticeably enough, to reconstruct some classical generally relativistic structures and to transcribe classification schemes for world curvature tensors. Notwithstanding the fact that the construction of curvature spinors is implicitly carried by the formalisms, the spin curvatures that occur in the classification schemes and some of the spinor structures mentioned above were obtained in an artificial way by carrying out straightforward spinor translations of Riemann and Weyl tensors. A fairly complete algebraic description of the affine and curvature structures tied in with the formalisms is supplied in Refs. [30-32].

The most striking physical feature of the $\gamma\varepsilon$-framework lies over the result that any curvature spinors are expressed as sums of purely gravitational and electromagnetic contributions which produce, in an inextricably geometric way, the occurrence of wave functions for gravitons and photons of both handednesses. The gravitational contributions for the ε-formalism were utilized in Refs. [18, 19] to support a spinor translation of Einstein's equations. It had been established a little earlier [20] that any of them should show up as a spinor pair which must be associated to the irreducible decomposition of a Riemann tensor. Any gravitational wave functions for either formalism are defined as totally symmetric curvature pieces that occur in spinor decompositions of Weyl tensors [18]. On the other hand, each electromagnetic curvature contribution emerges as a pair of suitably contracted pieces which enter the spinor representation of a locally defined Maxwell bivector [30]. The work of Ref. [4] gives a rough description of the propagation of gravitons for the ε-formalism together with the obtainment of the patterns for their interactions with external electromagnetic fields. In Refs. [30, 32], the full $\gamma\varepsilon$-description of the propagation of spin curvatures in vacuum is brought out. It thus appears that the couplings between gravitons and photons are strictly borne in both formalisms by the wave equations that govern the electromagnetic propagation. The propagation of gravitons in the presence of arbitray sources is described in Ref. [33] where a somewhat important condition on the first covariant derivative of energy-momentum tensors is deduced. A specialization of this description for the particular case of sources coming from eletromagnetic curvatures, is given in Ref. [26]. The work of Ref. [27] touches upon an interesting situation concerning the occurrence of geometric sources in the field equations for Infeld-van der Waerden photons. In Ref. [30], it was suggested for the first time that a description of some of the physical properties of the cosmic microwave background could be achieved by looking at the propagation in Friedmann-like conformally flat spacetimes of electromagnetic curvatures. A notable class of conformally flat spacetimes which admit decomposable Christoffel connexions, was considered in Ref. [21] in conjunction with a derivation of the corresponding spin-affine and curvature configurations for the γ-formalism. It was shown thereabout that the whole derivation can actually be implemented only if a specific constancy property is imposed on one of the spin densities borne by the expression for a characteristic γ-metric function. Explicit expressions for the gravitational spinors of those spacetimes were then derived. A detailed description of the interaction couplings that take place in the formulation of Dirac's theory in curved spacetimes has likewise been given [22]. This latter work has really made up the original description of Dirac fields as given by Infeld and van der Waerden.

In the present work, we exhibit the formulation of the theory of classical Proca fields within the framework of the γ-formalism. The theory of spin densities and the gauge transformations inherently borne by the ε-formalism [30, 31] will not be exhibited by this point since the ε-counterparts of our key developments just bear meaninglessness. The spinor field equations are obtained out of transcribing directly the statements that make out the world version of the theory. Some well-known calculational techniques are then utilized for deriving the wave equations that control the propagation of the fields taken into consideration. Indeed, these techniques are just the same as the ones employed in Refs. [23, 30] for obtaining the typical wave equations of the entire $\gamma\varepsilon$-framework. It thus appears that no interaction couplings between Proca fields and electromagnetic curvatures are ultimately carried by the resulting wave equations. What comes about is, in effect, that the only interactions which occur in the theoretical context being considered involve strictly Proca fields and wave functions for gravitons. One of our motivations for elaborating upon the situation entertained herein is related to the absence from the literature of any two-component description of the propagation of external massive bosons in generally relativistic spacetimes. In our view, it might be worthwhile to work out such a massive case.

We will allow for the notation adhered to in Ref. [30] except that spacetime components will now be labelled by lower-case Greek letters. Kernel letters for world and spin quantities will broadly appear as Greek and Latin letters. In particular, we denote as x^μ some local coordinates on a spacetime $\mathfrak{M}$ equipped with a torsionless covariant derivative operator ∇_μ. The partial derivative operator for x^μ is denoted by ∂_μ. A world metric tensor $g_{\mu\nu}$ on $\mathfrak{M}$ presumably bears the local signature $(+---)$. We require $g_{\mu\nu}$ to fulfill the metric compatibility condition

$$\nabla_\mu g_{\lambda\sigma} = 0.$$

The Riemann tensor of ∇_μ is written as $R_{\mu\nu\lambda\sigma}$. Our sign convention for the respective Ricci tensor $R_{\mu\nu}$ is the same as the one adopted in Ref. [4]. The determinant of $g_{\mu\nu}$ and the covariant Levi-Civita world density in $\mathfrak{M}$ will especially be denoted as $\mathfrak{g}$ and $\epsilon_{\mu\nu\lambda\sigma}$, respectively. We shall use the primed-unprimed index notation of Ref. [4] upon dealing with conjugate spinor components. World indices all range over the four values $0, 1, 2, 3$ whereas spinor indices take either the values $0, 1$ or $0', 1'$. We will utilize the convention according to which the effect on any index block of the actions of the symmetry and antisymmetry operators is indicated by surrounding the indices singled out with round and square brackets, respectively. A horizontal bar lying over a kernel letter will sometimes be used to denote the operation of complex conjugation. Further conventions will be explained in due course.

Our outline has been set as follows. In Section 2, the world version of the Proca theory is formulated on the basis of the standard least-action principle for classical fields in curved spacetimes [34]. Section 3 brings out the overall system of spinor field equations. In Section 4, we carry out the derivation of our wave equations. Some remarks on our work are made in Section 5. The calculational techniques referred to above shall be taken for granted from the beginning.

2 World theory

The least-action principle for the Proca theory in $\mathfrak{M}$ is written as

$$\delta S = \delta \int_{\Omega} \mathcal{L}\sqrt{-\mathfrak{g}}d^4x = 0, \tag{1}$$

where $\mathcal{L}$ denotes the Lagrangian density

$$\mathcal{L} = -\frac{1}{4}f^{\mu\nu}f_{\mu\nu} + \frac{1}{2}m^2 A^{\mu}A_{\mu}, \tag{2}$$

which carries the Proca bivector

$$f_{\mu\nu} = 2\partial_{[\mu}A_{\nu]} = 2\nabla_{[\mu}A_{\nu]}, \tag{3}$$

with A_μ and m being a Proca potential and the mass of $f_{\mu\nu}$, respectively. Usually, the variation δ bears linearity and obeys the Leibniz rule, in addition to being defined so as to commute with partial derivatives and integrations. The integral of Eq. (1) is taken over a volume Ω in $\mathfrak{M}$ whose closure is compact, and

$$d^4x = \frac{1}{4!}\epsilon_{\mu\nu\lambda\sigma}dx^{\mu} \wedge dx^{\nu} \wedge dx^{\lambda} \wedge dx^{\sigma} \tag{4}$$

defines an elementary volume density in Ω.

With the help of Eqs. (2) and (3), we can reinstate the statement (1) as

$$\delta S = \int_{\Omega} (-f^{\mu\nu}\partial_{\mu}\delta A_{\nu} + m^2 A^{\nu}\delta A_{\nu})\sqrt{-\mathfrak{g}}d^4x = 0, \tag{5}$$

where δA_ν is taken as an arbitrary covariant quantity in Ω that vanishes on the boundary $\partial\Omega$ of Ω. Hence, performing an integration by parts in (5), yields

$$\int_{\Omega} [\frac{1}{\sqrt{-\mathfrak{g}}}\partial_{\mu}(\sqrt{-\mathfrak{g}}f^{\mu\nu}) + m^2 A^{\nu}]\sqrt{-\mathfrak{g}}\delta A_{\nu}d^4x$$
$$- \int_{\partial\Omega} f^{\mu\nu}\sqrt{-\mathfrak{g}}\delta A_{\nu}d^3x_{\mu} = 0, \tag{6}$$

Journal of Physics: Conference Series **706** (2016) 052033 doi:10.1088/1742-6596/706/5/052033

with

$$d^3x_\mu = \frac{1}{3!}\epsilon_{\mu\nu\lambda\sigma}dx^\nu \wedge dx^\lambda \wedge dx^\sigma, \tag{7}$$

whence we can write down the field equations

$$\frac{1}{\sqrt{-\mathfrak{g}}}\partial_\mu(\sqrt{-\mathfrak{g}}f^{\mu\nu}) + m^2A^\nu = 0, \tag{8}$$

which amount to the same thing as

$$\nabla_\mu f^{\mu\nu} + m^2A^\nu = 0. \tag{9}$$

Equations (9) constitute the first world half of Proca's theory in $\mathfrak{M}$. The second half comes into play as the Bianchi identity

$$\nabla_{[\mu}f_{\lambda\sigma]} = 0, \tag{10}$$

which can be reexpressed as

$$\nabla^\mu \, {}^*f_{\mu\nu} = 0, \tag{11}$$

where

$${}^*f_{\mu\nu} = \frac{1}{2}\sqrt{-\mathfrak{g}}\epsilon_{\mu\nu\lambda\sigma}f^{\lambda\sigma} \tag{12}$$

is the dual bivector of $f_{\mu\nu}$. It follows that, by taking the covariant divergence of the left-hand side of Eq. (9), likewise combining together the relations

$$\nabla_\nu\nabla_\mu f^{\mu\nu} = -\nabla_{[\mu}\nabla_{\nu]}f^{\mu\nu} = -\frac{1}{2}[\nabla_\mu, \nabla_\nu]f^{\mu\nu} \tag{13}$$

and the commutator expansion

$$[\nabla_\mu, \nabla_\nu]f^{\mu\nu} = R_{\mu\nu\lambda}{}^\mu f^{\lambda\nu} + R_{\mu\nu\lambda}{}^\nu f^{\mu\lambda} = 2R_{\mu\nu}f^{\mu\nu} \equiv 0, \tag{14}$$

we promptly arrive at the condition

$$\nabla_\mu A^\mu = 0. \tag{15}$$

A similar procedure involves inserting (3) into (9) to get

$$(\Box + m^2)A_\mu - \nabla_\lambda\nabla_\mu A^\lambda = 0, \tag{16}$$

whence, making use of the equality

$$\nabla_\lambda\nabla_\mu A^\lambda = \nabla_\mu(\nabla_\lambda A^\lambda) + R_{\lambda\mu\sigma}{}^\lambda A^\sigma, \tag{17}$$

and calling upon (15), we end up with the wave equation

$$(\Box + m^2)A_\mu + R_\mu{}^\lambda A_\lambda = 0. \tag{18}$$

Needless to say, Eq. (15) may be thought of as a theoretically intrinsic subsidiary condition on A_μ.

3 Spinor field equations

By definition, any Proca wave functions enter formal bivector expansions like

$$\sigma^{\mu}_{AA'}\sigma^{\nu}_{BB'}f_{\mu\nu} = f_{AA'BB'} = \gamma_{A'B'}\psi_{AB} + \gamma_{AB}\psi_{A'B'}, \tag{19}$$

with

$$\psi_{AB} = \frac{1}{2}f_{ABC'}{}^{C'} = \psi_{(AB)},\ \psi_{A'B'} = \frac{1}{2}f_{A'B'C}{}^{C} = \psi_{(A'B')}, \tag{20}$$

and $(\gamma_{AB}, \gamma_{A'B'})$ being a pair of covariant metric spinors for the γ-formalism. The σ-symbols carried by Eq. (19) are some of the pertinent Hermitian connecting objects, which supposedly fulfill the covariant constancy requirement (see, for instance, Ref. [32])

$$\nabla_{\mu}\sigma^{\lambda}_{BB'} = 0. \tag{21}$$

Accordingly, either of ψ_{AB} and $\psi_{A'B'}$ is a massive spin-one uncharged field that represents locally the six degrees of freedom of $f_{\mu\nu}$ in $\mathfrak{M}$. The corresponding field-potential relationships are given by

$$\psi_{AB} = -\nabla^{C'}_{(A}A_{B)C'},\ \psi_{A'B'} = -\nabla^{C}_{(A'}A_{B')C} \tag{22}$$

and

$$\psi^{AB} = \nabla^{(A}_{C'}A^{B)C'},\ \psi^{A'B'} = \nabla^{(A'}_{C}A^{B')C}. \tag{23}$$

In passing, we point out that, in deriving Eqs. (22) and (23), it may be necessary to implement the Infeld-van der Waerden eigenvalue equations [6, 31]

$$\nabla_{\mu}\gamma_{BC} = i\beta_{\mu}\gamma_{BC},\ \nabla_{\mu}\gamma^{BC} = -i\beta_{\mu}\gamma^{BC}, \tag{24}$$

as well as their complex conjugates. The quantity β_{μ} amounts to a gauge invariant world vector, which is expressed by

$$\beta_{\mu} = \nabla_{\mu}\Phi + 2\Phi_{\mu}, \tag{25}$$

with Φ and Φ_{μ} being, respectively, the polar argument of the independent component of γ_{AB} and a γ-formalism electromagnetic potential. It is useful to introduce the Maxwell bivector associated to Φ_{μ}. We have, in effect,

$$F_{\mu\nu} = 2\partial_{[\mu}\Phi_{\nu]} = 2\nabla_{[\mu}\Phi_{\nu]}. \tag{26}$$

The spinor decomposition of $F_{\mu\nu}$ takes up the wave functions ϕ_{AB} and $\phi_{A'B'}$ which thus supply dynamical states for Infeld-van der Waerden photons in $\mathfrak{M}$. These wave functions essentially constitute the electromagnetic curvature of $\mathfrak{M}$, thereby being deeply rooted into the geometric structure of $\mathfrak{M}$. It is obvious that the geometric field-potential relationships may right away be attained from (22) and (23) by making trivial replacements of kernel letters. Such relationships shall be utilized later in Section 4.

Journal of Physics: Conference Series **706** (2016) 052033 doi:10.1088/1742-6596/706/5/052033

The first spinor half of Proca's theory arises here from the two-component transcription of Eq. (9) whence, by invoking (21), we obtain the field equations[1]

$$\nabla^{AA'}(\gamma_{A'B'}\psi_{AB} + \text{c.c.}) + m^2 A_{BB'} = 0. \tag{27}$$

The second half now consists of the statements

$$\nabla^{AA'} f^{*}_{AA'BB'} = i\nabla^{AA'}(\gamma_{AB}\psi_{A'B'} - \text{c.c.}) = 0, \tag{28}$$

which effectively account for the dual expansion

$$\sigma^{\mu}_{AA'}\sigma^{\nu}_{BB'} f^{*}_{\mu\nu} = i(\gamma_{AB}\psi_{A'B'} - \gamma_{A'B'}\psi_{AB}). \tag{29}$$

Of course, Eq. (28) may be reset as the Hermitian configuration

$$\nabla^{AA'}(\gamma_{A'B'}\psi_{AB}) = \nabla^{AA'}(\gamma_{AB}\psi_{A'B'}). \tag{30}$$

We should also observe that the pattern of Eq. (29) oftenly emerges from the combination of (19) with the alternating expansion

$$\sqrt{-\mathfrak{g}}\,\epsilon_{AA'BB'CC'DD'} = i(\gamma_{AC}\gamma_{BD}\gamma_{A'D'}\gamma_{B'C'} - \gamma_{AD}\gamma_{BC}\gamma_{A'C'}\gamma_{B'D'}). \tag{31}$$

Typically, the entire Proca theory in $\mathfrak{M}$ is written out explicitly as the field equations

$$\nabla^{A}_{B'}\psi_{AB} + \frac{1}{2}m^2 A_{BB'} - i\beta^{A}_{B'}\psi_{AB} = 0 \tag{32}$$

and

$$\nabla^{B'}_{A}\psi^{AB} - \frac{1}{2}m^2 A^{BB'} + i\beta^{B'}_{A}\psi^{AB} = 0, \tag{33}$$

together with the conjugates of (32) and (33). A formal simplification to it can be accomplished by utilizing Eqs. (24) along with metric prescriptions of the type

$$\nabla^{AB'}\psi_{AB} = \nabla^{AB'}(\psi^{C}_{A}\gamma_{CB}),\ \nabla_{AB'}\psi^{AB} = \nabla_{AB'}(\gamma^{BC}\psi^{A}_{C}). \tag{34}$$

For the unprimed wave functions, for instance, we thus have the equivalent statements

$$\nabla^{AB'}\psi^{B}_{A} + \frac{1}{2}m^2 A^{BB'} = 0,\ \nabla_{AB'}\psi^{A}_{B} - \frac{1}{2}m^2 A_{BB'} = 0. \tag{35}$$

Evidently, the symmetry borne by the wave functions makes it immaterial to stagger their indices.

[1] The symbol "c.c." will henceforth denote an overall complex conjugate piece.

4 Wave equations

At this stage, we shall follow up the procedure which amounts to implementing the calculational techniques mentioned in Section 1 towards deriving the wave equations for the fields that occur in the statements (32)-(35). We will initially work out the procedure for ψ^{AB} and ψ_A^B. The wave equation for ψ_{AB} will then be obtained by taking into effect a valence interchange rule that had been deduced originally [30] in connection with the presentation of the general description of $\gamma\varepsilon$-curvatures. We may certainly get the wave equations for any primed fields by taking complex conjugates. Equations (24) will be used so many times in what follows that we will no longer refer to them explicitly.

We start by operating with $\nabla_{CB'}$ on the configuration of Eq. (33). Hence, using the operator correlation

$$\nabla_{CB'}\nabla_A^{B'} = \gamma_{LC}\nabla_{B'}^L(\gamma_{MA}\nabla^{MB'}) = i\beta_{CB'}\nabla_A^{B'} + \gamma_{LC}\gamma_{MA}\nabla_{B'}^L\nabla^{MB'}, \tag{36}$$

together with the splitting

$$\nabla_{B'}^L\nabla^{MB'} = \Delta^{LM} + \frac{1}{2}\gamma^{LM}\Box \tag{37}$$

and the definition[2]

$$\Delta^{AB} = \nabla_{C'}^{(A}\nabla^{B)C'}, \tag{38}$$

we get the contribution

$$\nabla_{CB'}\nabla_A^{B'}\psi^{AB} = i\beta_{CB'}\nabla_A^{B'}\psi^{AB} + (\Delta_{AC} - \frac{1}{2}\gamma_{AC}\Box)\psi^{AB}. \tag{39}$$

The Δ-derivative of (39) reads

$$\Delta_{AC}\psi^{AB} = \frac{R}{6}\gamma_{CA}\psi^{AB} + \Psi_{AMC}{}^B\psi^{AM} - 2i\phi_{AC}\psi^{AB}, \tag{40}$$

where ϕ_{AB} stands for a wave function for Infeld-van der Waerden photons and Ψ_{ABCD} is a wave function for gravitons in $\mathfrak{M}$. For the β-term of (39), we have

$$i\beta_{CB'}\nabla_A^{B'}\psi^{AB} = i(\beta_{B'(A}\nabla_{C)}^{B'} - \frac{1}{2}\gamma_{AC}\beta^\mu\nabla_\mu)\psi^{AB}. \tag{41}$$

In addition, recalling the unprimed relation of (23), produces the following expansion for the differential kernel of the operated mass term of (33):

$$\nabla_{CB'}A^{BB'} = \gamma_{MC}(\psi^{MB} + \frac{1}{2}\gamma^{MB}\nabla_\mu A^\mu), \tag{42}$$

which, by virtue of the condition (15), may be simplified to

$$\nabla_{CB'}A^{BB'} = \gamma_{MC}\psi^{MB}. \tag{43}$$

[2]The object $\Box$ equals the covariant D'Alembertian operator $\nabla^\mu\nabla_\mu$ whilst Δ^{AB} is linear and enjoys the Leibniz rule property.

We have next to allow for the contribution

$$\nabla_{CB'}(i\beta_A^{B'}\psi^{AB}) = (i\nabla_{CB'}\beta_A^{B'})\psi^{AB} + i\beta_A^{B'}\nabla_{CB'}\psi^{AB}. \tag{44}$$

It is evident that the sum of the β-term of (39) with the second term lying on the right-hand side of (44), bears skewness in the indices A and C. For the other individual term of (44), we spell out the auxiliary configuration

$$\nabla_{CB'}\beta_A^{B'} = \nabla_{CB'}(\gamma^{B'D'}\beta_{AD'}), \tag{45}$$

such that, after performing some computations, we get

$$(i\nabla_{CB'}\beta_A^{B'})\psi^{AB} = i(\Delta_{AC}\Phi + 2\phi_{AC})\psi^{AB}, \tag{46}$$

where Φ is given by Eq. (25). The Δ-derivative of (46) vanishes identically[3] because of the torsionlessness of ∇_μ. It follows that, fitting pieces together, yields

$$(\Box + 2i\beta^\mu\nabla_\mu + \Theta + \frac{R}{3} + m^2)\psi^{AB} - 2\Psi^{AB}{}_{LM}\psi^{LM} = 0, \tag{47}$$

with

$$\Theta = -\beta^\mu\beta_\mu + i\nabla_\mu\beta^\mu. \tag{48}$$

The entire derivation of the wave equation for the field involved in Eqs. (35) does not produce any couplings other than a gravitational one which looks like that borne by (47). Roughly speaking, the only reason for this stems from the result that we can carry out the relevant derivation without having to call for any correlations like (36) or (45), with the valence pattern of ψ_A^B accordingly ensuring the absence of any $\phi\psi$-interactions.[4] For the first of Eqs. (35), say, we thus reexpress (37) as

$$\nabla_{CB'}\nabla^{AB'} = \gamma_{MC}(\Delta^{AM} - \frac{1}{2}\gamma^{AM}\Box), \tag{49}$$

and let the splitting (49) act on ψ_A^B such that the relation (43) still holds. Consequently, by taking account of the derivative

$$\Delta_C^A\psi_A^B = \frac{R}{6}\psi_C^B + \Psi^{AB}{}_{CD}\psi_A^D, \tag{50}$$

while resetting the kernel for the mass term as

$$\nabla_{CB'}A^{BB'} = \psi_C^B, \tag{51}$$

and making some index substitutions, we obtain

$$(\Box + \frac{R}{3} + m^2)\psi_A{}^B + 2\Psi_{AD}{}^{BC}\psi_C{}^D = 0. \tag{52}$$

[3] In fact, within the $\gamma\varepsilon$-framework, the quantity Φ is looked upon as a world scalar subject to a suitable gauge behaviour.

[4] The work of Ref. [30] describes in detail on the basis of the theory of spin densities the situation related to the eventual absence of electromagnetic contributions from Δ-derivatives.

It is worth pointing out that the Δ-derivative of (50) possesses the property

$$\Delta^{A[C}\psi_A{}^{B]} = 0. \tag{53}$$

The wave equation for ψ_{AB} can indeed be derived from (47) by applying to it the simultaneous interchanges

$$i\beta^\mu \nabla_\mu \leftrightarrow -i\beta^\mu \nabla_\mu, \ \Theta \leftrightarrow \overline{\Theta}, \tag{54}$$

which come naturally from the utilization of the differential devices

$$\Box\psi_{AB} = \Box(\psi^{CD}\gamma_{CA}\gamma_{DB}), \ \Box(\gamma_{CA}\gamma_{DB}) = -\overline{\Upsilon}\gamma_{CA}\gamma_{DB} \tag{55}$$

and

$$2(\nabla_\mu\psi^{CD})\nabla^\mu(\gamma_{CA}\gamma_{DB}) = 4(2\beta^\mu\beta_\mu + i\beta^\mu\nabla_\mu)\psi_{AB}, \tag{56}$$

with the definition

$$\Upsilon = 2(\beta^\mu\beta_\mu - \overline{\Theta}). \tag{57}$$

We thus have

$$(\Box - 2i\beta^\mu\nabla_\mu + \overline{\Theta} + \frac{R}{3} + m^2)\psi_{AB} - 2\Psi_{AB}{}^{LM}\psi_{LM} = 0. \tag{58}$$

5 Concluding remarks

We saw clearly that the coupling $2i\phi_{AC}\psi^{AB}$ occurs through the expansions (40) and (46) in the derivation that leads to the wave equation (47), but it nevertheless gets cancelled when the derivation is actually carried through. If we had instead worked out explicitly the derivation procedure for ψ_{AB}, then such a $\phi\psi$-coupling would also have arisen at some intermediate calculational steps, but it would once again be cancelled upon completion of the derivation. It should be obvious that both of the wave equations (47) and (58) could be readily derived from (52) by implementing the correlations

$$(\Box\psi_A{}^C)\gamma_{CB} = (\Box - 2i\beta^\mu\nabla_\mu + \overline{\Theta})\psi_{AB}$$

and

$$\gamma^{AC}(\Box\psi_C{}^B) = (\Box + 2i\beta^\mu\nabla_\mu + \Theta)\psi^{AB},$$

along with the eigenvalue equations

$$\Box\gamma_{AB} = \Theta\gamma_{AB}, \ \Box\gamma^{AB} = \overline{\Theta}\gamma^{AB}.$$

One of the most remarkable properties of the wave equations we have exhibited is, in fact, that the only interactions carried by them just involve Proca fields and gravitational wave functions. Hence, Proca fields propagate in $\mathfrak{M}$ as if electromagnetic curvatures were absent. Moreover, in the limiting case of external massless fields, our wave equations naively recover the form of some of the traditional statements for the γ-formalism. Therefore, we could say that the work of Sections 3 and 4 has filled in the gap associated to the absence of a two-component description of external massive spin-one fields in general relativity.

References

[1] Infeld L. and Van der Waerden B.L.: Sitzber. preuss. Akad. Wiss., Physik-math. Kl. (1933) 380.

[2] Geroch R.: Jour. Math. Phys. 9 (1968) 1739.

[3] Geroch R.: Jour. Math. Phys. 11 (1970) 343.

[4] Penrose R. and Rindler W.: *Spinors and Space-Time, Vol. 1* (Cambridge University Press, Cambridge 1984).

[5] Weyl H.: Z. Physik 56 (1929) 330.

[6] Bade W.L. and Jehle H.: Rev. Mod. Phys. Vol. 25, 3 (1953) 714.

[7] Laporte O. and Uhlenbeck G.E.: Phys. Rev. 37 (1931) 1380.

[8] Jehle H.: Phys. Rev. 75 (1949) 1609.

[9] Heller J. and Bergmann P.G.: Phys. Rev. 84 (1951) 665.

[10] Thompson A. and Bergmann P.G.: Phys. Rev. 88 (1952) 1215.

[11] Corson E.M.: *Introduction to Tensors, Spinors and Relativistic Wave Equations* (Blackie, Glasgow 1953).

[12] Bergmann P.G.: Phys. Rev. Vol. 107, 2 (1957) 624.

[13] Plebanski J.: Acta Phys. Polon. 27 (1965) 361.

[14] Papapetrou A.: Compt. Rend. Series A 272, 23 (1971) 1537.

[15] Ludwig G.: Classical and Quantum Gravity 3 (1986) 141.

[16] Kroon J.A.V.: Jour. Math. Phys. 2 (2000) 898.

[17] Ludwig G. and Edgon S.B.: Gen. Rel. Grav. 34 (2002) 807.

[18] R. Penrose, Ann. Phys. 10 (1960) 171.

[19] Penrose R. and Rindler W.: *Spinors and Space-Time, Vol. 2* (Cambridge University Press, Cambridge 1986).

[20] Witten L.: Phys. Rev. Vol. 113, 1 (1959) 357.

[21] Cardoso J.G.: Jour. Math. Phys. Vol. 46, 4 (2005) 042504.

[22] Cardoso J.G.: Classical and Quantum Gravity 23 (2006) 4151.

[23] Cardoso J.G.: Acta Physica Polonica B Vol. 38, 8 (2007) 1001.

[24] Cardoso J.G.: Jour. Math. Phys. Vol. 51, 2 (2010) 023513.

[25] Cardoso J.G.: Int. Jour. Theor. Phys. **49**, (2010) 1274.

[26] Albrecht R. and Cardoso J.G.: Eur. Phys. J. Plus **126**, (2011) 77.

[27] Kuerten A. and Cardoso J.G.: Int. Jour. Theor. Phys. **50**, (2011) 3007.

[28] Cardoso J.G.: Int. Jour. Theor. Phys. **51**, (2012) 2559.

[29] Weber K. and Cardoso J.G.: Adv. Appl. Clifford Algebras **24**, Number 3, (2014) 875.

[30] Cardoso, J.G.: Czech Journal of Physics, Vol. 55, 4 (2005) 401.

[31] Cardoso, J.G.: Adv. Appl. Clifford Algebras **22**, (2012) 955.

[32] Cardoso, J.G.: Adv. Appl. Clifford Algebras **22**, (2012) 985.

[33] Cardoso, J.G.: Il Nuovo Cimento **6** Vol. 124 B, (2009) 631.

[34] M. Burgess, *Classical Covariant Fields,* Cambridge Monographs on Mathematical Physics (Cambridge University Press, 2005).

Journal of Physics: Conference Series **706** (2016) 052034 doi:10.1088/1742-6596/706/5/052034

Block-diagonal similarity renormalization group and effective nucleon-nucleon interactions

S. Szpigel[1], V. S. Timóteo[2] and E. Ruiz Arriola[3]

[1] Centro de Rádio-Astronomia e Astrofísica Mackenzie, Universidade Presbiteriana Mackenzie 01302-907, São Paulo, SP, Brasil

[2] Grupo de Óptica e Modelagem Numérica - GOMNI, Universidade Estadual de Campinas 13484-332, Limeira, SP, Brasil

[3] Departments de Física Atómica, Molecular y Nuclear, Universidad de Granada E-18071, Granada, Andalucia, España

E-mail: E-mail: szpigel@mackenzie.br (SS)

E-mail: E-mail: varese@ft.unicamp.br (VST)

E-mail: E-mail: earriola@ugr.es (ERA)

Abstract. We apply the block-diagonal similarity renormalization group to a simple toy-model for the nucleon-nucleon (NN) interaction in the $^1S_0$ channel, aiming to analyze the complementarity between the explicit and the implicit renormalization approaches in nuclear physics. By *explicit* renormalization we mean the methods based on the wilsonian renormalization group in which high-energy modes above a given cutoff scale are integrated out while their effects are replaced by scale dependent effective interactions consistently generated in the process. We call *implicit* renormalization the usual procedure of cutoff effective theories in which the high-energy modes above the cutoff scale are simply removed and their effects are included through parametrized cutoff dependent counterterms whose strengths are fixed by fitting low-energy data. We compare the effective interactions obtained in both schemes and find a wide range of cutoff scales where they overlap. We further analyze the role played by the one-pion exchange (OPE) considering a δ-shell plus OPE representation for the NN interaction.

1. Introduction

The idea of effective interactions has been strongly pursued since the late 1950s after the pioneering works by Goldstone [1], Moshinsky [2] and Skyrme [3], who suggested to use this notion to overcome the complications which appear in the solution of the nuclear many-body problem due to the large short-range repulsive core. This allowed to take advantage of the much simpler mean field framework based on those effective interactions in the implementation of nuclear structure calculations [4]. The main problem of the effective interaction approach is both the proliferation of independent parameters as well as their huge numerical diversity [5]. This reflects both the lack of an unambiguous link to the fundamental nucleon-nucleon (NN) interaction as well as the quite disparate finite nuclei and nuclear matter observables which have been used to fix the effective theory parameters. In recent works [6, 7] an effort has been made in order to understand the origin of the NN effective interactions from free space NN scattering in a model independent way, without invoking finite nuclei nor nuclear matter properties. This approach corresponds to what will be called here as *implicit* renormalization.

Journal of Physics: Conference Series **706** (2016) 052034 doi:10.1088/1742-6596/706/5/052034

An intense reformulation of the nuclear many-body problem has taken place in the last ten years inspired by the wilsonian renormalization group ideas, providing an alternative approach to the determination of effective interactions directly from the NN *bare* potentials fitted to the scattering data (for reviews see e.g. [8, 9, 10] and references therein). The basis of the whole approach is to take advantage of choosing the proper resolution scale in the formulation of the problem, separating explicitly what degrees of freedom and interactions behave dynamically below that scale. This approach corresponds to what will be called here as *explicit* renormalization. Within this framework, the so called $V_{\text{low k}}$ approach [11] has been used to consistently integrate out the high-momentum components of both high-precision [12, 13] and chiral effective field theory (ChEFT) [14, 15] NN potentials in order to derive phase-shift equivalent softer forms. A more recent approach that has been applied in this context is the similarity renormalization group (SRG), which provides a great deal of simplification in the derivation of effective interactions for many-body calculations in nuclear physics [16]. The basic strategy underlying the application of the SRG methods to nuclear forces is to evolve the initial *bare* potential via a continuous unitary transformation that runs a cutoff λ on energy differences. Such a transformation generates a family of unitarily equivalent smooth interactions $H_\lambda = U_\lambda \; H \; U_\lambda^\dagger$ with a band-diagonal structure of a prescribed width roughly given by the SRG cutoff λ. One of the main advantages of the SRG method over the $V_{\text{low k}}$ is that it allows for a straightforward and consistent treatment of the scale dependence of induced as well as initially introduced many-body forces [17].

In this contribution we present a summary of previous works [18, 19, 20] where we have applied the block-diagonal similarity renormalization group (BD-SRG) scheme [21] to the NN system. We considered a simple toy-model for the NN interaction in the $^1S_0$ and $^3S_1$ partial-wave channels, which is constructed such that the phase-shifts at low-momenta and the deuteron binding-energy are reasonably described with a short-range interaction and makes the SRG evolution towards small values of the SRG cutoff λ much more practical. As we have shown, the BD-SRG allows to implement a block-diagonal separation of the Hilbert space in two orthogonal (decoupled) subspaces, $\mathcal{H} = \mathcal{H}_P \oplus \mathcal{H}_Q$, which are respectively below or above a cutoff scale Λ, which will be referred as the block-diagonal (BD) cutoff. The unitary evolution runs the SRG cutoff from $\lambda \to \infty$ (the ultraviolet limit) to $\lambda \to 0$ (the infrared limit) and interpolates between the initial *bare* hamiltonian $H \equiv H_{\lambda\to\infty}$ and the block-diagonal one $H_{\lambda\to 0}$. This corresponds to a unitary implementation to all energies of the $V_{\text{low k}}$ approach in which the higher energy states are missing and in practice a free theory is assumed above the energy determined by the BD cutoff Λ. We compared the effective interactions obtained in the explicit and implicit renormalization approaches and analyzed to what extent in terms of the BD cutoff Λ the two approaches overlap. Here we restrict the analysis to the case of the $^1S_0$ channel NN interaction. We also discuss the results of a preliminary study on the role played by the one-pion exchange (OPE) interaction which is carried out by implementing the BD-SRG evolution of the δ-shell plus OPE representation for the NN interaction described in Ref. [22].

2. Toy-model separable gaussian potential

In the applications of the SRG method to nuclear physics, realistic potentials which fit NN scattering data up to the pion-production threshold ($\sqrt{m_\pi M_N} \sim 400$ MeV) are usually taken as the initial NN *bare* interaction. Due to the short-range repulsive core, such potentials exhibit a long high-momentum tail which complicates the numerical convergence when solving the SRG flow-equations. Therefore, for illustration purposes, we will consider here as the NN *bare* interaction a simple separable gaussian potential in the $^1S_0$ channel, given by

$$V(p,p') = C \; g_L(p) g_L(p') = C \; \exp\left[-\left(p^2 + p'^2\right)/L^2\right] \; . \tag{1}$$

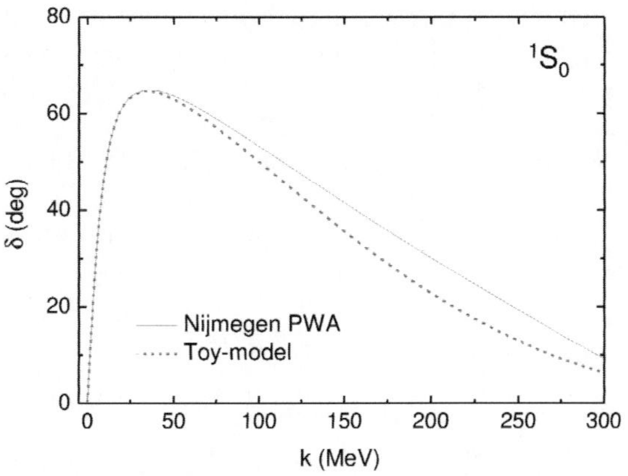

Figure 1. Phase-shifts for the toy-model potential in the $^1S_0$ channel compared to the results obtained from the Nijmegen PWA [13].

The parameters C and L are determined from the solution of the Lippmann-Schwinger (LS) equation for the on-shell reactance matrix K by fitting the experimental values of the Effective Range Expansion (ERE) parameters. Namely, we solve the partial-wave LS equation for the K-matrix,

$$K(p,p';E) = V(p,p') + \frac{2}{\pi}\,\mathcal{P}\int_0^\infty dq\, q^2\, \frac{V(p,q)}{E-q^2}\, K(q,p';E)\ , \tag{2}$$

where E is the scattering energy, and match the resulting on-shell K-matrix to the ERE expansion,

$$K^{-1}(k,k;k^2) = -\left[-\frac{1}{a_0} + \frac{1}{2}\, r_e\, k^2 + \mathcal{O}(k^4)\right] = -k\, \cot\, \delta(k)\ , \tag{3}$$

where $k = \sqrt{E}$ is the on-shell momentum and $\delta(k)$ stands for the phase-shifts. Adjusting the parameters of the toy model potential in the $^1S_0$ channels to fit the ERE parameters to second order in the on-shell momentum k, i.e. the scattering length $a_0 = -23.74$ fm and the effective range $r_e = 2.77$ fm we obtain $C = -1.915884$ fm and $(1/L^2) = 0.691269$ fm^2.

In the case of the toy-model separable potential, given by Eq. (1), it is straightforward to evaluate the phase-shifts $\delta(k)$ from the solution of the partial-wave LS equation for the K-matrix using the *ansatz*

$$K(p,p';k^2) = g_L(p)\ \tilde{K}(k)\ g_L(p')\ , \tag{4}$$

where $\tilde{K}(k)$ is called the reduced on-shell K-matrix. This leads to the relation

$$k\cot\delta(k) = -\frac{1}{V(k,k)}\left[1 - \frac{2}{\pi}\,\mathcal{P}\int_0^\infty dq\, q^2 \frac{1}{k^2-q^2} V(q,q)\right]\ . \tag{5}$$

The phase-shifts for the toy-model potential in the $^1S_0$ channel evaluated from this equation are shown in Fig. 1, compared to the results obtained from the Nijmegen partial-wave analysis (PWA) [13]. As one can see, despite its simplicity, our toy model for the NN interaction provides a reasonable qualitative description of the $^1S_0$ phase-shifts.

3. Explicit renormalization: BD-SRG evolution

The similarity renormalization group (SRG) approach, developed by Glazek and Wilson [23] and independently by Wegner [24], is a renormalization method based on a series of continuous unitary transformations that evolve hamiltonians with a cutoff on energy differences, driving the original hamiltonian towards a band-diagonal form. Here we employ the formulation for the SRG developed by Wegner, which is based on a non-perturbative flow-equation that governs the unitary evolution of a hamiltonian $H = T_{\rm rel} + V$ with a flow parameter s that ranges from zero to infinity. Namely, assuming that $T_{\rm rel}$ is independent of s, we have

$$\frac{dH_s}{ds} = \frac{dV_s}{ds} = [\eta_s, H_s] \, , \tag{6}$$

where η_s is an anti-hermitian operator. The flow parameter s has dimensions of $[\text{energy}]^{-2}$ and in terms of the SRG cutoff λ with dimension of momentum is given by the relation $s = \lambda^{-4}$. The flow equation is to be solved with the boundary condition $H_s|_{s\to 0} \equiv H_0$, where H_0 $(\equiv H_{\lambda\to\infty})$ is the hamiltonian corresponding to the initial *bare* interaction.

The anti-hermitian operator η_s is usually chosen as $\eta_s = [G_s, H_s]$, where G_s is a hermitian operator which we call the SRG generator since it defines η_s and so the flow of the hamiltonian. Here, we take the block-diagonal (BD) SRG generator [21], given by

$$G_s = H_s^{\rm BD} \equiv \begin{pmatrix} PH_sP & 0 \\ 0 & QH_sQ \end{pmatrix} \, , \tag{7}$$

where P and $Q = 1 - P$ are projection operators. In a partial-wave momentum-space basis, the projection operators are determined in terms of the BD cutoff Λ that divides the momentum space into a low-momentum P-space $(p < \Lambda)$ and a high-momentum Q-space $(p > \Lambda)$. Here we define the projection operators just as step functions, $P \equiv \theta(\Lambda - p)$; $Q \equiv \theta(p - \Lambda)$.

Thus, the flow-equation with the BD-SRG generator can be written in matrix-form as

$$\begin{pmatrix} \frac{d}{ds}[PV_sP] & \frac{d}{ds}[PV_sQ] \\ \frac{d}{ds}[QV_sP] & \frac{d}{ds}[QV_sQ] \end{pmatrix} = \begin{pmatrix} P\eta_sQH_sP - PH_sQ\eta_sP & P\eta_sQH_sQ - PH_sP\eta_sQ \\ Q\eta_sPH_sP - QH_sQ\eta_sP & Q\eta_sPH_sQ - QH_sP\eta_sQ \end{pmatrix} . \tag{8}$$

This equation has to be solved numerically on a finite momentum grid with N points p_n and weights w_n $(n = 1, \dots N)$ by implementing a high-momentum ultraviolet cutoff $P_{\rm max}$ and an infrared momentum cutoff $P_{\rm min}$. The discretization of the momentum space on a grid with N points leads to a system of $4N^2$ non-linear first-order coupled differential equations. By choosing the BD-SRG generator $G_s = H_s^{\rm BD}$, the matrix-elements inside the off-diagonal blocks PV_sQ and QV_sP are suppressed as the flow parameter s increases (or as the similarity cutoff λ decreases), such that the hamiltonian is driven to a block-diagonal form. In the infrared limit $s \to \infty$ $(\lambda \to 0)$ the P-space and the Q-space become completely decoupled, i.e.

$$\lim_{s\to 0} \begin{pmatrix} PV_sP & PV_sQ \\ QV_sP & QV_sQ \end{pmatrix} = \begin{pmatrix} PV_{\rm low\,k}P & 0 \\ 0 & QV_{\rm high\,k}Q \end{pmatrix} . \tag{9}$$

Thus, while unitarity implies that the phase-shifts evaluated from the solution of the LS equation remain invariant along the SRG evolution, i.e. $\delta_\lambda(p) = \delta(p)$ for any λ, one has

$$\lim_{\lambda\to 0} \delta_\lambda(p) = \delta_{\rm low\,k}(p) + \delta_{\rm high\,k}(p) \, , \tag{10}$$

where $\delta_{\rm low\,k}(p) = \delta(p)\theta(\Lambda - p)$ and $\delta_{\rm high\,k}(p) = \delta(p)\theta(p - \Lambda)$ are the phase-shifts of the $V_{\rm low\,k}$ and $V_{\rm high\,k}$ potentials respectively.

4. Implicit renormalization

Implicit renormalization can be defined in simple way by looking for an effective NN interaction regulated by a sharp or smooth momentum cutoff Λ which reproduces NN scattering data up to a given center-of-mass (CM) momentum $k \leq \Lambda$. The requirement that observables should be cutoff independent determines the implicit Λ-dependence of the effective interaction.

At low cutoffs Λ, we may approximate the toy-model potential by an effective field theory (EFT) with only contact interactions regulated by a smooth exponential momentum cutoff,

$$V_\Lambda(p,p') = \exp[-(p/\Lambda)^{2n})]\left[C_0 + C_2(p^2+p'^2)+\ldots\right]\exp[-(p'/\Lambda)^{2n}], \tag{11}$$

where $C_0, C_2, \ldots$ are Λ-dependent coefficients to be determined through a renormalization procedure and $n = 1, 2, \ldots$ defines the sharpness of the cutoff regulating function.

The running of the coefficients C_i with the cutoff Λ is determined from the solution of the LS equation for the on-shell K-matrix by fitting the experimental values of the ERE parameters. For the contact theory potential at next-to-leading order (NLO) the coefficients $C_0^{(2)}$ and $C_2^{(2)}$ are fixed at a given cutoff Λ by fitting the scattering length a_0 and the effective range r_e. One should note that, as a consequence of the Wigner causality bound, there is a maximum value $\Lambda_{\rm WB}$ for the cutoff Λ above which we cannot fix the NLO potential coefficients by fitting the experimental values of both a_0 and r_e while keeping the renormalized potential hermitian [25].

5. Comparison between the explicit and the implicit renormalization approaches

We solve the BD-SRG flow equation for the toy-model potential in the ${}^1S_0$ channel on a gaussian grid with $N = 50$ points and $P_{\rm max} = 5$ fm^{-1}, using an adaptative variable-step fifth-order Runge-Kutta algorithm. Then, we compare the running of the coefficients $C_0^{(2)}$ and $C_2^{(2)}$ with the cutoff Λ in the effective NLO contact theory potential on the same grid to the running of the corresponding coefficients $\tilde{C}_0^{(2)}$ and $\tilde{C}_2^{(2)}$ with the BD cutoff ($\equiv \Lambda$) extracted from a polynomial fit of the BD-SRG evolved toy-model potential,

$$V_{\lambda,\Lambda}(p,p') = \tilde{C}_0^{(2)} + \tilde{C}_2^{(2)}\,(p^2+p'^2)+\cdots\,. \tag{12}$$

The parameters C and L in the initial toy-model potential and the coefficients $C_0^{(2)}$ and $C_2^{(2)}$ in the NLO contact theory potential are determined from the numerical solution of the LS equation for the K-matrix by fitting the experimental values of a_0 and r_e. The coefficients $\tilde{C}_0^{(2)}$ and $\tilde{C}_2^{(2)}$ are determined by fitting the diagonal matrix-elements of the BD-SRG evolved toy-model potential for the lowest grid momenta with the polynomial form.

In Fig. 2 we show the density plots for the BD-SRG evolution of the toy-model potential in the ${}^1S_0$ channel for a BD cutoff $\Lambda = 0.5$ fm^{-1} [18]. As expected, the P-space and the Q-space become decoupled as the SRG cutoff λ decreases towards the infrared limit $\lambda \to 0$. In Fig. 3 we show the results for the coefficients $\tilde{C}_0^{(2)}$ and $\Lambda^2\tilde{C}_2^{(2)}$ extracted from the BD-SRG evolved toy-model potential in the ${}^1S_0$ channel compared to the corresponding coefficients $C_0^{(2)}$ and $\Lambda^2 C_2^{(2)}$ obtained for the NLO contact theory potential regulated by a smooth exponential momentum cutoff with a sharpness parameter $n = 16$. As one can see, there is a remarkably good agreement between the coefficients extracted from the BD-SRG evolved potential and those obtained for the NLO contact theory potential as the SRG cutoff λ decreases, which can be traced to the decoupling between the P-space and the Q-space. Thus, in the infrared limit $\lambda \to 0$ we expect to achieve a high degree of agreement between the effective interactions obtained in the explicit and the implicit renormalization approaches for BD cutoffs Λ up to $\Lambda_{\rm WB}$ determined by the Wigner causality bound for the NLO contact theory potential. Indeed, for the lowest SRG cutoff considered in the calculations, $\lambda = 0.1$ fm^{-1}, the overlap between the two approaches is verified within a range of BD cutoffs Λ from 0.5 fm^{-1} to 1.5 fm^{-1}.

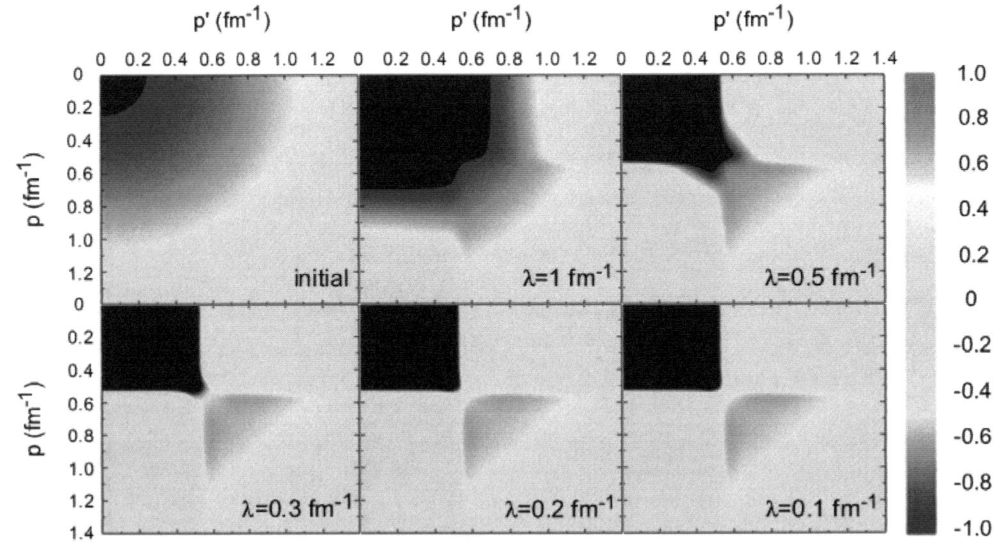

Figure 2. Density plots for the BD-SRG evolution of the toy-model potential in the $^1S_0$ channel for a BD cutoff $\Lambda = 0.5$ fm^{-1} ($N = 50$ points and $P_{\max} = 5$ fm^{-1}) [18].

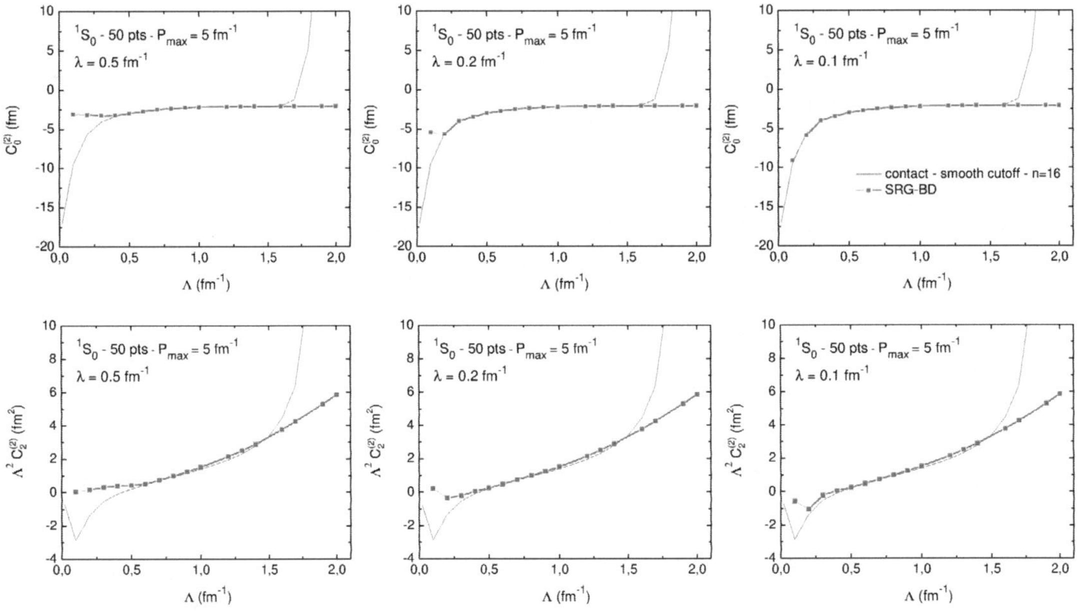

Figure 3. Coefficients $\tilde{C}_0^{(2)}$ and $\Lambda^2\tilde{C}_2^{(2)}$ extracted from the BD-SRG evolved toy-model potential in the $^1S_0$ channel compared to the corresponding coefficients $C_0^{(2)}$ and $\Lambda^2 C_2^{(2)}$ for the NLO contact theory potential regulated by a smooth exponential momentum cutoff with $n = 16$.

6. Role of the one-pion exchange

As we have pointed out, the agreement between the effective interactions obtained in the explicit and implicit renormalization approaches over a wide range of relatively low cutoff scales Λ is due to the decoupling between the low-momentum P-space and the high-momentum Q-space. This motivates an analysis of the role played by the OPE interaction in the implicit approach.

We consider the δ-shell (DS) plus OPE representation for the NN interaction described in Ref. [22], built with basis on a PWA of about 8000 pp and np data. According to this analysis, OPE is the only needed contribution for $r > 3$ fm such that the NN interaction can be split as

$$V = V(r \leq 3 \text{ fm}) + V(1\pi, r \geq 3 \text{ fm}) \,, \tag{13}$$

where $V(r \leq 3 \text{ fm})$ corresponds to the short- and intermediate-range interactions parametrized through the δ-shells and $V(1\pi, r \geq 3 \text{ fm})$ corresponds to the long-range OPE interaction. In the top-panels of Fig. 4 we show the diagonal and fully off-diagonal matrix-elements of the DS + OPE potential in the $^1S_0$ channel, together with the corresponding contributions from $V(r \leq 3 \text{ fm})$ and $V(1\pi, r \geq 3 \text{ fm})$. As one can see, $V(1\pi, r \geq 3 \text{ fm}) \ll V(r \leq 3 \text{ fm})$ and hence it might be possible to implement a perturbative expansion of the BD-SRG evolved DS + OPE potential in which only the $V(r \leq 3 \text{ fm})$ piece is evolved and the corrections from the evolved OPE are included order by order, i.e.

$$V_{\lambda,\Lambda} \equiv V_{\lambda,\Lambda}(r \leq 3 \text{ fm}) + V(1\pi, r \geq 3 \text{ fm}) + \mathcal{O}\left[V^2_{\lambda,\Lambda}(1\pi, r \geq 3 \text{ fm})\right] \,. \tag{14}$$

In the bottom panels of Fig. 4 we show the diagonal and fully off-diagonal matrix-elements of the difference between the full DS + OPE potential in the $^1S_0$ channel evolved with the BD-SRG up to $\lambda = 1 \text{ fm}^{-1}$ and the zeroth-order perturbative approximation for several values of the BD cutoff Λ. We get that the corrections from the BD-SRG evolved OPE piece are indeed small, namely $\mathcal{O}[V^2_{\lambda,\Lambda}(1\pi, r \geq 3 \text{ fm})] \leq 10^{-2}$ fm for $0.5 \text{ fm}^{-1} \leq \Lambda \leq 3.0 \text{ fm}^{-1}$. This result suggests that the contribution from the BD-SRG evolved OPE interaction may be treated perturbatively.

7. Summary and Outlook

We have presented a summary of our previous works on the application of the block-diagonal similarity renormalization group (BD-SRG) to the NN interaction, whose main purpose was to investigate the complementarity between the implicit and the explicit renormalization approaches in nuclear physics. In order to simplify the analysis and reduce the computational effort, we have considered a separable gaussian potential toy-model for the NN interaction in S-wave channels. Here we have focused on the analysis of the NN system in the $^1S_0$ channel.

By comparing the NN effective interactions obtained both in the explicit and the implicit renormalization schemes, we verify that the complementarity between these two approaches holds for a wide range of cutoff scales Λ. This suggests that the bulk of the effective NN interaction and its scale dependence can be extracted directly from low-energy NN data. We have also presented a preliminary analysis of the role played by the OPE interaction, based on a perturbative expansion of the BD-SRG evolved $^1S_0$ channel NN potential in a δ-shell plus OPE representation. As we have shown, the contribution from the OPE interaction remains small after the BD-SRG evolution, suggesting that it may be treated perturbatively in calculations of light nuclei structure. In a forthcoming publication we will present a detailed study on this issue, including also an analysis of the relevance of the two-pion exchange (TPE) interaction.

Acknowledgements

E.R.A. would like to thank the Spanish DGI (Grant FIS2011-24149) and Junta de Andalucia (Grant FQM225). S.S. is partially supported by FAPESP and V.S.T. thanks FAEPEX, FAPESP and CNPq for financial support.

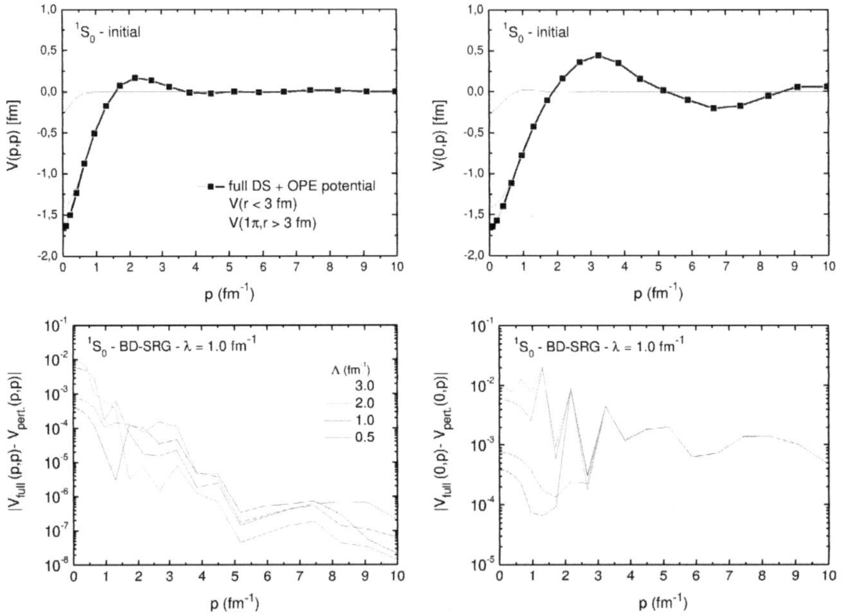

Figure 4. Top-panels: diagonal and fully off-diagonal matrix-elements of the DS + OPE potential in the $^1S_0$ channel, together with the corresponding contributions from $V(r \leq 3$ fm) and $V(1\pi, r \geq 3$ fm); Bottom panels: diagonal and fully off-diagonal matrix-elements of the difference between the full DS + OPE potential evolved with the BD-SRG up to $\lambda = 1$ fm^{-1} and the zeroth-order perturbative approximation for several values of the BD cutoff Λ.

References

[1] Goldstone J 1957 *Proc. Roy. Soc. Lond. A Math. Phys. Eng. Sci.* **239** 267–279
[2] Moshinsky M 1958 *Nuclear Physics* **8** 19 – 40 ISSN 0029-5582
[3] Skyrme T 1959 *Nucl. Phys.* **9** 615–634
[4] Bender M, Heenen P H and Reinhard P G 2003 *Rev. Mod. Phys.* **75** 121–180
[5] Dutra M, Lourenco O, Sa Martins J S, Delfino A, Stone J R *et al.* 2012 *Phys. Rev.* **C85** 035201
[6] Ruiz Arriola E 2010 (*Preprint* `1009.4161`)
[7] Navarro Perez R, Amaro J E and Ruiz Arriola E 2013 *Phys. Rev.* **C88** 064002
[8] Bogner S K, Furnstahl R J and Schwenk A 2010 *Prog. Part. Nucl. Phys.* **65** 94–147
[9] Furnstahl R J 2012 *Nucl.Phys.Proc.Suppl.* **228** 139–175
[10] Furnstahl R J and Hebeler K 2013 *Rept. Prog. Phys.* **76** 126301
[11] Bogner S K, Furnstahl R J, Ramanan S and Schwenk A 2007 *Nucl. Phys.* **A784** 79–103
[12] Wiringa R B, Stoks V G J and Schiavilla R 1995 *Phys. Rev.* **C51** 38–51
[13] Stoks V G J, Klomp R A M, Terheggen C P F and de Swart J J 1994 *Phys. Rev.* **C49** 2950–2962
[14] Entem D R and Machleidt R 2003 *Phys. Rev.* **C68** 041001
[15] Epelbaum E, Glockle W and Meissner U G 2005 *Nucl.Phys.* **A747** 362–424
[16] Bogner S K, Furnstahl R J and Perry R J 2007 *Phys. Rev.* **C75** 061001
[17] Jurgenson E D, Navratil P and Furnstahl R J 2011 *Phys. Rev.* **C83** 034301
[18] Ruiz Arriola E, Szpigel S and Timoteo V S 2014 *Few Body Syst.* **55** 989–992
[19] Ruiz Arriola E, Szpigel S and Timoteo V S 2014 *Phys. Lett.* **B728** 596–601
[20] Ruiz Arriola E, Szpigel S and Timoteo V S 2015 *Annals of Phys.* **353** 129–149
[21] Anderson E *et al.* 2008 *Phys. Rev.* **C77** 037001
[22] Navarro Perez R, Amaro J E and Ruiz Arriola E 2013 *Phys. Rev.* **C88** 024002
[23] Glazek S D and Wilson K G 1993 *Phys. Rev.* **D48** 5863–5872
[24] Wegner F 1994 *Annalen der physik* **506** 77–91 ISSN 1521-3889
[25] Szpigel S, Timoteo V S and Duraes F O 2011 *Annals of Phys.* **326** 364–405

Jet Fragmentation and Energy Loss in Ultrarelativistic Heavy Ion Collisions

[1]S. Martins[†], [2]C. Brenner Mariotto[⋆]
[†]Instituto de Física, Universidade Federal do Rio Grande do Sul,
Caixa Postal 15051 - CEP 91501-970 - Porto Alegre, RS, Brazil
[⋆]Instituto de Matemática, Estatística e Física, Universidade Federal do Rio Grande,
Av. Itália, km 8, Campus Carreiros, CEP 96203-900, Rio Grande, RS, Brazil
E-mail: [1]sony.fisica@gmail.com, [2]cristianomariotto@furg.br

Abstract. In this work we study the suppression of charged hadrons ($\pi^{\pm}$, $K^{\pm}$, p and $\bar{p}$) in ultrarelativistic heavy ion (UHI) collisions, where the quark-gluon plasma (QGP) is believed to be produced. Jet energy loss in the QGP is the main explanation for jet quenching, which may suppress the number of hadrons produced while the jets traverse the dense and colorful medium created. By taking into account the energy loss of the partons traversing the medium and the nuclear shadowing of the parton distributions depending on the impact-parameter, we calculate the nuclear modification factor R_{AA} for lead-lead collisions at the LHC. Finally, we estimate the opacity of the QGP in several centrality classes.

1. Introduction

The suppression of high-p_T jets is one of the signals of the formation of the quark-gluon plasma (QGP) in ultrarelativistic heavy-ion (UHI) collisions at RHIC and LHC. The medium-induced gluon radiation is claimed to be the dominant mechanism underlying the jet quenching phenomena observed in UHI collisions. In this contribution, we study the production of charged hadrons in UHI collisions at the LHC - a good scenary to test the properties of the hot and dense medium created. This study is motivated by recent LHC data [1] on the production of charged particles in several centrality intervals. Our approach is to consider nuclear parton distributions dependent on the impact-parameter [2] and an energy loss mechanism [3] which affects the jet fragmentation processes, in order to calculate the nuclear modification factor R_{AA} in several centrality classes.

2. Particle Production in pp Collisions

The baseline calculation is the cross section for charged hadron production in proton-proton (pp) collisions, which can be written as

$$\frac{d\sigma^{pp\rightarrow h+X}}{dp_T^2dy} = \sum_{abcd}\int dx_a dx_b f_a^p\left(x_a, Q^2\right) f_b^p\left(x_b, Q^2\right) \frac{2}{z_c}\frac{d\hat{\sigma}}{d\hat{t}}(ab\rightarrow cd) D_{h/c}\left(z_c, \mu_f^2\right) \tag{1}$$

where $\frac{d\hat{\sigma}}{dt}$ are the partonic cross sections (all $2\rightarrow 2$ processes), $f^p_{a,b}\left(x, Q^2\right)$ are the parton distribution functions (PDF's) in the proton and $D_{h/c}\left(z_c, {\mu_f}^2\right)$ are the fragmentation functions relating the partons produced (c) and the formed hadrons (h). The CTEQ6L [4] parametrization is used for the parton distributions $f^p_{a,b}(x,Q^2)$ and the KKP [5] parametrization is used for the fragmentation functions $D_{h/c}(z_c,{\mu_f}^2)$. We use $Q = q_T$ and $\mu_f = q_T/z_c$, where q_T is the transverse momentum of the parton produced in the hard scattering and z_c is the momentum fraction of this parton carried by the hadron produced via fragmentation. The other kinematic variables are

$$x_{a,b} = \frac{q_T}{\sqrt{s}}\left(e^{\pm y_c} + e^{\pm y_d}\right) \qquad z_c = \frac{x_T e^{-y}}{2x_b} + \frac{x_T e^{y}}{2x_a}$$

$$x_{a,min} = \frac{x_T e^{y}}{2 - x_T e^{-y}} \qquad x_{b,min} = \frac{x_a x_T e^{-y}}{2x_a - x_T e^{y}}$$

with $x_T = \frac{2m_T}{\sqrt{s}}$ and $m_T = \sqrt{m_c^2 + q_T^2}$. Also, y_c and y_d are the rapidities of the produced partons, whereas p_T and y are the transverse momentum and rapidity of the produced hadron.

3. Particle Production in UHI Collisions

In the eikonal limit, the inelastic cross-section σ_{AA} can be derived from the corresponding inelastic nucleon-nucleon (NN) cross-section σ_{NN}. In the Glauber multiple collision model [6], this cross-section reads

$$\sigma_{AB} = \int d^2b\left[1 - e^{-\sigma_{NN}(s)T_{AB}(b)}\right] \tag{2}$$

$$\approx \int d^2b\,\sigma_{NN}^{hard} T_{AB}(b) \qquad \text{for a hard scattering.} \tag{3}$$

Here, $T_{AB}(b) = \int d^2\vec{s}\,T_A(\vec{s})\,T_B(|\vec{b} - \vec{s}|)$ is the nuclear overlap function of the nuclei A and B separated by the impact parameter b. The vectors $\vec{s}$ and $\vec{b} - \vec{s}$ are impact parameter variables in the transverse plane.
For a given centrality class $C_1 - C_2$, it is useful to define the geometric fraction [7]

$$f_{geo}(b_1 < b < b_2) = \left[2\pi \int_{b_1}^{b_2} b\,db\left(1 - e^{-\sigma_{NN}(s)T_{AB}(b)}\right)\right] / \sigma_{AB}^{geo}\,, \tag{4}$$

where f_{geo} simply corresponds to a 0.X (e.g. 0.1) factor for the X%(10%) centrality and σ_{AB}^{geo} is given by Eq. (2). In the case of a hard scaterring,

$$f_{geo}(b_1 < b < b_2) \rightarrow f_{hard}(b_1 < b < b_2) = \frac{2\pi}{AB}\int_{b_1}^{b_2} b\,db T_{AB}(b)\,. \tag{5}$$

From this, we can define the nuclear modification factor for a given centrality class

$$R_{AA}^{C_1-C_2} = \left(\frac{d\sigma_{C_1-C_2}^{AA\rightarrow h+X}}{dp_T^2 dy}\right) \Big/ \left(A^2 f_{hard}\frac{d\sigma^{pp\rightarrow h+X}}{dp_T^2 dy}\right) \tag{6}$$

In nuclear collisions, the cross section for particle production can be obtained from the Glauber model and the equation (1) with some modifications,

$$\frac{d\sigma^{AA\rightarrow h+X}}{dp_T^2 dy} = \int d^2b\,d\vec{s}\,T_A(\vec{s})T_B\left(|\vec{b} - \vec{s}\,|\right)\sum_{abcd}\int dx_1 dx_2 f_1^A\left(x_1, Q^2, b\right) f_2^A\left(x_2, Q^2, b\right) \times$$

$$\times \frac{d\hat{\sigma}}{d\hat{t}}\frac{z_c^*}{z_c}\frac{D_{h/c}\left(z_c^*, \mu_f^2\right)}{\pi z_c} \tag{7}$$

Here, $f_{1,2}^A(x, Q^2, b)$ are nuclear parton distributions depending on the impact-parameter b, given by the parametrization FGS [2], which is based on the leading twist theory of nuclear shadowing [2], and it is valid in the range 0.58 fm < b < 9.98 fm. Also, z_c^* is the momentum fraction of the fragmenting parton, modified by energy loss effects [8],

$$z_c^* = z_c/(1 - \Delta E/p_c)\,. \tag{8}$$

In UHI collisions at RHIC and LHC, one expects the production of the quark-gluon plasma (QGP). Considering the medium created, we take into account the energy loss of the partons traversing the medium via a shift in the momentum fraction, given by eq. (8), of the fragmenting parton carried by the produced hadrons. The energy loss ΔE is calculated via GLV formalism, based on the reaction operator approach to opacity expansion [3]. Assuming a thin plasma, the mean number of jet scatterings $\bar{n} = L/\lambda$ is small, allowing the opacity expansion. The induced radiation intensity in first order in opacity is given by [3],

$$\frac{dI^{(1)}}{dx} = \frac{C_R\alpha_s}{\pi}\left(1 - x + \frac{x^2}{2}\right)\frac{L}{\lambda_g}E\int_{k_{min}^2}^{k_{max}^2}\frac{dk_\perp^2}{k_\perp^2}\int_0^{q_{max}^2} d^2q_\perp \frac{\mu_{eff}^2}{\pi\,(q_\perp^2+\mu^2)}\,^2\,\frac{2k_\perp\cdot q_\perp(k-q)_\perp^2 L^2}{16x^2E^2+(k-q)_\perp^4L^2}$$

where μ is a screening scale from color Yukawa potential, L is the medium lenght and L/λ is the medium opacity. The kinematical limits for the transfer momentum q and transverse momentum k are

$$q_{max}^2 = s/4 \simeq 3E\mu \qquad k_{min}^2 = \mu^2 \qquad k_{max}^2 = min\left[4E^2x^2, 4E^2x(1-x)\right]\,.$$

Applying the asymptotical limit $k_{min} \to 0$, $k_{max} \to \infty$ and $q_{max} \to \infty$, and integrating in the light-cone momentum fraction x of the gluon, one obtains the approximation for induced energy loss (for more details, see [3])

$$\Delta E^{(1)} \approx \frac{C_R\alpha_s}{4}\frac{L^2\mu^2}{\lambda_g}\log\frac{E}{\mu}\,. \tag{9}$$

This expression will be used in eq. (8) in order to calculate the parton energy loss effects.

4. Results and Conclusions

Our results for the nuclear modification factor R_{AA} of charged particles in several centrality classes are shown in Figure 1, compared with LHC data [1] from lead-lead collisions at $\sqrt{s} = 2.76\,\text{TeV}$. Our calculation is expected to be applicable for p_T higher than about 8 TeV. As expected, the suppression is stronger for the more central collisions, and it decreases with increasing p_T. From this comparison we obtain (via best fit to data) the opacity parameter L/λ of the medium for each centrality class. These results are summarized in Table 1 and should be considered as rough estimates of this property of the medium formed in the intermediate stages of ultrarelativistic heavy-ion collisions at the LHC.

Considering the Glauber model, b-dependent nuclear parton distributions and parton energy loss effects, we estimate the nuclear modification factor R_{AA} and obtain values for the opacity parameter of the medium created in UHI collisions at the LHC, for different centrality classes. We obtain that when we move from more central (0-5%) to more peripheral (30-40%) centrality classes, the opacity parameter of the medium decreases from 3.7 to 2.8, which indicates that the scattering power of the medium created (QGP) depends on the number of participant nucleons. Our next steps include the study of other energy loss mechanisms and their implications on the transport coefficient $\hat{q}$ of the QGP.

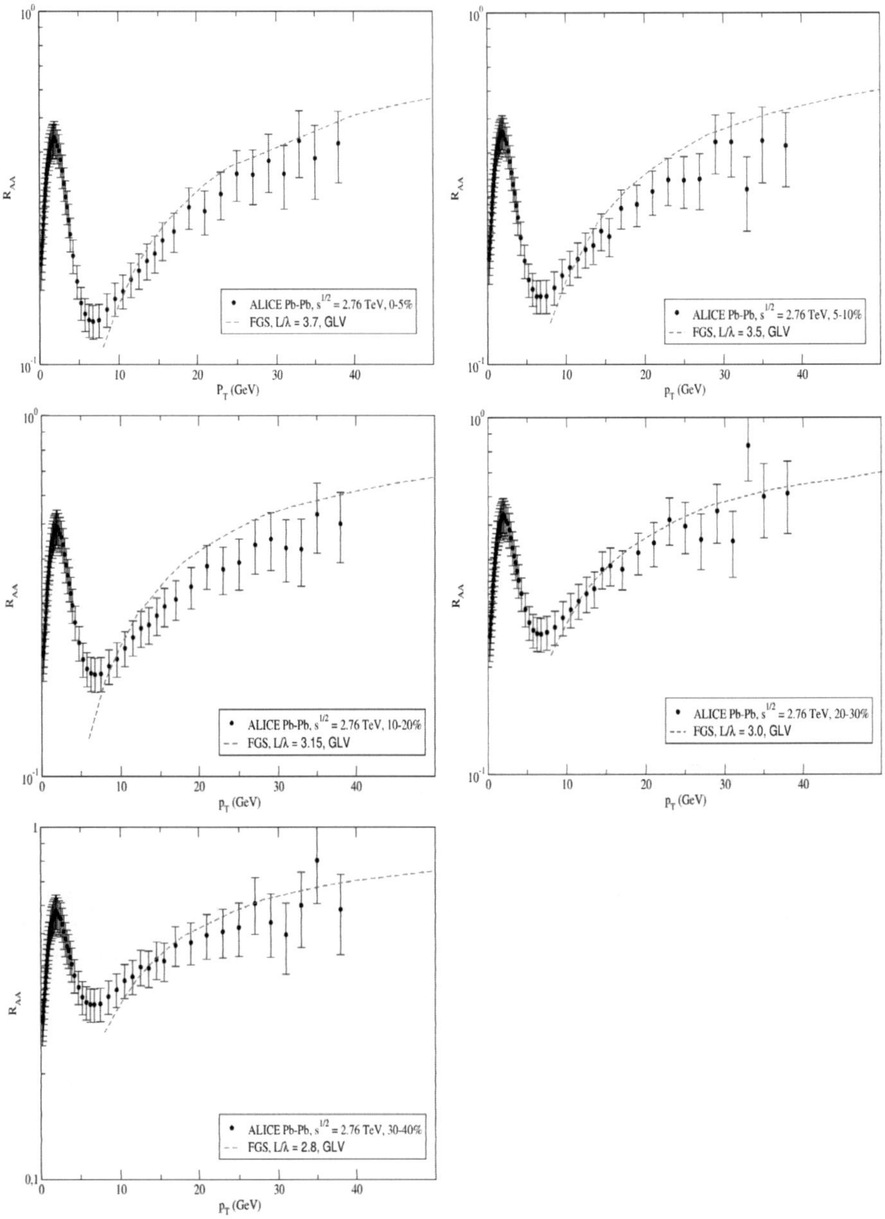

Figure 1. Nuclear modification factor R_{AA} of charged particles produced in $Pb - Pb$ collisions at the LHC ($\sqrt{s} = 2.76\,\mathrm{TeV}$), in different centrality classes, compared with ALICE data [1].

4.1. Acknowledgments

This work is partially financed by CAPES, CNPQ and by the Programa de Pós-Graduação em Física FURG, Brazil.

Journal of Physics: Conference Series **706** (2016) 052035 doi:10.1088/1742-6596/706/5/052035

Table 1. Values obtained for the opacity parameter of the medium, in different centrality classes.

Centrality class	L/λ
$0-5\%$	3.7
$5-10\%$	3.5
$10-20\%$	3.15
$20-30\%$	3.0
$30-40\%$	2.8

5. References

[1] Abelev B *et al.* (ALICE Collaboration) 2013 *Phys. Lett.* B **720** 52
[2] Frankfurt L, Guzey V and Strikman M 2012 *Phys. Rept.* **512** 255
[3] Gyulassy M, Levai P and Vitev I 2001 *Nucl. Phys.* B **594** 371
[4] Pumplin J *et al.* 2002 *JHEP* **07** 012
[5] Kniehl B A, Kramer G and Potter B 2000 *Nucl. Phys.* B **582** 514
[6] Glauber R J and Matthiae G 1970 *Nucl. Phys.* B **21** 135
[7] D. G. d'Enterria, nucl-ex/0302016
[8] Levai P 2011 *Nucl. Phys.* A **862** 146

Subtractive Renormalization of the NJL model *

V. S. Timóteo[1], E. F. Batista[2], R. L. S. Farias[3,4] and S. Szpigel[5]

[1]Grupo de Óptica e Modelagem Numérica - GOMNI, Faculdade de Tecnologia - FT
Universidade Estadual de Campinas - UNICAMP, 13484-332, Limeira, SP, Brasil
[2]Departamento de Ciências Exatas e Naturais
Universidade Estadual do Sudoeste da Bahia, 45700-000, Itapetinga, BA, Brasil
[3]Physics Department, Kent State University, Kent, OH 44242, United States
[4]Departamento de Física, Universidade Federal de Santa Maria
97105-900, Santa Maria, RS, Brasil
[5]Centro de Rádio-Astronomia e Astrofísica Mackenzie
Universidade Presbiteriana Mackenzie, 01302-907, São Paulo, SP, Brasil

E-mail: `varese@ft.unicamp.br (VST)`

E-mail: `efbatista@uesb.edu.br (EFB)`

E-mail: `rfarias@kent.edu (RLSF)`

E-mail: `szpigel@mackenzie.br (SS)`

Abstract.
In this work, we apply a subtractive renormalization method to the $SU(2)$ Nambu–Jona-Lasinio (NJL) model in Born approximation and compare the results obtained here with those achieved by the standard momentum cutoff regularization. We have computed the dynamical quark mass, the chiral condensate, the pion mass and the quark-antiquark T-matrix as a function of the cutoff Λ and the subtraction scale ξ. We have shown that both approaches give similar results and the dependence of the physical quantities on the subtraction point ξ is much weaker than the dependence on the cutoff Λ.

1. Introduction

The Nambu–Jona-Lasinio (NJL) model has originally been proposed to describe the nucleon-nucleon (NN) interaction via contact forces[1]. Later, this model was reinterpreted as an effective field theory for quantum chromodynamics (QCD), the fundamental theory of strong interactions between quarks and gluons. Moreover, the NJL model became the most popular model for QCD and inspired a whole program of applying it to hadron structure [2] and quark matter problems [3]. More recently, the model has been used to study inhomogeneous quark condensates [4] and quark matter under strong magnetic fields [5, 6].

Since the NJL model is based on fermion contact interactions, it seems reasonable that renormalization methods which are applied to the nucleon-nucleon interaction may also be applied to the NJL model for quarks. We have shown that the NN interaction at leading order in chiral perturbation theory can be renormalized by performing one subtraction in the kernel of the NN scattering equation [7], since in this case the divergency comes from a pure contact interaction which is just a constant in momentum space.

*Presented by V.S.T. at XIII Hadron Physics, March 22 - 27, 2014, Angra dos Reis, RJ, Brazil

When higher order interactions are included in the chiral expansion additional subtractions are required [8] so that NLO interactions need three subtractions [9], N2LO interactions need four subtractions [10, 11] and N3LO interactions need five subtractions [12]. In this work, we apply one subtraction to the NJL model in the Born approximation and compare the results to those is obtained in the standard cutoff approach.

2. The NJL model

The Lagrangian for the NJL model is written as

$$\mathcal{L}_{\rm NJL} = \bar{\psi}(i\gamma_\mu\partial^\mu - m_0)\psi + G\,\left[(\bar{\psi}\psi)^2 + (\bar{\psi}i\gamma_5\vec{\tau}\psi)^2\right]\,. \tag{1}$$

Linearizing the interaction in a mean field approximation yields to the well known gap equation of the NJL model

$$m = -2G\,\langle\bar{\psi}\psi\rangle\,, \tag{2}$$

where $\langle\bar{\psi}\psi\rangle$ is the scalar density corresponding to the quark loop,

$$\langle\bar{\psi}\psi\rangle = -i\ {\rm tr}\ S_F(0)\,, \tag{3}$$

with

$$S_F(x-y) = -i\ \langle T\left[\psi(x)\bar{\psi}(y)\right]\rangle = \int\frac{d^4p}{(2\pi)^4}\frac{e^{ip\cdot(x-y)}}{\not{p}-m+i\epsilon}\,. \tag{4}$$

Evaluating the trace of equation (3) we have

$$\begin{aligned}{\rm tr}\ S_F(0) &= \int\frac{d^4p}{(2\pi)^4}\frac{4m}{p_0^2-p^2-m^2}\,,\\ &= -\int\frac{d^3p}{(2\pi)^3}\frac{4}{(2\pi)}\frac{\pi}{2}\frac{i\ m}{\sqrt{p^2+m^2}}\,,\\ &= -\frac{4i}{8\pi^2}\int dp\ p^2\frac{m}{\sqrt{p^2+m^2}}\,.\end{aligned} \tag{5}$$

3. The cutoff scheme

At this point, a momentum cutoff $p_{\rm max}=\Lambda$ is usually applied to the integral in order to ensure that the gap equation, equation (2), has a finite solution. In figure 1 we show the solutions of the gap equation (upper left panel) and the result obtained for the chiral condensate (upper right panel), both as a function of the cutoff Λ.

Both the mass and the condensate depend strongly on the cutoff and they grow very rapidly as Λ increases so that this scale is normally kept in the range $0.6 < \Lambda < 0.7$ GeV. In figure 1 we also show the dependence of the pion mass on the cutoff as obtained from the Gell-Mann–Oakes–Renner (GOR) relation (lower left panel) and its dependence on the current quark mass within the usual cutoff range (lower right panel), for a current quark mass of 5.5 MeV and a pion decay constant of 93 MeV.

The quark-antiquark T-matrix is given by

$$\begin{aligned}T &= K + KJK + \cdots\,,\\ &= \frac{K}{1-KJ}\,,\end{aligned} \tag{6}$$

where

$$J(q^2) = i\ {\rm tr}\int\frac{d^4p}{(2\pi)^4}\left(i\gamma_5\frac{1}{\not{p}+\frac{\not{q}}{2}-m+i\epsilon}i\gamma_5\frac{1}{\not{p}-\frac{\not{q}}{2}-m+i\epsilon}\right)\,, \tag{7}$$

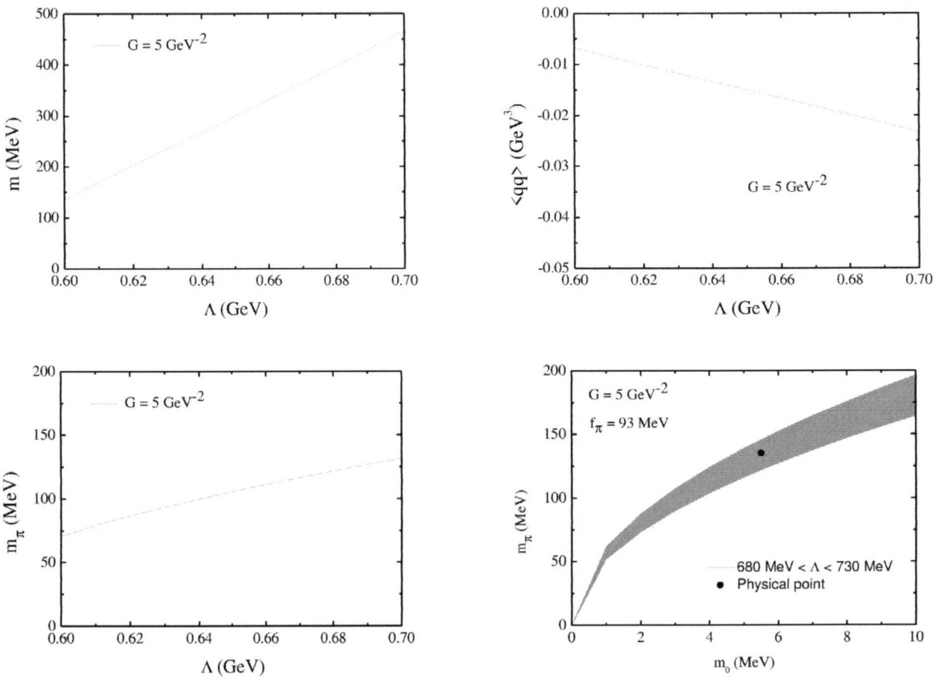

Figure 1. Top panels: Dynamical mass for the quarks (left) and chiral condensate (right) as a function of the momentum cutoff. Bottom panels: Pion mass as a function of the momentum cutoff (left) and as a function of the current quark mass (right) for $G = 5.5\ \mathrm{GeV}^{-2}$ and $f_\pi = 93$ MeV. The square represents the physical point $(m_{u,d}\ ,\ m_\pi) = (5.5\ ,\ 140)$ MeV.

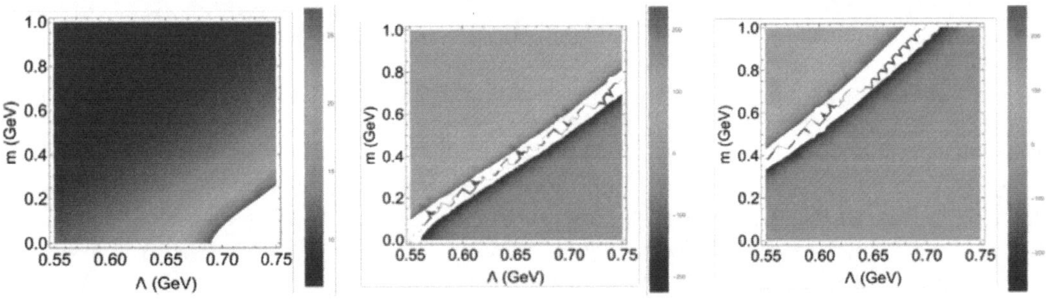

Figure 2. The quark-antiquark T-matrix, in GeV^{-2}, as a function of the dynamical mass and the cutoff for $G = 0.5\,G_\Lambda$ (left), $G = 1.0\,G_\Lambda$ (middle) and $G = 1.5\,G_\Lambda$ (right). The reference coupling in the cutoff scheme is $G_\Lambda = 5.0\ \mathrm{GeV}^{-2}$.

and $K = 2G$ in the Born approximation. In the cutoff scheme we obtain

$$\begin{aligned} T &= \frac{2G}{1 - 2G\,J}\,, \\ &= \frac{2G}{1 - \frac{N_c N_f}{m\pi^2}\int_0^\Lambda dp p^2 \frac{2Gm}{\sqrt{p^2+m^2}}}\,, \\ &= \frac{2G}{1 - 2G\frac{N_c N_f}{\pi^2}\left[\Lambda\sqrt{m^2+\Lambda^2} - m^2\,\text{arcsinh}\left(\frac{\text{m}}{\Lambda}\right)\right]}\,. \end{aligned} \tag{8}$$

Note that the gap equation, equation (2), can be obtained by setting the denominator of equation (8) equal to zero, which corresponds to the poles of the T-matrix:

$$2G\frac{N_c N_f}{\pi^2}\left[\Lambda\sqrt{m^2+\Lambda^2} - m^2\,\text{arcsinh}\left(\frac{\text{m}}{\Lambda}\right)\right] = 1\,. \tag{9}$$

For a given value of the coupling G, the pole structure can be observed in figure 2 where the T-matrix is shown in a density plot as a function of the dynamical mass and the cutoff.

4. $q\bar{q}$ subtracted T-matrix at zero temperature

We want to derive the subtracted T-matrix inspired in the two-nucleon case. This requires an approximation of equation (5) with the replacement $\sqrt{p^2+m^2} \approx m + \frac{p^2}{2m} + \cdots$, namely

$$\text{tr}\, S_F(0) \sim -\frac{i}{2\pi^2}\int dp p^2 \frac{2m^2}{2m^2+p^2}\,. \tag{10}$$

Now the fermion propagator has a non-relativistic form and we can derive a subtracted T-matrix for the NJL model with the exact same formalism that has been applied to the NN interaction [7]. In the Born approximation, the NJL model can be treated in the same way as a pionless effective field theory for the two-nucleon system at leading order.

The scattering amplitude for two nucleons is given by the Lippman-Schwinger (LS) equation

$$\begin{aligned} T_{\text{NN}} &= V_{\text{NN}} + V_{\text{NN}} G_0 T_{\text{NN}}\,, \\ &= \frac{V_{\text{NN}}}{1 - V_{\text{NN}} G_0}\,, \end{aligned} \tag{11}$$

where

$$G_0 \equiv G_0(q;k) = \frac{1}{k^2 - q^2}\,, \tag{12}$$

and $V_{\text{NN}} = C_0$ is the two-nucleon potential at leading order in a pionless EFT (C_0 is usually fitted to the NN s-wave scattering length). Performing subtractions in the kernel of the LS equation has proven to be effective to renormalize the nuclear two-body force [9, 10, 11]. With only a pure contact interaction, one subtraction is enough to renormalize the two-body T-matrix and the kernel G_0 is subtracted at a scale ξ:

$$\begin{aligned} T_{\text{NN}} &= V_{\text{NN}} + V_{\text{NN}} G_1 T_{\text{NN}}\,, \\ &= \frac{V_{\text{NN}}}{1 - V_{\text{NN}} G_1}\,, \end{aligned} \tag{13}$$

where G_1 is the Green's function with one subtraction

$$G_1 \equiv G_1(q;k,\xi) = G_0(q;k) - G_0(q;-\xi^2) = \frac{\xi^2 + k^2}{\xi^2 + q^2} G_0(q;k)\,. \tag{14}$$

We then make the non-relativistic approximation on the quark propagator and introduce the subtraction in the same way as the subtracted Green's function $G_1(q;k,\xi)$ was obtained in Ref. [7]. The result is

$$J(q^2=0) = \frac{1}{2m\pi^2}\int dpp^2 \frac{2m^2}{p^2+2m^2}\frac{\xi^2+m^2}{\xi^2+p^2} \; . \tag{15}$$

In the born approximation, $K = 2G$ and one subtraction is enough to provide a finite amplitude

$$\begin{aligned} T &= \frac{2G}{1-2G\,J}\,, \\ &= \frac{2G}{1-2G\,\frac{N_cN_f}{2m\pi^2}\int dpp^2\frac{2m^2}{p^2+2m^2}\frac{\xi^2+m^2}{\xi^2+p^2}}\,, \\ &= \frac{2G}{1-\frac{2N_cN_fGm}{\pi}\frac{\xi^2+m^2}{\xi+m\sqrt{2}}}\; . \end{aligned} \tag{16}$$

The gap equation for the mass is then

$$2G\frac{N_cN_f}{\pi}\left[\frac{\xi^2+m^2}{\xi/m+\sqrt{2}}\right] = 1\; . \tag{17}$$

In figure 3 the solutions of equation (17) (upper left panel) and the chiral condensate (upper right panel) are shown as functions of the subtraction point ξ.

Again, fixing the pion decay constant f_π we can obtain the pion mass from the GOR relation as a function of the subtraction point for fixed current quark mass or as a function of the current quark mass for fixed subtraction point. This is shown in the lower panels of figure 3. The pole structure of the T-matix is shown in the density plot of figure 4, where T is shown as a function of the mass m and the subtraction point ξ.

The dynamical mass depends on both the coupling G and the renormalization scale Λ or ξ. Then, the running of the coupling G with the renormalization scales can be obtained by fixing the dynamical mass. Graphically, the running is given by the intersection of the $m(G,\Lambda)$ surface with the physical mass plane. Setting the mass at $m = 325$ MeV, the running of the coupling G with both Λ and ξ can be observed in figure 5.

While the coupling G diverges at $\Lambda = 0$, it is finite at $\xi = 0$. Also, the mass doesn't increase with the subtraction point ξ as much as it grows with the cutoff Λ. The dependence on the renormalization scale is clearly much weaker in the subtraction scheme when compared to the standard cutoff approach.

5. Final remarks

So far we have compared the cutoff and the subtractive renormalization procedures in the NJL model at zero temperature. The subtractive renormalization procedure has been applied to the NN interaction and here we extended it to quark degrees of freedom.

We found that the subtraction scheme gives good results when the strength of the coupling G_ξ is half the strength G_Λ. Furthermore, the dependence of the dynamical mass on the subtraction point is much weaker than in the cutoff approach.

The subtractive renormalization may then be useful in the description of mesons with one heavy quark, where an additional (and larger) cutoff is required in order to describe the mass of the D meson [13].

Journal of Physics: Conference Series **706** (2016) 052036 doi:10.1088/1742-6596/706/5/052036

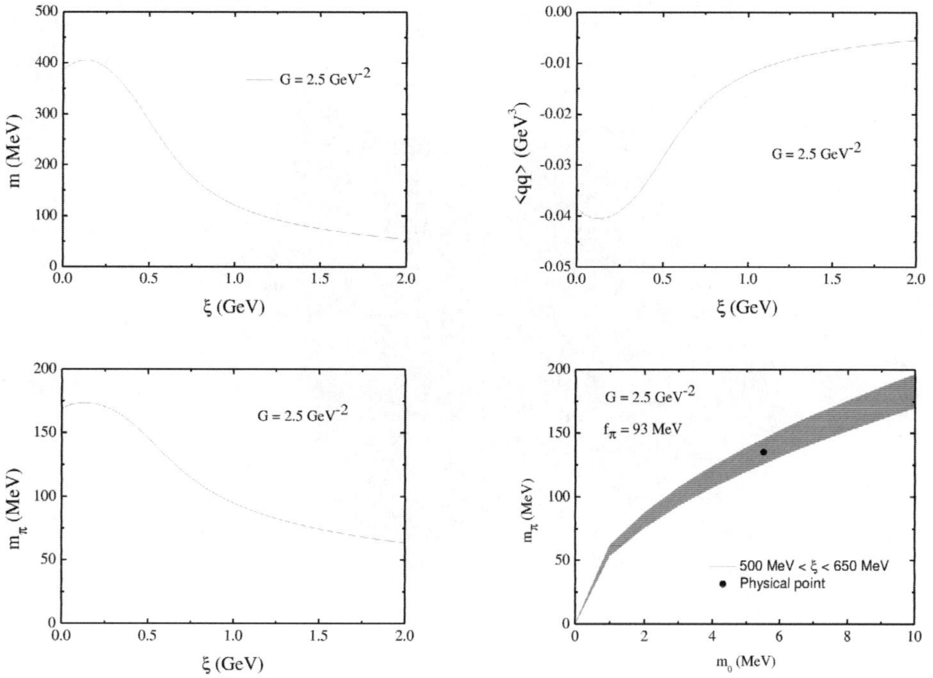

Figure 3. Top panels: Dynamical mass for the quarks (left) and chiral condensate (right) as a function of the subtraction point ξ. Bottom panels: Pion mass as a function of the subtraction point (left) and of the current quark mass (right) for $G = 2.5\ \mathrm{GeV}^{-2}$ and $f_\pi = 93$ MeV. The square represents the physical point $(m_{u,d}\ ,\ f_\pi) = (5.5\ ,\ 140)$ MeV.

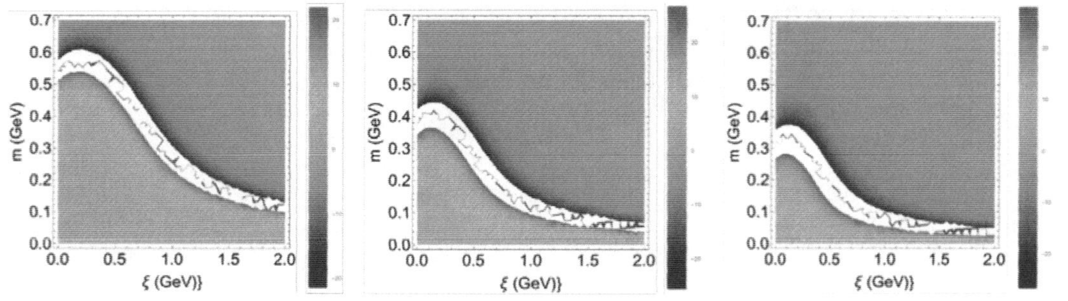

Figure 4. The quark-antiquark T-matrix, in GeV^{-2}, as a function of the dynamical mass and the subtraction point for $G = 0.5\,G_\xi$ (left), $G = 1.0\,G_\xi$ (middle) and $G = 1.5\,G_\xi$ (right). The reference coupling in the subtraction scheme is $G_\xi = 2.5\ \mathrm{GeV}^{-2}$.

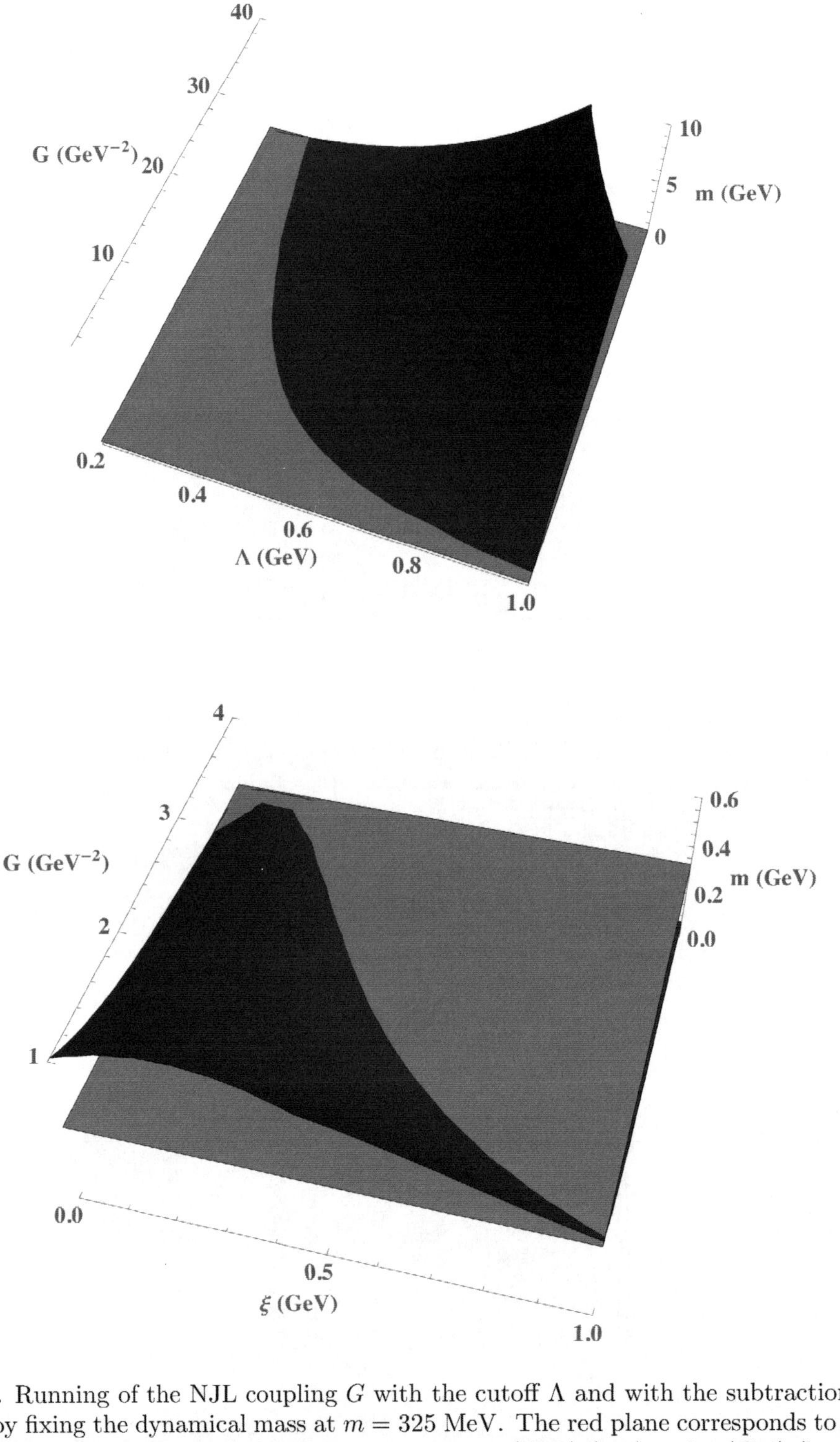

Figure 5. Running of the NJL coupling G with the cutoff Λ and with the subtraction point ξ obtained by fixing the dynamical mass at $m = 325$ MeV. The red plane corresponds to the fixed value of the mass and the blue surface corresponds to $m(G, \Lambda)$ (top) and $m(G, \xi)$ (bottom).

Acknowledgements
The authors would like to thank FAEPEX, FAPESP and CNPq for financial support.

References
[1] Nambu Y and Jona-Lasinio G 1961 *Phys. Rev.* **124** 146.
[2] Vogl U and Weise W 1991 *Prog. Part. Nucl. Phys.* **27** 195.
[3] Klevansky S P 1992 *Rev. Mod. Phys.* **64** 649.
[4] Buballa M 2005 *Phys. Rep.* **407** 205.
[5] Menezes D P, Pinto M B, Avancini S S, Martínez A P and Providência C 2009 *Phys. Rev. C* **79** 035807.
[6] Avancini S S, Menezes D P, Pinto M B and Providência C 2012 *Phys. Rev. D* **85** 091901(R).
[7] Frederico T, Timóteo V S and Tomio L 1999 *Nucl. Phys. A* **653** 209.
[8] Frederico T, Delfino A and Tomio L 2000 *Phys. Lett. B* **481** 143.
[9] Timóteo V S, Frederico T, Delfino A, Tomio L 2005 *Phys. Lett. B* **621** 109.
[10] Timóteo V S, Frederico T, Delfino A, Tomio L 2011 *Phys. Rev. C* **83** 064005.
[11] Szpigel S and Timóteo V S 2012 *J. Phys. G: Nucl. Part. Phys.* **39** 105102.
[12] Batista E F, Szpigel S and Timóteo V S 2015 *Journal of Physics: Conf. Series* **630** 012056.
[13] Gottfried F O and Klevansky S P 1992 *Phys. Lett. B* **286** 221.

The LAUE project: latest developments

V. Liccardo[1], **E. Virgilli**[2], **F. Frontera**[2] **and P. Rosati**[2]

[1] ITA-Instituto Tecnológico de Aeronáutica, São José dos Campos - SP - Brasil

[2] Physics and Earth Sciences Department, University of Ferrara - Italy

E-mail: liccardo@fe.infn.it

Abstract. We present the status of the LAUE project devoted to develop a technology for building long focal length Laue lenses for hard X-/soft gamma-ray astronomy (80-600 keV). The Laue lens will be composed of bent crystals of Gallium Arsenide (GaAs, 220) and Germanium (Ge, 111), and, for the first time, the focusing property of bent crystals will be exploited for this field of applications. At the present the goal of the project is the building of a 20 m focal length Laue lens petal capable of focusing X-ray in the energy range 90-300 keV.

1. Introduction

The LAUE project, supported by the Italian Space Agency (ASI), is a follow-up of the HAXTEL project [1] and is devoted to create a technology to build a Laue lens with long focal length (20-100 m) able to focus photons in the 80-600 keV energy range, unreachable with multilayer telescopes. The experience acquired from the HAXTEL project has shown that, for focal length longer than 10 m a new assembling technique is needed. Furthermore, a new generation of crystals must be developed for Laue lenses, given that the Point Spread Function (PSF) of a flat crystal is related to the size of the crystal itself [2]. Simulations have shown that a Laue lenses made of Ge (111) bent crystals can overcome by far the sensitivity of the current X-ray astrophysics missions [3].

On the basis of these requirements, the LAUE project was conceived to face the following challenging goals:

- to realize a technology to produce a large number of diffractive crystals with high efficiency;
- to find a method for assembling the crystals and fixing them in place with high accuracy in a relatively short time.

Regarding the first item, the development of bent crystals was considered of key importance, in order to increase the reflection efficiency and to obtain a better focusing of the photons. As of the last two items in the LAUE project, an industrial accommodation study of a modular lens was included.

Concerning the second goal, in order to decrease the cumulative error budget, we decided to develop an apparatus to correctly orient and fix each crystal to the lens frame under the control of a gamma-ray beam. For the same reason, we also decided to keep the lens petal fixed and to move the gamma-ray beam parallel to the lens axis [4].

2. The project apparatus

The entire apparatus for both assembling and testing the lens was designed to be installed in the LArge Italian X-ray laboratory (LARIX) located at the Physics and Earth Sciences Department of the University of Ferrara. The laboratory includes an experimental room (LARIX A) with a 12 m long facility and an assembling facility consisting of a 100 m long tunnel. The length of the tunnel (Fig. 1) will allow to obtain a small divergence of the beam impinging on each crystal and to build Laue lenses with long focal length [4].

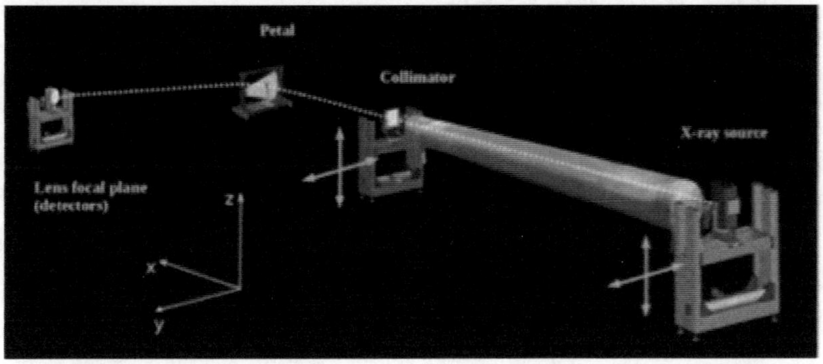

Figure 1. Layout of the LAUE apparatus at the University of Ferrara.

In the LAUE project a flat frame is used as a support for the crystal tiles composing the single petal. During the entire building phase the petal frame is kept fixed while a pencil beam of radiation is moved in front of the petal frame to mimic the presence of an astronomical gamma-ray source. The source of radiation is also directly used to control the correct positioning of the crystals. The total number of crystals that composes the designed petal is 275 and the tiles are distributed in 18 concentric ring sections [5]. Positioning each crystal with the correct diffraction angle and keeping this orientation unchanged is the most challenging phase.
The layout of the developed apparatus is shown in Figure 1, with details reported in another paper [4]. The main components of the facility are:

- A gamma-ray source produced by an X-ray tube with maximum voltage of 320 kV and a very small spot size (0.4 mm diameter) from which the photons are emitted. The source photons are mechanically collimated to get an output beam of about 20 arcmin. The source can be translated in a plane (y, z) perpendicular to the beam axis (called x axis) and tilted around axes z (vertical axis) and y (Fig. 2);
- A 21 m beam-line (60 cm inner diameter) within which the gamma-ray beam travels under vacuum to avoid absorption and scattering interactions;
- At the exit of the beam-line, a final shield of the gamma-ray beam with a square hole in the center, where a slit with the aperture defined by 4 independently movable Tungsten Carbide blades is set up (Fig. 2);
- Lens-petal frame, which is held by a pedestal that can be manually rotated and translated. It has the shape of a circular sector, 18 degree wide as seen from the lens axis;.
- A fine six-axes motorized robot (hexapod) that allows the correct positioning of each crystal tile on the lens frame under the control of the gamma-ray pencil beam. Once the diffracted photons are focused on the lens focus, the crystal tile is fixed to the lens frame (Fig. 3);
- Two focal plane detectors (a gamma-ray imager and a spectrometer), one above the other, held by a pedestal. The pedestal can be moved along the x axis using a rail oriented along

this axis. For a fixed value of x, the detectors can be translated in the (y, z) plane and rotated around each of the three axes x, y, and z (Fig. 3);

- Both the final slit and the petal frame are located in the already mentioned clean room to guarantee the crystal positioning and fixing to the lens frame under constant cleanliness and environment conditions.

Figure 2. The X-ray tube (*left*) and the Tungsten collimator (*Right*).

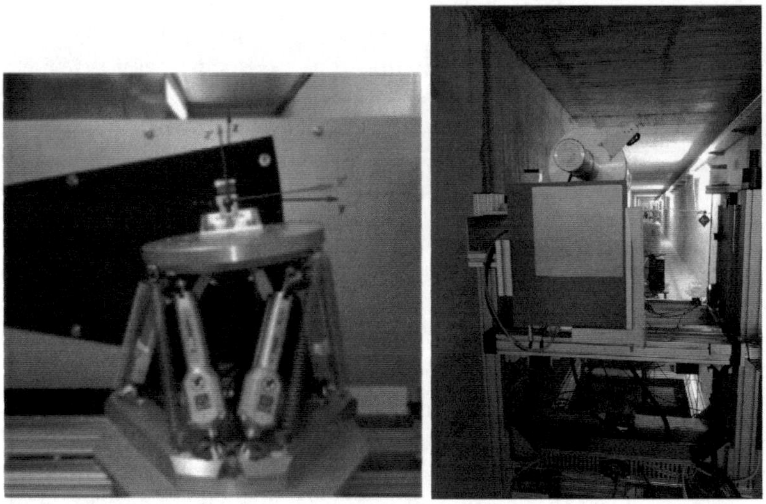

Figure 3. The hexapod system with the petal lens frame (*left*) and the X-ray detectors (*Right*).

3. Petal features and crystals selection

The lens petal is designed in such a way as to diffract photons in the 90-300 keV energy band. The upper threshold is due to the maximum energy of the photons available with the X-ray tube adopted. The lens petal passband is defined by the inner and outer radius of the lens petal.

The crystals will be positioned and glued on the frame so that to focus the diffracted beam at a distance of 20 m. The width of the designed lens petal subtends an angle of 18°. It is being covered with bent crystal tiles of Ge (111) and GaAs (220) as shown in Figure 4.

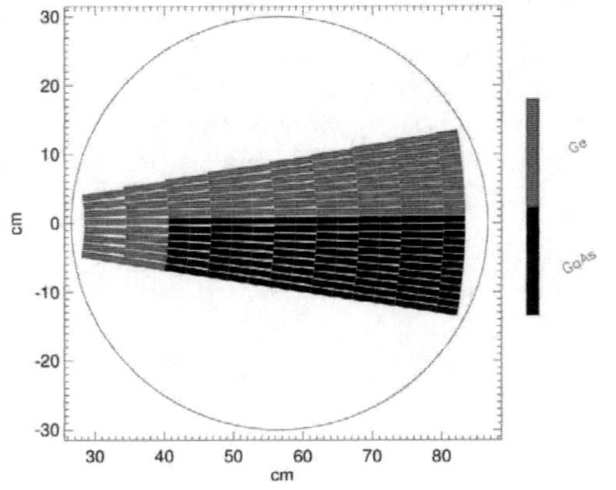

Figure 4. The upper part of the petal (*red*) will be filled with Ge (111) tiles while the bottom part (*black*) will be filled with GaAs (220) tiles.

The crystal tile size of both crystal types as required for the Laue project is 30 x 10 mm^2, with a thickness of 2 mm (Fig. 5). For the lens petal prototype, which is being assembled, all the bent Ge (111) crystal tiles (about 150) have been provided by LSS (Sensor and Semiconductor Laboratory of Ferrara) with a nominal angular spread of 4 arcsec. A self-standing uniform spherical curvature of perfect crystals has been achieved by making indentation on the main external surface of the crystal. The produced grooves yield an external curvature (primary curvature) which for some crystallographic orientations, induces a quasi-mosaic (QM) deformation of the internal diffracting planes [6]. The curvature is preliminary measured by the LSS staff by means of a profilometer, obtaining a curvature radius of about 40m.
IMEM (Istituto dei Materiali per l' Elettronica ed il Magnetismo) has supplied bent crystal tiles of GaAs (220). The bending technique adopted in this case relies instead on a controlled surface damaging, obtained by means of a mechanical lapping process on one side of the sample [7].

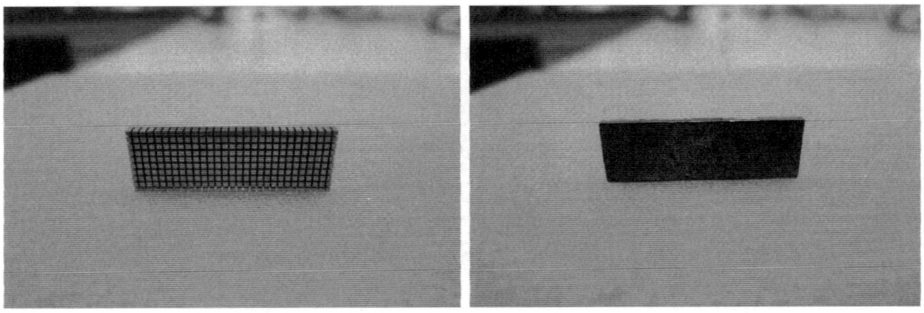

Figure 5. Ge (111) crystal tile (*Left*) and GaAs (220) crystal tile (*Right*).

Journal of Physics: Conference Series **706** (2016) 052037 doi:10.1088/1742-6596/706/5/052037

4. Last results of the project

Various different methods to set the crystals at the correct diffraction angle have been considered. All the procedures were based on setting the crystals over a frame and the match between crystal and substrate was made with a structural adhesive. The most critical effect while positioning a crystal is the rotation around the z axis. Once the crystal is positioned, the curing process induces a differential adhesive shrinkage that results in a non negligible tilt around that axis. The petal frame is a 2 mm thick carbon fiber with an hole for each crystal. The procedure foresees the coupling between the tile and the support by injecting the adhesive through the hole, with the crystal set at the correct position. The crystal tile is positioned in such a way to have the diffracted image barycentre in the center of the detector. After 60 minutes from the glue injection and at regular time intervals thereafter, the focused point is monitored with an optical camera (Fig. 6).

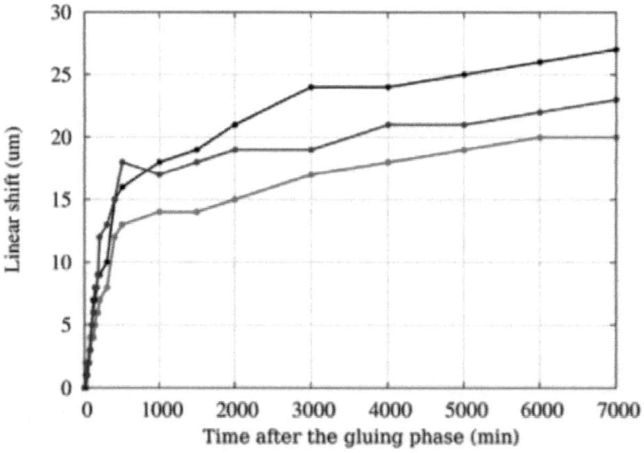

Figure 6. Linear shift, as a function of time, of a glued crystal obtained with the optical camera.

As shown in Figure 6 the total shift (20-25 μm) corresponds to a combination of rotation and translation of the crystal caused by the glue shrinkage. For the first 60 minutes the sample was set into the clamp and no motion was observed. After the release of the clamp, a common trend was observed which corresponded to a decrease of the Bragg angle. Even though the trend is monotonically decreasing for all curves, the effect is not systematic. The procedure adopted turned out to be limited for our goals due to errors introduced by the glue shrinkage which are not negligible.

A new method has been developed in order to reduce the error budget and to have a more control over the total assembling procedure (gluing + paste polymerization). It relies on the employment of a new adhesive (DYMAX OP-61-LS) which is low-shrink and allow us to control the curing by using an UV light. A new transparent support made of polymethyl methacrylate has been also adopted in order to better exploit the new curing paste. The main advantage of the UV adhesive is that the crystal tile position can be fine adjusted even if the glue has been already applied, since the polymerization is controlled by the application of the UV light.

The new technique has shown good improvements and the results obtained can be found elsewhere [8].

Journal of Physics: Conference Series **706** (2016) 052037 doi:10.1088/1742-6596/706/5/052037

5. Conclusions

The latest LAUE project results are very promising for the prospects of the hard X-ray astronomical observations (up to 600 keV). The accuracy with the new adopted crystal assembling method has turned out to be what we expected, allowing the building of a Laue lens petal with a precision in the crystal positioning of 15-20 arcsec. The new generation of bent crystals Ge (111) and GaAS (220) have shown good results in terms of curvature, efficiency and focusing capability, and represent good candidates for an astrophysics mission based on a Laue lens. After a preliminary study of adhesives, the new UV curing paste showed be satisfactory and is a perfect candidate to fulfill the requirement of a short polymerization time and a good positioning accuracy (15-20 arcsec). Improvements in the mounting process chain are still possible.

5.1. Acknowledgments

The authors wish acknowledge the financial support by the Italian Space Agency (ASI) through the project "LAUE - Una Lente per i raggi Gamma" under contract I/068/09/0.

References

[1] E. Virgilli, F. Frontera, V. Valsan, V. Liccardo, V. Carassiti, F. Evangelisti, S. Sqerzanti: Laue lenses for hard X-/soft Gamma-rays: new prototype results, in *Optics for EUV, X-Ray, and Gamma-Ray Astronomy V*, **8147** (SPIE Conference Series, 2011);

[2] V. Liccardo, E. Virgilli, F. Frontera, V. Valsan, E. Buffagni, C. Ferrari, E. Bonnini, A. Zappettini, V. Guidi, V. Bellucci, R. Camattari: Study and characterization of bent crystals for Laue lenses, Experimental Astronomy **38**, 3 (2014);

[3] V. Valsan, E. Virgilli, V. Liccardo, F. Frontera, Ezio Caroli, J. B. Stephen: Simulation of a Laue lens with bent Ge(111) crystals, Submitted to Experimental Astronomy (2015);

[4] E. Virgilli, F. Frontera, V. Liccardo, V. Valsan, V. Carassiti, S. Chiozzi, F. Evangelisti, M. Parise, S. Squerzanti, M. Statera, V. Guidi, V. Bellucci, R. Camattari, I. Neri, E. Caroli, N. Auricchio, A. Basili, S. Silvestri, J. Stephen, F. Cassese, L. Recanatesi, C. Ferrari, A. Zappettini, E, Buffagni, S. Mottini, M. Pecora, B. Negrii: The LAUE project and its main results, in *Optics for EUV, X-Ray, and Gamma-Ray Astronomy V*, **8661** (SPIE Conference Series, 2013);

[5] V. Valsan, F. Frontera, E. Virgilli, V. Liccardo: Results versus expectations obtained with the LAUE project, in *Optics for EUV, X-Ray, and Gamma-Ray Astronomy V*, **8661** (SPIE Conference Series, 2013);

[6] V. Bellucci, R. Camattari, V. Guidi, I. Neri, N. Barriere: Self-standing bent silicon crystals for very high efficiency Laue lens, Experimental Astronomy **31**, 45 (2011);

[7] E. Buffagni, E. Bonnini, A. Zappettini, G.M. Guadalupi, F. Rossi, C. Ferrari: Crystal bending by surface damaging in mosaic GaAs crystals for the LAUE project, in *Optics for EUV, X-Ray, and Gamma-Ray Astronomy V*, **8661** (SPIE Conference Series, 2013);

[8] E. Virgilli, F. Frontera, P. Rosati, V. Liccardo, V. Carassiti, S. Squerzanti, E. Caroli, N. Auricchio, S. Silvestri, J.B. Stephen: Hard X-ray broad band Laue lenses (80-600 keV): building methods and performances, in *Optics for EUV, X-Ray, and Gamma-Ray Astronomy V*, **9603** (SPIE Conference Series, 2015);

Journal of Physics: Conference Series **706** (2016) 052038 doi:10.1088/1742-6596/706/5/052038

Heavy-Quarkonium Potential from Lattice Gluon Propagator

W M Serenone, A Cucchieri and T Mendes

Instituto de Física de São Carlos, Universidade de São Paulo
Caixa Postal 369, CEP 13560-970, São Carlos SP, Brazil

E-mail: `willian.serenone@usp.br`

Abstract. The study of heavy quarks is of great interest for the search of physics beyond the Standard Model and the understanding of nonperturbative aspects of QCD. One of the early attempts to study these systems was the potential model approach. The Cornell potential is perhaps the most successful of these potentials. However, the use of perturbation theory in its building process implies that it is unable to model confinement without the *ad-hoc* addition of a linear term. In this paper, we modify the Cornell potential by using a (nonperturbative) lattice gluon propagator. This approach allowed us to verify that the use of perturbation theory washes away confinement. We were able to use this modified potential in the Schrödinger equation to obtain numerically the spectrum of heavy quarkonia (charmonium and bottomonium). We use the Cornell-potential spectrum as a benchmark of our potential. The result shows that our potential was able to describe better the spin-average of the experimental states than the Cornell potential. We also computed interquark distances for the quarkonia states.

1. Introduction

After the J/Ψ discovery, its non-relativistic character become evident and the idea of describing heavy quarkonia with potential models was proposed [1, 2]. One of the first proposed models was the Cornell potential. It is to date very popular, due to its simplicity and yet good description of the spectrum. Corroborating it, one can fit it to lattice calculations of the potential between static quarks [3].

The Cornell potential is characterized by the sum of two terms

$$V = -\frac{4}{3}\frac{\alpha_s}{r} + \sigma r\,. \tag{1}$$

The first term resembles the Coulomb potential from electromagnetism, the most notable difference being the 4/3 factor, called "color factor". The similarity of this term with the Coulomb potential is not a coincidence, since its derivation is done by the Fourier transform of the scattering amplitude computed perturbatively — the same process that one can use to derive the Coulomb potential. The second term is added in an "*ad-hoc*" way to reproduce confinement. The motivation for it comes from discretizing QCD on the lattice and computing the Wilson loop in the strong-coupling limit. Also, lattice data support this shape as well, as seen in [3, 4, 5].

The derivation of the Cornell potential uses the free-gluon propagator, thus it is natural to investigate the effects of using a non-perturbative one. Our choice was to use the propagator from Ref. [6], which was obtained by simulating a pure gauge theory with $SU(2)$ symmetry in

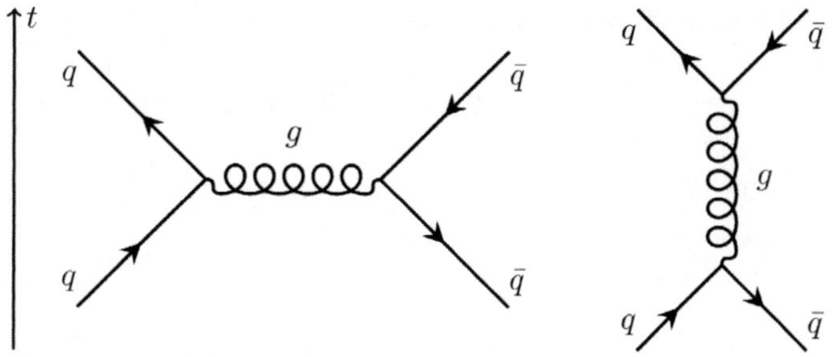

Figure 1. Feynman graph contributing at tree-level to the scattering amplitude T_{fi}. The diagram on the left is a gluon exchange and the one on the right is the brief annihilation of quarks. The latter yields a null contribution to T_{fi}.

Landau gauge. We normalized the propagator so that it would match the perturbative one in the limit $k^2 \to \infty$. We found that it shares the general shape of the Coulomb-like term of the Cornell potential, but presenting some additional features. It is still a non-confining potential and thus needs the addition of the linear term.

Both models have as free parameters the quark masses (m_c and m_b) and the string force σ. We fit these parameters to the spin-averaged spectrum of charmonium and bottomonium. Doing so gives us a way to decide if the obtained potential is indeed an improvement over the Cornell potential. The method allowed us to obtain the wave functions as well and thus we provide estimates of the average particle diameters. This information may be useful to Effective Field Theories, such as potential Non-Relativistic QCD (pNRQCD) [7].

Preliminary results were presented in Refs. [8, 9]. A more complete study can be found in [10].

2. Potential from Lattice Gluon Propagator

For completeness, we will briefly review the method to obtain the Cornell potential. A more detailed review on the subject can be found in Ref. [11].

The scattering-matrix S_{fi} can be expressed, in a first-order approximation, as

$$S_{fi} \equiv \langle f|i\rangle = \delta_{fi} + i(2\pi)^4\, \delta^{(4)}(Q-P)\, T_{fi}\,, \tag{2}$$

where Q and P correspond respectively to the final and initial total momentum and T_{fi} is the scattering amplitude. The two Feynman diagrams that contribute to T_{fi} at tree-level are presented in Fig. 1. We assume that the propagator will be of the form $P^{ab}_{\mu\nu}(k^2) = \delta^{ab} D_{\mu\nu}(k^2)$ and impose that the quark color-state is given by

$$c_q = \frac{(|r\rangle + |g\rangle + |b\rangle)}{\sqrt{3}}\,, \tag{3}$$

while the antiquark is in the color-state $c_{\bar{q}} = c_q^\dagger$. Under these conditions, we obtain that just the left diagram in Fig. 1 actually contributes at tree-level. The same calculation for the remaining term shows that it gets multiplied by a factor 4/3.

In the next step, we impose the non-relativistic approximation $p^2 \ll m^2$, i.e. we drop terms proportional to $1/m^2$. As a consequence, only the component $D_{00}(0, \boldsymbol{k}^2)$ contributes to the scattering amplitude

$$T_{fi} = \frac{4}{3}\frac{g_0^2}{(2\pi)^6} D_{00}(\boldsymbol{k}^2)\,. \tag{4}$$

The potential is then obtained by performing the Fourier transform

$$V(\boldsymbol{r}) = -(2\pi)^3 \int \exp(-i\boldsymbol{k}\cdot\boldsymbol{r})\, T_{fi}(\boldsymbol{k}^2)\, d^3k = -\frac{4}{3}\frac{g_0^2}{(2\pi)^3}\int \exp(-i\boldsymbol{k}\cdot\boldsymbol{r})\, D_{00}(0,\boldsymbol{k}^2)\, d^3k\,. \quad (5)$$

When we choose the free-gluon propagator $D_{\mu\nu}(k^2) = -g_{\mu\nu}/k^2$, Eq. (5) results in the Coulomb-like term of the Cornell potential

$$V(\boldsymbol{r}) = -\frac{4}{3}\frac{\alpha_s}{r}\,, \quad (6)$$

where $\alpha_s = g_0^2/(4\pi)$. As said in Section 1, the linear term is added in an *ad-hoc* way, motivated by lattice computations of the Wilson loop, both numerical [3, 4, 5] and analytical (by expanding e^{-S} in the strong limit coupling [12]).

2.1. Gluon Propagator from Lattice Simulations

In our work, we redo the calculation above using a propagator computed non-perturbatively, using a lattice simulation of a pure SU(2) gauge theory in Landau gauge. It is given by (see Ref. [6, 13, 14])

$$P^{ab}_{\mu\nu}(k) = \frac{C\,(s+k^2)}{t^2+u^2k^2+k^4}\left(\delta_{\mu\nu} - \frac{k_\mu k_\nu}{k^2}\right)\delta^{ab}\,,$$
$$C = 0.784, \qquad s = 2.508\,\mathrm{GeV}^2\,,$$
$$u = 0.768\,\mathrm{GeV}, \quad t = 0.720\,\mathrm{GeV}^2\,. \quad (7)$$

Although the simulated theory has $SU(2)$ gauge symmetry, we expect that the only difference with the $SU(3)$ case will be a multiplicative constant [15]. We adjust the multiplicative constant C by requiring that, in the limit $k^2 \to \infty$, the propagator in Eq. (7) behaves like the perturbative one, i.e. $D_{\mu\nu} \sim 1/k^2$. In the non-relativistic approximation, the energy carried by the exchanged gluon is approximately zero ($k_0 \cong 0$) and therefore the contribution of the term $k_\mu k_\nu/k^2$ to the component D_{00} can be neglected. Lastly, we rotate from Euclidean space to Minkwoski space by performing the substitution $\delta_{\mu\nu} \to -g_{\mu\nu}$.

After the above considerations, we insert T_{fi} in Eq. (5) to obtain the potential

$$V(\boldsymbol{r}) = -\frac{4}{3}\frac{g_0^2}{(2\pi)^3}\int \frac{C\,(s+\boldsymbol{k}^2)e^{-i\boldsymbol{k}\cdot\boldsymbol{r}}}{t^2+u^2\boldsymbol{k}^2+\boldsymbol{k}^4}\,d^3k = -\frac{4}{3}\frac{g_0^2 C}{2i(2\pi)^2 r}\int_{-\infty}^{\infty}\frac{(s+k^2)\left(e^{ikr}-e^{-ikr}\right)}{t^2+u^2k^2+k^4}k\,dk\,, \quad (8)$$

where we already performed the trivial angular integration. The remaining integral can be evaluated using residues. More precisely, we have one simple pole in each quadrant of the complex plane, distributed symmetrically. We consider a semicircular contour closed above (below) to integrate the first (last) term. The result is a sum of four terms (one for each pole). However, since the poles are symmetrically distributed, it is possible to relate the four poles by complex conjugations and changes in signal. This allows us to write the Lattice-Gluon-Propagator potential (V_{LGP}) as

$$V_{LGP}(r) = -\frac{4}{3}\frac{2\alpha_s}{r}\Re\left[\frac{C(s+k_{0,0}^2)\,e^{ik_{0,0}r}}{u^2+2k_{0,0}^2}\right], \quad (9)$$

where

$$k_{0,0} = i\sqrt{t}\,\exp\left[i\frac{\theta}{2}\right]. \quad (10)$$

Both V_{LGP} and the Cornell potential depend on the strong coupling constant α_s. As it is known, the QCD coupling has a dependence on the energy scale of the process. We use as the energy scale the mass of the 1S state of the system considered [J/ψ for charmonium and $\Upsilon(1S)$ for bottomonium]. Ref. [16, Section 9] presents us with with the four-loop calculation for α_s as well as values for Λ_{QCD}, which we use in computing the two different values of α_s.

The result in Eq. (9) is a non-confining potential, just as in the Cornell-potential case. But we notice that it contains qualitative differences from the Coulomb-like one. For instance, it does rise above zero, showing a behavior that would be expected for a confining potential until around 1.5 fm. After this region, the potential is damped. For the calculation of the eigenstates (see Section 3), we thus add a linear term σr to model confinement, as it is done for the Cornell potential

$$V_{LGP+L}(r) = V_{LGP}(r) + \sigma r \, . \tag{11}$$

3. Finding Eigenstates and Fitting

The V_{LGP+L} and the Cornell potential are very similar and thus it is expected that V_{LGP+L} will yield a spectrum similar to the Cornell potential, but perhaps with an improved description of the bound-state energies. Since we are in a non-relativistic regime, it is possible to perform this check by solving the Schrödinger equation. We perform the usual variable separation, i.e. the wave function is given by[1] $\psi_{l,n}^m(\boldsymbol{r}) = f_n(r) Y_l^m(\theta, \varphi)/r$. The function $f_n(r)$ has to obey the boundary conditions $f_n(0) = 0$ and $f_n(r \to \infty) = 0$. Due to the complexity of the potentials involved, we choose to solve the ordinary differential equation for $f_n(r)$ numerically.

For the numerical solution, we use the *shooting method* [17]. It consists in discretizing a likely range to find the eigenenergies and, for each one of these points, we compute the wave function. We know that the boundary conditions will only be met for the exact value of the eigenenergy. Therefore, if we determine the energy value for which the wave function most closely matches the boundary conditions, we know that this energy will be a good approximation for an eigenenergy.

The way we implement the strategy above is to start the integration of $f_n(r)$ for a large value of r, where we can consider only the linearly rising term of the potential. In this regime, we know that the solution will be an Airy function of the first kind and thus we have analytic values for the initial point and its derivative. We integrate towards $r = 0$ using a second-order Runge-Kutta method (more precise methods showed little improvements). If we compare the value of $f_n(0)$ for two consecutive energies and find that it changed sign, we know that the eigenenergy must be inside the range delimited by these energies. We then adapt the bisection method to further refine the precision of the found eigenenergy.

We remark that our model has as free parameters the charm and bottom quark mass (m_c and m_b) as well as the string force σ. To fit the eigenenergies obtained numerically, to the experimental spectrum, we set up a grid of points m_c, m_b and σ and compute the spectrum for each one of the points. It is then possible to compute the χ^2 value

$$\chi^2(m_c, m_b, \sigma) = \sum_i \left(\frac{E_i - E_{i,\text{experimental}}}{\sigma_i} \right)^2 \tag{12}$$

and select the parameters that minimize it. We also compute the uncertainty associated with this calculation by delimiting the area for which the χ^2 rises by no more than one unit. The errors in the parameters are then obtained by projecting the area on the parameter axis. We propagate these errors to the spectrum by generating random numbers around the central values of the fit, following a Gaussian distribution with standard deviation equal to the established errors.

The experimental states found are spin dependent, while our spectrum is not. Therefore, we average over states with same angular momentum and principal quantum number, but different

[1] We notice that the potential in Eq. (11) is central.

Figure 2. The charmonium spectrum obtained using V_{LGP+L} and the Cornell potential, along a comparison with the spin-averaged states.

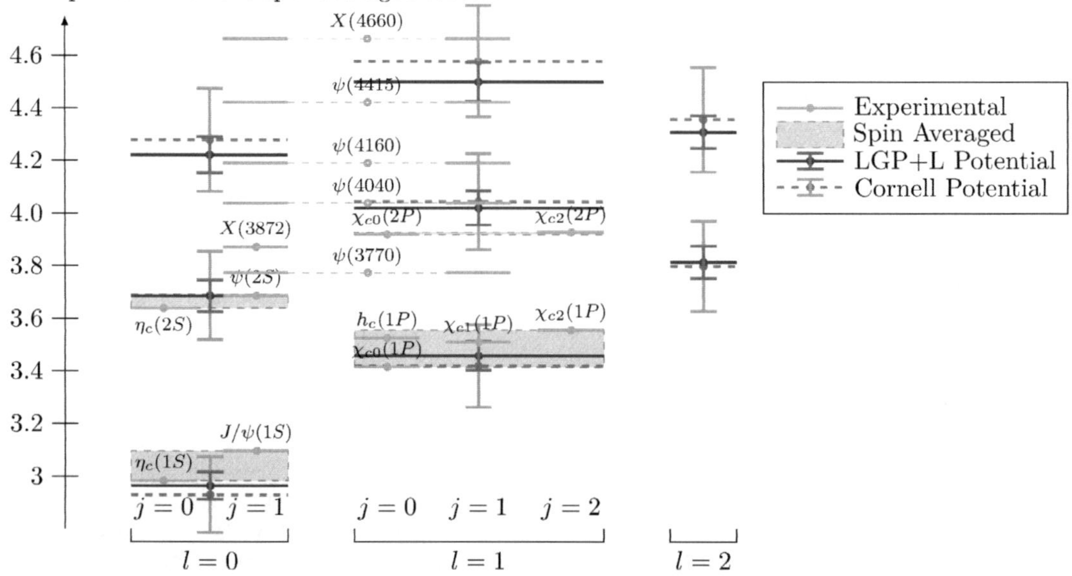

spins. Our choice was to simply take the average of the minimum and maximum values of mass for each case and assign the uncertainty σ_i as half the distance between them. The values were extracted from Ref. [16].

4. Results

The numerical results below were obtained using the following parameters: The wave function was integrated from $r = 20.0\,\text{GeV}^{-1}$ until the origin (as described in the Section 3) in steps of $5.0 \times 10^{-3}\,\text{GeV}^{-1}$. We searched the charm quark mass in the range between 1.00 GeV and 2.00 GeV in steps of 1×10^{-3} GeV. The search for the bottom quark mass used the same step, but in the range between 4.15 GeV and 4.85 GeV. The last parameter, i.e. the string force σ, is searched in the range from $0.10\,\text{GeV}^2$ to $0.50\,\text{GeV}^2$ in steps of $0.01\,\text{GeV}^2$. We used $\alpha_s = 0.2663$ when solving the differential equation for charmonium and $\alpha_s = 0.1843$ for bottomonium. Lastly, we present the discretization used in the search for eigenenergies. For simplicity, our search was made for the total particle mass (by replacing in the differential equation the eigenenergy with $M_{\text{Tot}} - 2m_q$). For the charmonium, we searched for masses between 2.00 GeV and 6.00 GeV in steps of 0.04 GeV. For the bottomonium, we searched from 8.50 GeV until 12.50 GeV using the same step as for charmonium.

The spectrum obtained with both potentials can be found in Tables 1 and 2 as well in Figures 2 and 3. The parameters that minimized χ^2 can be found in Table 3. As it is possible to see from these tables and figures, the spectrum obtained from V_{LGP+L} is similar to the one from the Cornell potential, as expected. But we must point out that V_{LGP+L} shows a small but observable improvement over the Cornell potential. This can also be seen by the smaller value of χ^2 obtained from the fit. We notice as well that the error associated to each state in the spectrum is smaller in the V_{LGP+L} case. We can see in Table 3 that we are able to determine the model parameters more precisely for the V_{LGP+L} as well.

As described in Section 3, the process of determining the spectrum goes through finding the wave function. The raw output yields non-normalized wave functions. We normalize them and

Table 1. Results for the charmonium eigenstates using V_{LGP+L} and the Cornell potential.

V_{LGP+L}			Cornell Potential		
State	Predicted mass (GeV)	Deviation from averaged spin (GeV)	State	Predicted mass (GeV)	Deviation from averaged spin (GeV)
1S	2.96(5)	−0.10	1S	2.93(14)	−0.14
2S	3.69(6)	0.01	2S	3.69(17)	0.01
3S	4.22(7)	-	3S	4.28(20)	-
1P	3.46(6)	−0.07	1P	3.42(16)	−0.11
2P	4.02(6)	0.09	2P	4.04(18)	0.12
3P	4.50(7)	-	3P	4.58(21)	-
1D	3.81(6)	-	1D	3.80(17)	-
2D	4.31(6)	-	2D	4.36(20)	-

Table 2. Results for the bottomonium eigenstates using V_{LGP+L} and the Cornell potential.

V_{LGP+L}			Cornell Potential		
State	Predicted mass (GeV)	Deviation from averaged spin (GeV)	State	Predicted mass (GeV)	Deviation from averaged spin (GeV)
1S	9.47(4)	0.04	1S	9.49(8)	0.06
2S	10.00(4)	−0.01	2S	10.00(10)	−0.01
3S	10.37(5)	0.01	3S	10.39(12)	0.03
4S	10.67(6)	0.10	4S	10.72(14)	0.14
5S	10.95(7)	-	5S	11.01(16)	-
1P	9.86(4)	−0.03	1P	9.84(10)	−0.04
2P	10.24(5)	−0.01	2P	10.25(11)	0.00
3P	10.56(5)	0.03	3P	10.59(13)	0.05
4P	10.84(6)	-	4P	10.89(15)	-
1D	10.11(5)	−0.05	1D	10.10(11)	−0.06
2D	10.44(5)	-	2D	10.45(12)	-
3D	10.73(6)	-	3D	10.77(14)	-

Table 3. Quark masses and string tension obtained from our fit. These parameters are used to obtain the spectrum in Tables 1 and 2.

V_{LGP+L}	Cornell Potential	Quark Mass in Ref. [16]
$m_c = 1.16(3)$ GeV	$m_c = 1.11^{+0.08}_{-0.02}$ GeV	$m_c = 1.275(25)$ GeV
$m_b = 4.61^{+0.02}_{-0.01}$ GeV	$m_b = 4.58^{+0.04}_{-0.01}$ GeV	$m_b(\overline{MS}) = 4.18(3)$ GeV
$\sigma = 0.23(1)$ GeV	$\sigma = 0.26^{+0.01}_{-0.03}$ GeV	$m_b(1S) = 4.66(3)$ GeV
$\chi^2 = 6.20$	$\chi^2 = 12.13$	

Figure 3. The bottomonium spectrum obtained using V_{LGP+L} and the Cornell potential, along a comparison with the spin-averaged states.

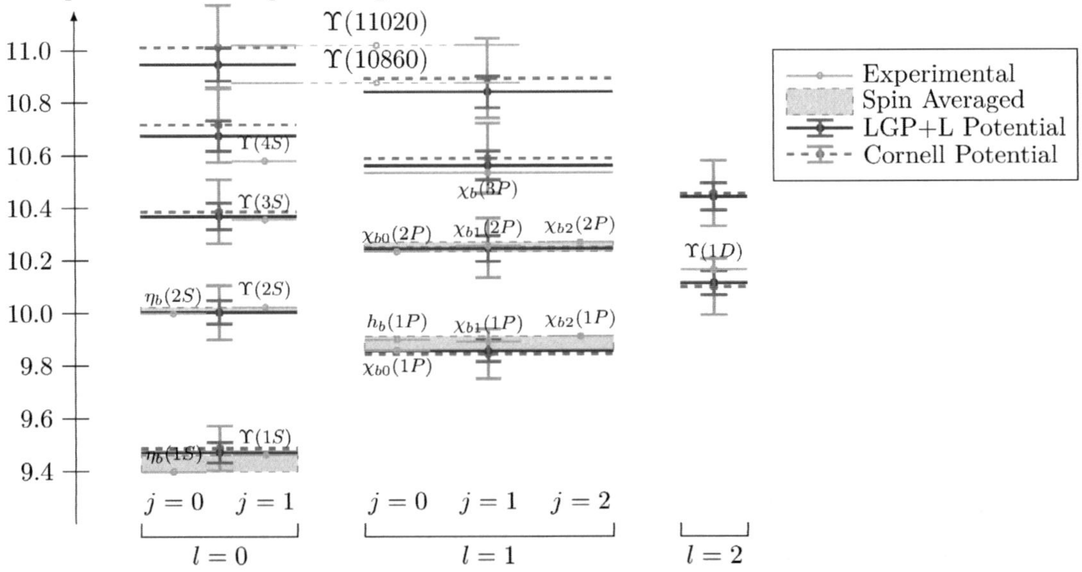

Table 4. Typical inter-quark distances for charmonium and bottomonium. Errors are expected to be negligible.

Charmonium			Bottomonium		
State	V_{LGP+L} distance (fm)	Cornell Potential distance (fm)	State	V_{LGP+L} distance (fm)	Cornell Potential distance (fm)
1S	0.80	0.83	1S	0.45	0.47
2S	1.59	1.56	2S	0.94	0.94
3S	2.20	2.14	3S	1.35	1.31
1P	1.29	1.28	1P	0.76	0.77
2P	1.94	1.90	2P	1.18	1.16
3P	2.50	2.43	3P	1.55	1.49

compute the mean diameter of particles

$$\langle d\rangle = 2\int_0^\infty |f(r)|^2\, dr\,. \tag{13}$$

The results are found in Table 4. These results may be useful in Effective Field Theories, such as pNRQCD [7].

5. Summary

In this work we employed a potential model for heavy-quarkonia states using a non-perturbative potential obtained through lattice simulations. We found that the potential is still non-confining, which is expected, since perturbation theory is employed for its derivation. Despite this, we find that it shows small qualitative differences from the Coulomb-like term of Cornell potential. The most interesting one is that it rises above zero, showing a behavior similar to the one expected for a confining potential up to around $\sim 1.5\,\text{fm}$. It is damped however for larger distances.

We proceeded to compute the spectrum using the Cornell and V_{LGP+L} for charmonium and bottomonium. The results showed that the corrections introduced in V_{LGP+L} through the use of the lattice gluon propagator improved the spectrum and narrowed the errors. Our interpretation is that the propagator in Eq. (7) contains information about QCD that is missing from the free propagator, even in the perturbative regime. We also obtained the wave functions of the quarkonia states and were able to compute the mean diameter of such states, which may be useful to pNRQCD [7].

Acknowledgments
The authors thank B. Blossier and F. Navarra for useful comments. W.S. thanks the Brazilian funding agencies FAPESP and CNPq for financial support. A.C. and T.M. thank CNPq for partial support. For drawing Fig. 1, we used the `feynman` LaTeXpackage available at `http://feynman.aivazis.com/`.

References
[1] Eichten E, Gottfried K, Kinoshita T, Lane K D and Yan T M 1978 *Phys. Rev.* D **17** 3090 [Erratum: Phys. Rev.D21,313(1980)]
[2] Eichten E, Gottfried K, Kinoshita T, Lane K D and Yan T M 1980 *Phys. Rev.* D **21** 203
[3] Bali G S 2001 *Phys. Rept.* **343** 1–136 (*Preprint* hep-ph/0001312)
[4] Koma Y, Koma M and Wittig H 2006 *Phys. Rev. Lett.* **97** 122003 (*Preprint* hep-lat/0607009)
[5] Koma M, Koma Y and Wittig H 2008 *PoS* **CONFINEMENT8** 105
[6] Cucchieri A, Dudal D, Mendes T and Vandersickel N 2012 *Phys. Rev.* D **85** 094513 (*Preprint* 1111.2327)
[7] Brambilla N *et al.* 2014 *Eur. Phys. J.* C **74** 2981 (*Preprint* 1404.3723)
[8] Serenone W M and Mendes T 2013 *AIP Conf. Proc.* **1520** 364–366
[9] Serenone W M, Cucchieri A and Mendes T 2014 *PoS* **LATTICE2013** 434 (*Preprint* 1404.7436)
[10] Serenone W M, Cucchieri A and Mendes T 2015 (*Preprint* 1505.06720)
[11] Lucha W, Schoberl F F and Gromes D 1991 *Phys. Rept.* **200** 127–240
[12] Creutz M 1983 *Quarks, Gluons and Lattices* Cambridge Monographs on Mathematical Physics (Cambridge University Press) ISBN 9780521315357
[13] Cucchieri A and Mendes T 2007 *PoS* **LAT2007** 297 (*Preprint* 0710.0412)
[14] Cucchieri A and Mendes T 2008 *Phys. Rev. Lett.* **100** 241601 (*Preprint* 0712.3517)
[15] Cucchieri A, Mendes T, Oliveira O and Silva P J 2007 *Phys. Rev.* D **76** 114507 (*Preprint* 0705.3367)
[16] Olive K A *et al.* (Particle Data Group) 2014 *Chin. Phys.* C **38** 090001
[17] Press W H, Teukolsky S A, Vetterling W T and Flannery B P 2007 *Numerical Recipes 3rd Edition: The Art of Scientific Computing* (Cambridge University Press) ISBN 9780521880688

AUTHOR INDEX